About the Author

Roger Daniels is Charles Phelps Taft Professor of History Emeritus at the University of Cincinnati. He received his Ph.D. from UCLA in 1961 and is a past president of both the Immigration and Ethnic History Society and the Society for Historians of the Gilded Age and the Progressive Era. He has written widely about Asian Americans and immigration. Among his most recent books are *Not Like Us: Immigrants and Minorities in America, 1890–1924; Debating American Immigration, 1882–Present* (with Otis Graham); and *American Immigration: A Student Companion.*

COMING TO AMERICA

A HISTORY OF IMMIGRATION AND ETHNICITY IN AMERICAN LIFE

SECOND EDITION

Roger Daniels

NEW YORK • LONDON • TORONTO • SYDNEY

Developed by Visual Education Corporation, Princeton, N.J.

A previous edition of this book was published in 1991 by HarperPerennial, an imprint of HarperCollins Publishers.

First Perennial edition published 2002.

Designed by Rhea Braunstein

Library of Congress Cataloging-in-Publication Data

Daniels, Roger.

Coming to America : a history of immigration and ethnicity in American life / Roger Daniels.—2nd ed.

p. cm.

Includes bibliographical references and index.

ISBN 0-06-050577-X

1. Ethnology—United States—History. 2. Minorities—United States—History. 3. United States—Ethnic relations. 4. United States—Emigration and immigration—History. 5. Immigrants—United States—History. I. Title.

E184.A1 D26 2002

305.8'00973—dc21 2002072436

12 RRD 20

For

Günter and Joetta Moltmann

Gracious hosts and good friends to a *Gastarbeiter und Frau*

Contents

Part III: Modern Times

Tables, Charts, and Maps

Tables

Charts

Maps

Preface to the Second Edition

When I first undertook the revision of this book, a dozen years after writing it, my notion was that I would rewrite parts of Chapter 16 substantially to reflect the changes that had taken place up to 1990, and then add a final chapter on events and developments since then. But it soon became apparent that amending chapter sixteen was not to be easily accomplished, so, as is the case for the earlier chapters, I am leaving the text as it was except for correcting typographical and other errors and updating tables. In the new seventeenth chapter, "Immigration in an Age of Globalization," I look again at the legislation of the late 1980s, provide an overview of twentieth-century immigration, and then analyze the convoluted history of immigration and immigration legislation in the 1990s.

Were I writing the book from scratch now, there would be alterations in almost every chapter to reflect the changes in scholarship in the intervening years. This would be nowhere more true than in the chapters dealing with the colonial era. There has been a great deal of monographic scholarship about immigration in those years: important works by Marianne S. Wokeck, *Trade in Strangers: The Beginnings of Mass Migration to North America* (1999) and Hans-Jürgen Grabbe, *Vor der Grossen Flut: Die Europäische Migration in die Vereinigten Staaten von Amerika 1783–1820* (2001) are merely two of several books that would cause me to write somewhat differently. And one broadly interpretative work, Jon Butler, *Becoming America: The Revolution Before 1776* (2000), makes points and raises fundamental questions that would have been taken notice of in the text itself. But, all in all, I believe that the text has stood up reasonably well. I have added to the original bibliography a separate list of some of the more important publications of the intervening years and include appendixes about the spate of recent immigration legislation.

As this text went to the publisher at the end of September 2001, the terrorist atrocities against New York's World Trade Center and the Pentagon had just taken place. There seemed to be no reason to believe that, in the long run, the established patterns of immigration would change, although it would not be surprising if it became more difficult for immigrants from most of the so-called Middle Eastern nations to gain admittance, and certainly there would be increased surveillance and interrogation of foreigners already here. There also was sporadic violence against presumed Arabs/Muslims—the first fatal victim of such post–September 11 violence seems to have been Balbir Singh Sodhi, a forty-nine-year-old Sikh American gas station owner of Mesa, Arizona, shot down in his place of business by a gunman in broad daylight on September 15.

PART I

COLONIAL AMERICA

1

Overseas Migration from Europe

Migration is a fundamental human activity. The very first person of whom we are aware—that African Eve whom her discoverer, Donald Johanson, decided to call Lucy—was in the process of moving (migrating or returning "home"?) as her footprints, still legible after three million years, show.[1] In the ensuing millennia her descendants settled, somehow, every continent save Antarctica. The New World, last to be settled, was apparently uninhabited by man until some thirty thousand years ago, when Asian migrants crossed the then-existing land bridge between Siberia and Alaska and deployed throughout the two Americas, creating cultures of great variety and complexity. To cite but one example of that variety, consider the Inuit, one of the native peoples of Canada's Arctic regions. (Americans tend to call them Eskimos.) Though the Inuit number but several thousand, their two language families, specialists assure us, are as dissimilar as English and Russian.

In any event, by prehistoric times, the human race had peopled almost the entire globe by migration. In historic times we make a distinction between the term *migration,* which simply means moving, and *immigration* which means moving across national frontier. If we posit two unemployed auto workers in Detroit leaving the motor city to seek a job elsewhere—one of them taking a trolley across the bridge to Windsor, Ontario, and the other flying to San Diego—only the first is immigrating even though his is the shorter journey. Thus an immigrant is simply a migrant whose move has involved crossing at least one international frontier.

The struggle to impose global European hegemony began after the voyages of Columbus and others in the so-called Age of Discovery ("so-called" because everywhere Europeans went they found people there before them). In the course of that struggle Europeans' attitudes toward the peoples of the rest of the world changed. That change can most easily be seen by comparing the notions of one of the earlier European travelers, Marco Polo (1254?–1324?), with those of his successors. His famous book about his travels to China is written from the perspective of a person from a relatively underdeveloped region who has visited a more-developed one. Two centuries later, however, in the Age of Discovery, the attitudes of Europeans were quite different. Not only did they arrogantly assume their own superiority, but they also disregarded the legitimacy of the civilizations and cultures they encountered. Although still ready, like Marco Polo, to report wonders, and often capable of believing almost anything, European travelers in the new age exhibited an almost universal self-confidence of the superiority of their own culture. Europeans as different as Hernán Cortés (1485–1547) and John Winthrop (1588–1649) justified their usurpations as promoting the greater glory of God and/or empire. Similarly, the great Portuguese poet Luis de Camoëns (1524–80), who had himself been in India and China, begins his epic, *The Lusiads* (1572):

> *This is the story of Heroes who, leaving their native*
> *Portugal behind them, opened a way to Ceylon and*
> *farther, across seas no man had ever sailed before. . . .*
> *It is a story, too, of a line of kings who kept*
> *advancing the boundaries of faith and reason.*

Scholars disagree about the reasons for this confidence: Some ascribe it to European culture generally, others to Christianity, and still others to technology. One of my own teachers, the late Lynn White, Jr., ascribed it to a combination of religion and technology.[2] But, whatever the reasons, the growth of the notion of European superiority is quite clear.

Most Europeans also assumed that they and their stock were inherently superior to the various peoples they subjugated. With some notable exceptions, Europeans had a contempt for the cultures and peoples they encountered, an attitude that would soon evolve into modern racism. We who write of the triumphs of modern immigration to America must never forget the societies and peoples whom the newcomers conquered and sometimes exterminated. Although few contemporary writers are quite as crude as Theodore Roosevelt—in his vivid *Winning of the West* (1889–96) he could complain that the British Proclamation Line of 1763 was an attempt to keep "two thirds of a splendid continent as a hunting preserve for squalid savages"—too many historians of America still write as if the New World before the coming of the whites had been a tabula rasa or a virgin land for them to conquer and manipulate as they would. This theme, explicitly or implicitly, is dominant in most of the writing about what one Eurocentric historian called the "transit of civilization from the Old World to the New."[3] From John Winthrop's *Journal* (1630–44), through Frederick Jackson Turner's frontier thesis (1893), to Bernard Bailyn's *Voyages to the West* (1987), the theme of "an errand into the wilderness" or of "peopling" an empty land has dominated the discussion of the first centuries of modern immigration to America.

Faith and empire were, as Camoëns noted, prime forces in the process of European migration overseas, but so too was the desire for economic gain. From an imperial point of view—one that will be largely ignored in this book, which is focused on the experience of the human beings who came—the settlement of America north of Mexico was essentially a failure, an investment whose bottom line was written in red ink. No single fact illustrates this better than the willingness of some British policymakers during the negotiations preceding the Treaty of Paris of 1763 to take the profitable sugar island of Martinique (385 square miles) instead of all of Canada as spoils of war. Individuals, of course, did make money even out of North America. But whether they succeeded or failed, all the European powers who were engaged in colonization hoped to profit, either from primitive accumulation—stealing already mined precious metals and jewels from their owners, such as Aztec and Inca rulers—mining, or procuring low-bulk, high-value exports such as spices and furs for which there was an established market in Europe. In the New World the most prized exports in the period before 1800 were gold, silver, sugar, and furs. Only the last was found in North America in that period; the first of the many North American gold rushes did not begin until 1849.

The several hundred thousand Europeans who came to the New World in the sixteenth and seventeenth centuries came largely because of what we would now call the population and/or labor policies developed by the various colonizing powers. These policies were shaped both by the existing socioeconomic conditions in the home countries and by the opportunities for gain, real and imagined, that existed in the new colonies. While Anglo-American writers since at least the late sixteenth century have stressed the so-called "black legend"—the special cruelty exhibited by Spaniards—the modern social historian finds no moral difference between the ways that the European nations behaved in the

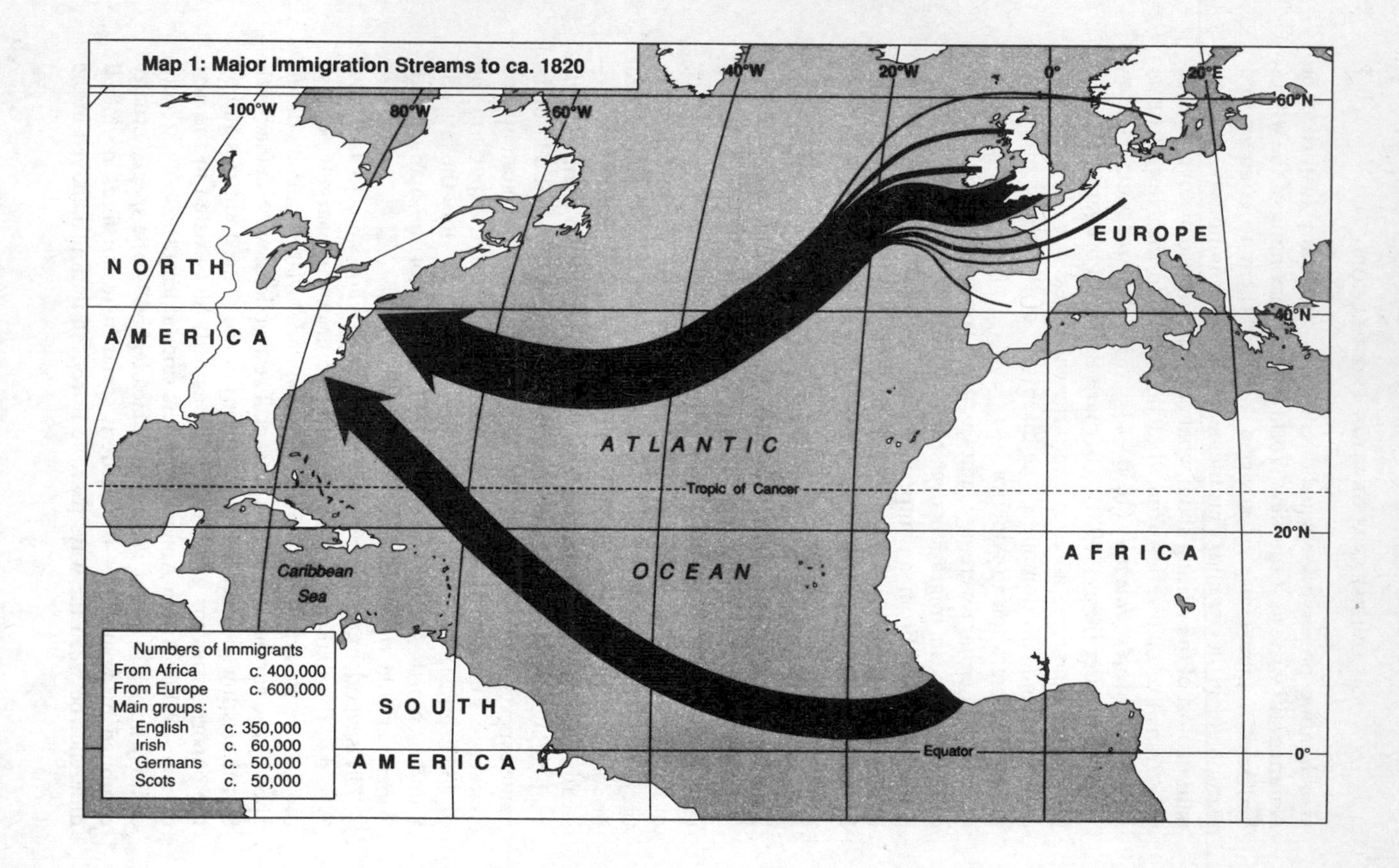
Map 1: Major Immigration Streams to ca. 1820
100°W
80°W
60°W
40°W
20°W
0°
20°E
60°N
40°N
20°N
0°
NORTH AMERICA
EUROPE
AFRICA
ATLANTIC OCEAN
Tropic of Cancer
Equator
Caribbean Sea
SOUTH AMERICA
Numbers of Immigrants
From Africa c. 400,000
From Europe c. 600,000
Main groups:
English c. 350,000
Irish c. 60,000
Germans c. 50,000
Scots c. 50,000

New World. Conquistadores hacking their way to riches through walls of human flesh are no more—or less—morally reprehensible than Puritan fathers rejoicing at the deaths of Indians from European diseases. John Winthrop, for example, wrote in 1634 that "the natives, they are neare all dead of the small Poxe, so as the Lord hathe cleared our title to what we possess." And, as Alfred Crosby has shown, European and African diseases to which the Amerindians had little or no natural immunity killed many more of them than did European bullets and blades, while the destruction of their culture may have caused even more fatalities.[4]

The numbers of deaths involved are staggering even though the experts' guesses vary wildly. The Argentinean scholar Angel Rosenblatt estimated in 1954 that the pre-Columbian population of the Americans had been about 13.3 million and that eight decades of conquest reduced this to some 10.8 million, a reduction of nearly 20 percent. More dramatic—and much more controversial—are the calculations of a group of historians at the University of California, Berkeley. For the central plateau of Mexico alone, these scholars have calculated from 1519 (the date of Cortés's arrival) a population decline as follows:

Table 1.1

"Berkeley School" Estimates of Central Mexican Population, 1519–1605

Year	Population
1519	25,200,000
1532	16,800,000
1548	6,300,000
1568	2,650,000
1580	1,900,000
1595	1,375,000
1605	1,075,000

If these figures even approximate what happened, the peoples of Central Mexico suffered one of the worst demographic disasters in human history. Even critics of the Berkeley school agree that its figures from 1568 on are probably good approximations: These show a decline of about 60 percent—about a million and a half human beings—in two generations. North of the Rio Grande the numbers are smaller, but here too the experts' investigations are constantly raising the total. When I was in graduate school in the late 1950s, "about a million" was an acceptable answer to the question, How many Indians lived in what is now the United States and Canada in 1492? Today an answer of "about two and a half or three million" would be acceptable.[5]

It would be a mistake, however, to assume that all Amerindians suffered equally or at all from the white invasion. The mounted Indians of the Great Plains, who because of Hollywood Westerns seem to be the archetypical North American Indians, had a way of life totally dependent on contributions taken from the whites: the horse and the gun. The Navajo, one of the few success stories in Amerindian demography, have grown from perhaps eight thousand in 1680 to some one hundred fifty thousand today. The continuing and changing relationships between Amerindians and whites, and sometimes between Amerindians and blacks, are ignored in most American histories, except to record the occasional battle or massacre. The first such battle—really a skirmish—is described in the Icelandic Eddas. The Viking colonists in North America—who called the Indians *skrellings*—had no overwhelming superiority in weaponry, so the result was essentially a standoff, although successful Indian opposition was clearly a factor in the Viking abandonment of America. But other kinds of relationships with later settlers were more significant, relationships symbolized by two almost-mythological real persons, Pocahontas and Squanto, each a lifesaver or sustainer. This supportive role was repeated, if less dramatically, in each area of settlement as the first newcomers learned survival techniques from those who had lived there for generations. Similarly, each increment of newcomers learned survival techniques from those already there, whether on the "wild" frontier or on the urban and suburban ones. But the newcomers, although learning from observation if in no other way, did not melt into the existing society any more than the Pilgrims became Indians.

Historians of America and American ethnic groups are all too ready to ignore the preexisting cultures and groups as they recount their version of life in the New World. Whether one is talking about the sixteenth century or the twentieth, few metaphors are more inappropriate than the one of the melting pot. Canadians have recently adopted a more satisfactory metaphor for their society, calling it a mosaic, in which groups tend to retain certain characteristics yet at the same time become Canadian.

This problem of identity has been a serious one since our earliest history and also troubles other "settler societies," particularly as long as they continue to receive significant numbers of additional immigrants. The more uniform the immigration and the more closely it corresponds in ethnicity, nationality, culture, and class to the existing society, the less severe the identity problem is likely to be. But even in relatively homogeneous settler societies such problems do exist. Scholars have recently described the problem of colonial identity as follows: "The process by which colonizing groups come to perceive themselves as a people distinct both from those resident in their country of origin and from those indigenous to their place of settlement." In Spanish-American colonial society, for example, where there were few ethnic differences within the colonizing population, the crucial distinction soon came to be that between *peninsulares,* those born in Iberia, and "creoles," those born in the New World. Similar distinctions were made elsewhere in the European colonial world.[6] In addition, in all colonial societies, distinctions are made between the colonizers and the colonized.

In the New World all the colonizing powers had to develop population-labor policies, policies which differed greatly between nations and often within the colonies of one nation. The differences between the policies of the three chief colonizing powers in the New World—Spain, France, and England—are to be explained by objective circumstances, not national character. (The best evidence for this is the virtually identical policies each—along with the Dutch and the Danes—adopted on their separate Caribbean islands.)

Spain found what all colonizers wanted: instant riches. As Cortés fancifully explained to Montezuma: "We Spaniards have a disease of the heart which only gold can cure." Once they had confiscated the already gathered treasure, the colonizers set about exploiting established mines and finding new ones. Thus they needed large amounts of labor. For this they first levied local Indian populations and then, probably because of declining Indian numbers, they began to import slaves from Africa. (Some historians have held that Bartolomé de las Casas [1474–1566], the Dominican "Apostle to the Indians," encouraged or even caused the introduction of Negro slavery to spare his beloved Indians, but most recent authorities deny this.) Spanish society thus became triracial. Up to the beginning of the nineteenth century, perhaps half a million *peninsulares* came to the New World, not very many people for nearly three centuries, especially considering the extent of Spanish dominion and the number of persons to be governed. In addition many quickly returned, and the mortality rates in the colonies were quite high. Add to this a very heavily male immigration and the extensive sexual relationships that existed between conquering males and subordinate females, and it is easy to understand why Spanish colonial society quickly found a place for persons of mixed ancestry, or mestizos. This was especially true if those persons were acculturated, more or less, to European standards in language, dress, diet, and so on.

And although language and law soon developed terms for all of the mixtures possible in a triracial society, practice soon defined most such persons by culture rather than by ancestry. Thus, an individual who spoke Spanish, wore European clothes, and ate European style food was considered, if not Spanish, not any longer an Indian. Such mestizos were crucial in helping to administer the empire.

In New France (Canada) an entirely different situation prevailed. French immigration was heavily male but much smaller than that from Spain: perhaps forty thousand persons at most during the colonial period. Since the fur trade was the key to wealth in New France, the fur traders, or *coureurs de bois,* became, after the soldiers and the priests, the most vital individuals. Unlike the Spaniards, who exacted various kinds of forced labor from Indians, the French wanted the Indians to become even more proficient at one of their traditional occupations: fur trapping. To this end they supplied them with tools to make them more efficient: metal traps and rifles. The Spanish policy forbade Indians, culturally defined, either to bear modern arms or ride horses, on pain of death. On the northern borders of Mexico, however, some horses—which, according to tradition, came initially from the expeditions of Francisco Vásquez de Coronado (1510?–1554)—escaped, found the environment to their liking, and multiplied. The Indians of northern Mexico and what is now the American Southwest soon captured some and learned how to ride, creating what anthropologists call the "horse and no gun culture." Meanwhile, on the swampy plains and forests of Canada, as noted, Indians were developing the "gun and no horse" culture. (The French would have supplied horses had they been useful, but trappers in eastern Canada needed canoes, not horses.) Eventually each culture expanded until they met, producing the previously mentioned "horse and gun" culture of the Plains Indians, whose great war/sport was the stealing of horses.

At the same time small settlements of French families settled in the Saint Lawrence Valley in what are now Quebec, eastern Ontario, and the Maritimes and tried to scratch out a living from relatively poor soil and a short growing season. Starting in the last third of the nineteenth century, descendants of these Quebecois would migrate to the United States in large numbers, as will be seen in chapter 9.

Unlike the Spaniards and the French, the English had no long-term use for the Indians. Although, the English, too, wished quick riches, agriculture and fishing quickly became their chief economic activities. As late as 1790, the first census showed that 95 percent of all Americans lived on farms, and only in 1920 would it record an "urban" majority: That is, just over half the population lived in places of 2,500 or more persons. As Richard Hofstadter put it, the United States was born in the country and only later moved to the city. Since forests were prevalent in the eastern third of the United States, the major social task—however individually performed—was clearing the land of timber and animal life. To the vast majority of those who settled what became the United States, Indians were simply another form of animal life, to be killed off, pushed west, and, eventually, to be placed in vast cages known as reservations.

European Migration and Exploration before Columbus

Since the beginning of modern times—for our purposes the end of the fifteenth century—some sixty million Europeans have migrated overseas, the majority of them to the United States. (The extracontinental migration of Africans, Asians, and Latin Americans will be discussed in later chapters.) But prior to the Age of Discovery and throughout most of recorded history, Europe has been a target rather than a source of mass immigration. As he tells us at the beginning of his history of the Peloponnesian War, Thucydides (471?–400? B.C.) believed:

> The country now called Hellas had in ancient times no settled population. On the contrary, migrations were of frequent occurrence, the several tribes readily abandoning their homes under the pressure of superior numbers.

Thucydides was trying to rationalize Greek migration myths, of which the *Iliad* and the *Odyssey* are only the most famous. Such myths abound in many cultures: The invasion of Canaan led by Moses and Joshua and the *Aeneid* are the two best known in Western civilization. All are attempts to explain how a given people got where it was. Thucydides was committing the historian's most typical sin, telling more than he knew. Even now, twenty-five centuries later, scholars are not clear about where the Greeks came from and their relation to and/or identity with the mysterious Sea Peoples—so described in an Egyptian document—who were clearly disturbers of the peace of the eastern Mediterranean some twelve centuries before the beginning of the Christian era.

In historic times, the Greeks themselves became colonizers, settling, outside Europe, first the coasts of Asia Minor and later the whole Mediterranean littoral, and then, in the time of Alexander the Great (356–323 B.C.) and his successors, they planted Greek cities as far east as India. They, with the Vandals, who crossed from Spain into North Africa in the fourth and fifth centuries, are the major exceptions to the generalization about Europe being a target rather than a source of mass migration in premodern times. The Romans, to be sure, deposited military garrisons across North Africa and Southwest Asia, but they did not migrate in large numbers outside Europe, while imperial Rome attracted immigrants from many parts of the Old World. Captives, soldiers, traders, and other voluntary migrants, the most famous of whom were the Apostles Peter and Paul—all made Rome a multiethnic metropolis.

The mass migrations into Europe chiefly took three routes: across the narrow straits at either end of the Mediterranean—the Bosporus and the Strait of Gibraltar—and, most important, westward across the eastern European plain. Successive mass migrations of whole peoples—what the Germans call *Völkerwanderungen*—were sometimes set off by repeated incursions into eastern Asia and Europe itself by armies of mounted horsemen from the grasslands of central Asia. Lasting until the final Mongol invasion in the early thirteenth century, these conquests and raids created a dominolike movement of so-called barbarian peoples throughout the Eurasian landmass and into North Africa. Tacitus (55?–117?), the first writer to examine this phenomenon, speaks of a *furor Teutonicus*—a rage of Germans—but modern scholarship has modified his views and sees these Germanic invasions more as a regularized process of settlement.[7]

From the south the chief invaders represented Islamic civilizations. First came the Moors, who reversed the Vandals' route by coming north across the Strait of Gibraltar. Their advance was stopped only at Tours in west-central France in 732. Islam remained in Iberia, or part of it, for more than seven centuries. Its presence helped create a tricultural civilization in Spain—Islamic, Christian, and Jewish—which, at its peak, was the most advanced in Europe.[8] When, in that famous year of 1492, the Christian crusade the Spanish call the *reconquista* was complete and the Moors were driven from their final footholds in Granada, much Spanish energy could be directed to the New World. The impact that the *reconquista* had on the Spanish imagination is hard to overestimate, but one of its symbolic hallmarks was the dotting of the New World landscape with the place-name Matamoros (death to the Moors), even though the nearest Moor was an ocean away.

The final invasion of Europe—until 1943, that is—came at the other end of the Mediterranean. After the fall of Constantinople in 1453, the Ottoman Turks, one of the peoples originally from Central Asia, overran the Balkans and reached into central Europe as far as the gates of Vienna. Although their advance was stopped in 1683, the Turks dominated much of southeastern Europe for centuries, and contemporary Turkey, a member of NATO, still has European footholds on the northern shores of the Bosporus and on Cyprus. Although cultivated eighteenth- and nineteenth-century Europeans like Mozart and Rossini could, in *Die Entführung aus dem Serail* and *L'Italiana in Algeri,* poke fun at Islamic rulers, for more than a millennium Islam was no laughing matter.

It was competition with Islam plus religious fervor that produced one of the first European attempts at overseas colonization, although, to be sure, the sea was the landlocked Mediterranean. These, the Crusades, the papally inspired attempts of western Christendom to regain the Holy Land, occurred, intermittently, from the late eleventh to the late thirteenth century. They succeeded only in establishing four ephemeral western Christian kingdoms at the eastern end of the Mediterranean. The most important of these, the Latin Kingdom of Jerusalem, was called Outremer—overseas—by its French suzerains. These tiny realms can be called the first European colonies overseas. The wars that were fought over them were the first European—as opposed to Greek and Roman—colonial wars. There was one intrinsic difference between these wars and their successors: Only in the Holy Land and its vicinity did European kings—for example, Louis IX (1215–70) of France and Richard I (1157–99) of England—personally lead armies in the colonies. And, in most subsequent colonial encounters, Europeans would enjoy superiority in both armament and techniques. In the Holy Land, after initial Crusader successes, this was not the case. From the time of the sultan Saladin (1137–93) most of the military initiative was with the crescent rather than the cross.

But even more important from the point of view of colonization and immigration history were the slow but steady steps which various groups of maritime Europeans began to take west, out into the Atlantic. For centuries before Columbus's epochal expeditions, traders and fishermen had been venturing far from European shores. Vilhjalmur Stefansson (1879–1962), himself an explorer of note, wrote that the discovery of Iceland by Irish monks in the sixth century was probably the effective discovery of the New World, since from Iceland, when the weather conditions are right, one can see Greenland, and from the southern tip of Greenland one can sail southwest to the North American mainland without ever being out of sight of land.[9]

That the Norsemen or Vikings—the word means "men of the *viks,*" or fjords—reached and tried to colonize North America, hardly any scholars now doubt. These hardy sailors and fierce fighters, who were the terror of maritime Europe for centuries, seem to have made several settlement-founding expeditions to the North American continent around the year 1000. These voyages are still generating scholarly controversy not to speak of outright fraud and ethnic jealousies. The late-nineteenth-century "Kensington Stone," which purported to show that Vikings reached at least as far south and inland as Douglas County, Minnesota, presumably after marching overland from southern Hudson Bay, and the celebrated "Vineland Map," announced by Yale University with great fanfare a few years ago, are only the most prominent and "successful" fakes. Most of those who have supported extreme claims for the Vikings have been of Scandinavian ancestry, and most of those who most vehemently deny the significance or even the existence of Viking settlements have been of Italian ancestry. (On the other hand, the most successful debunkers of the fakes, such as UCLA's Erik Wahlgren, whose demolition of the Kensington Stone is a classic, have also been of Scandinavian ethnicity.)[10]

No one knows exactly where "Vinland the Good" and the other Viking settlements were located, and guesses have ranged as far south as Virginia. Almost certainly, none was in the United States. The chief argument for a more southern site is the name that Leif Eriksson gave to his colony and the fact that grapes don't grow in the Far North. But when one remembers that his father named a snowy, treeless island Greenland, and that both men were promoters, it need not disturb us that Vinland was probably on Newfoundland. The most likely site has been identified: L'Anse aux Meadows on its northern shore. The site was first investigated by Helge Instad, a Norwegian archaeologist, in 1960, and subsequent digs have unearthed suggestive but not-quite-clinching evidence. As noted, Eriksson's settlements were short-lived, and Greenland ceased being inhabited by Europeans some time in the fifteenth century. Iceland has remained inhabited.

In more southerly waters the historical record is clearer. The Canary Islands, just beyond the Pillars of Hercules, had been known since classical times. Other islands were discovered by Portuguese navigators in the course of the fifteenth century, when Portugal became the major European seafaring power. Its Infante Henrique Navegador (1394–

1460)—Prince Henry the Navigator in English—established a center for maritime technology and discovery at Sagres (on Cape Saint Vincent) at Portugal's southwest corner. From this seagoing "mission control" were launched the successive probes of Africa's western coast: a Portuguese ship first rounded Cape Bojador in 1434; in 1441 the first cargo of gold and slaves was brought back from West Africa; and so on, until finally Bartholomeu Dias (1450?–1500) rounded the Cape of Good Hope in 1488. That paved the way for Vasco da Gama's (1469?–1524) epochal voyage that reached the west coast of India in 1498. Six years earlier, of course, the Italian navigator had, unwittingly, "discovered" the New World. (In all fairness, that greatest of "discoveries" should have been credited to the systematic successors of Prince Henry—but, as John Kennedy liked to say, life isn't fair.)

The Portuguese had also pushed the frontiers of European navigation farther west. In 1419, they discovered the Madeira group, some 350 miles off the coast of North Africa, and, beginning in 1431, discovered and colonized the previously uninhabited Azores nearly 750 miles out in the Atlantic. Samuel Eliot Morison describes the extensive Portuguese search for what he calls "flyaway islands" of both ancient myth and false reports by mariners. Almost certainly the Portuguese would eventually have reached some part of the New World on their own, as, after Columbus, they did when in 1500 Pedro Alvares Cabral (1460?–1526?) stumbled on that part of Brazil that juts toward Africa.[11]

Eight years previously Christopher Columbus (1451–1506), the Genoese sailing in the service of Their Most Catholic Majesties, Ferdinand and Isabella of Spain, made the first of his transatlantic voyages. His voyages represent the effective European discovery of the New World (although he himself never acknowledged that he had not reached parts of Asia) and catapulted Spain, which had chiefly been concerned with internal affairs and the Mediterranean, into the New World and the leadership of Europe. Cynics have sneered at Columbus as a man who, when he started, didn't know where he was going, when he arrived didn't know where he was, and when he returned, didn't know where he had been. The important fact to remember is that it was Columbus who planted the first permanent European colony in the New World—and, it should also be noted, he was a superb navigator.[12]

For nearly half a millennium after Columbus, Europeans migrated overseas in ever-growing numbers. Only in the years since World War II have migrations into Europe again become numerically more important than migrations out of it. During most of that period, European population soared, as the following table indicates:

Table 1.2

European Population, 1600–2000 (in millions)

YEAR	POPULATION	% INCREASE
1600	95	—
1650	100	5%
1700	120	20%
1750	140	17%
1800	190	36%
1850	270	42%
1900	400	48%
1950	580	48%
2000	729	26%

The causes of this population rise are complex and often quite different from one European country to another. (See, for example, the discussion of Irish population growth in chapter 4.) Among the chief causes of the early growth are increased agricultural production and a lessening in intensity of the plague: After 1750 or so, the industrial revolution was a major factor and, in more recent years, dramatic increases in the life span have been most important. Whatever the causes, this growth in population is clearly one of the factors that impelled Europeans to migrate: During that same period some sixty million Europeans left Europe to go overseas—most of them to the New World, and most of those to the United States.

The Laws of Migration

Why did sixty million leave Europe? Why did so many others stay? These are the most important questions a historian of immigration can ask—and among the most difficult to answer. One therefore needs a kind of generalized conceptual framework within which the experiences of groups and individuals can be structured, compared, and contrasted. A hundred years ago a British social scientist, E. G. Ravenstein, tried to propound what he, in good nineteenth-century fashion, called the "Laws of Migration."[13] Today we would call them *tendencies* or some such more flexible term, but, remarkably, most of Ravenstein's generalizations seem valid to contemporary scholars.

In addition, modern students of migration have developed a few special words to facilitate description of some of the major factors in migration. The chief of these are *push, pull,* and *means. Push* refers to those forces existing in the place of origin that encourage or impel persons to emigrate. These forces may be catastrophic, such as the Irish potato famine or an earthquake in the Azores; political, such as the revocation of the Edict of Nantes or the failure of the revolutions of 1848; or, most often, economic pressures of one kind or another, including the pressures of a growing population. *Pull* refers to those attractive forces emanating from the migrants' goal that draw migrants. Pull forces may be partially false, like the famous handbills that drew the Joads to California in *The Grapes of Wrath.* Noneconomic attractive forces include promises of political and/or religious freedom and such factors as climate and freedom from military service. As was the case with push forces, most pull forces are, in the final analysis, economic. *Means* is shorthand for the ability to migrate: This includes the availability of affordable transportation, the lack of restraints on mobility in the place of origin, and the absence of effective barriers at the destination.

Before turning to a discussion of Ravenstein's laws, I want to talk about the major immigration myths that most Americans believe. I call these the myths of Plymouth Rock, the Statue of Liberty, and the Melting Pot. The first holds that most immigrants came for religious or political liberty; the second holds that most immigrants were dirt poor, or as the poet Emma Lazarus (1849–87) put it, "the wretched refuse of [Europe's] teeming shore"; while the third argues that, in the phrase of another Jewish writer, Israel Zangwill (1864–1926), America was a "melting pot" in which nationalities and ethnic groups (but not races as we understand the term) would fuse into one. Although Zangwill apparently invented that usage of the term in his 1909 play of the same name, the idea had been around for more than a century. J. Hector St. John (1735–1813), the pseudonym of a Frenchman who sojourned here for a few years before returning to France, had written romantically of the American, the "new man," being "melted into a new race of men."

Like most enduring myths, each of the three contains a part of the truth. Persons did—and do—come to the United States seeking liberty; some have found—and will find—the way to wealth from poverty-stricken conditions; and while there has been a continuous genetic mixture of ethnic groups in the United States, most individuals are still aware of their ethnic background. As Nathan Glazer and Daniel P.

Moynihan put it, the melting pot "simply did not happen."[14] Not only have ethnic groups and, even more important, awareness of ethnicity, persisted, but in the United States and Canada (as opposed to Latin America) relatively little amalgamation has taken place. If one had to make a single sweeping generalization about immigration to the United States it would be that most immigrants were not the poorest persons in the countries from which they came and that economic betterment was the major motivating force. The myths do not square with the actual American experience.

Ravenstein's "laws," on the other hand, do so square, at least in their basic thrust. I have divided them into three general categories: characteristics of migrants, patterns of migration, and volume of migration. The examples I use to illustrate them are my own.

A basic characteristic of migration is that it is selective. Under all but catastrophic conditions, only a minority of any population is likely to migrate. The kinds of migrants from a given society will be determined, in large part, by the conditions in that society. Although most nineteenth-century migration to the United States was heavily male, that from Ireland after about 1851 was almost evenly divided by gender, and, after 1890, females predominated, for reasons peculiar to Irish society. During the same period and on into the twentieth century, two-thirds of the Poles and almost nine-tenths of the Greeks were males.

Pull migrants, according to Ravenstein, tend to be positively selected, while push migrants are negatively selected. Ravenstein means that migrants for whom conditions "at home" are not intolerable will come because they wish to and because their talents seem to fit the available opportunities. Conversely, those who are "pushed out," like most of the recent refugees from Cambodia, are persons who, except for the "push," would never have left. Similarly, the degree of positive selection tends to increase with the difficulty of the means. Ravenstein argues that the more difficult the migration is to accomplish, the more calculated is the decision to leave. Until late 1989, for a contemporary East German to decide to cross the *Grenze*—the border between the Germanies—was a much more consequential decision than that of a border crosser from Mexico into Texas. Both were committing illegal acts, but the consequences of being caught were much less serious for the Mexican.

Another "law" stressed by Ravenstein is the heightened propensity to migrate at certain stages of the life cycle. Most migrants tend to be young. Some 40 percent of Irish immigrants in the late nineteenth and early twentieth centuries were between the ages of twenty and twenty-four. From Greece and Italy, at the same time, large numbers of migrants were child laborers, part of the so-called padrone system that flourished in both the Old World and the New. In contemporary society much of the "brain drain" from Third World countries is composed of recent graduates of colleges and professional schools. Despite this, the iconography of American immigration—the way in which it is depicted in visual images—stresses women and children and family groups even though they were a minority—in some periods a very small minority—of all immigrants. Even today, when our immigration laws favor family reunification, families most often arrive serially. This is now usually called chain migration, as the migrants, whether members of a nuclear or an extended family, follow one another as links in a chain.

Ravenstein posited that the measurable social characteristics of migrants tended to be intermediate between the characteristics of the population of origin and the population of destination. Although much contemporary migration—particularly brain drain and some refugee migration—does not conform to this notion, what Ravenstein was pointing out was that most migrants were not at the very lowest levels of the societies they left but were usually below the average attainments in *measurable* social characteristics—education, income, and the like—of the societies to which they came.

Insofar as patterns of migration were concerned, Ravenstein pointed out that most migration takes place within well-defined streams. Whereas one generalizes about migration from Europe, from England, and from Italy going to the New World, to the American colonies, and to the cities of the northeastern United States, the fact of the matter is that migration often follows more precise patterns, often from a particular region, city, or village in the sending country to specific regions, cities, or even specific city blocks in the receiving nation. Sometimes, as in the case of the first considerable German migration to the New World in 1683, the pattern is set because a whole group of villagers with a pastor comes at once, in this instance from Krefeld to what became Germantown, Pennsylvania. This pattern was followed during the colonial period by thousands of other German immigrants who settled much of southeast Pennsylvania, becoming what their neighbors called the Pennsylvania "Dutch," apparently from the word *Deutsche.* The phenomenon of stream migration continues. Few have described it better than Robert Harney, writing about Italian migration:

> It was almost as if not just individuals and families but whole towns and villages formed an electric arc over the Atlantic. The poles at either end were the *paese* in Italy and its *colònia* in some

> eastern port or Chicago. The particles moving in the arc might be young bachelors or struggling fathers, *ritornati,* wives and children, remittances and steamship tickets, but they were all part of a single economy and an enclosed world. The *paese* for those who followed the arc became a psychic state, not a location.

He might have added that the arc could stretch to very small cities as well. For instance, several thousand Italians settled in and around Dunkirk, New York, in the late nineteenth and early twentieth centuries. Almost all of them came from one town in Italy, Valledolmo.[15] Such arcs continue today, mostly from places other than Europe. In Los Angeles, for example, if one travels along the seedier end of Sunset Boulevard, in the area north and east of Dodger Stadium, one will find enclaves of Central and South Americans—Salvadorans, Nicaraguans, Colombians—many of whom came not to "El Norte," or to California, or to LA, but to a precise area, perhaps a few blocks square, where they knew that their relatives, friends, and compatriots were established.

For every major migration stream, Ravenstein insisted, a counterstream develops. Most of those who have written about migration to America, from the earliest times to the numbers game currently being played by the Immigration and Naturalization Service about "illegal aliens," have ignored the fact that many immigrants return. That return is the counterstream. It is difficult to quantify this phenomenon, partly because of the way that the United States kept its immigration statistics. Congress ordered that incoming migrants be counted in 1819; no regular count was made of returning aliens until 1908; and, even today, no systematic count is made of American citizens who emigrate. Scholars were slow to investigate this, partly, I think, because returning was a kind of rejection of America. The serious study of return migration from America begins with the work of my mentor, Theodore Saloutos, who studied returned Greeks in *They Remember America* (1958). One of the more vexing problems in interpreting the available data is that more than a few immigrants came, went back, and came again, with each entry being counted as another immigrant. I interviewed one Issei—first-generation Japanese immigrant—who could document six trans-Pacific round trips before final passage to America. Just as enthusiastic "America letters"—letters written by or for immigrants for the folks back home—could be a spur to migration, the firsthand experience of disillusioned returnees could be a retardant.

Ravenstein liked to calculate what he called the "efficiency" of migratory streams: Stream minus counterstream equals net balance of migration or efficiency. He argued, in a common-sense way, that efficiency would be higher for immigrants who were pushed rather than pulled, for those who had a long and difficult migration rather than a short and easy one, and that efficiency would be greater in times of prosperity in the host country and lesser in times of economic hardship. This last point, although apparent to close students of the nineteenth century even without quantitative data, was underlined by the experience during the great depression. In the years 1932–36 (July 1, 1931–June 30, 1936) more than twice as many immigrants left as were admitted, some 260,000 to 120,000.

The rates of remigration vary widely between ethnic groups who migrated contemporaneously. Thomas Archdeacon has constructed a useful if too-precise chart that calculates remigration rates (what Ravenstein called efficiency) to the first decimal place for the period before 1930. The range runs from a miniscule 4.3 percent for "Hebrew" (only the Jews and the Irish have remigration rates below one in eight) to a high of 87.4 percent for "Bulgarian/Montenegrin/Serbian."[16] It is also clear that those who established families here were less likely to return than those who did not; however, family groups did travel both ways. My father's family, for example, first migrated to the United States in the 1880s, returned to Great Britain about 1890 in time for him to be born there, but returned, to stay, with him as a babe in arms. Thus he was an immigrant with both older and younger siblings who were native-born Americans.

Ravenstein's hypotheses about the volume of migration need not detain us long. He held that the volume of migration would vary according to the degree of diversity of the territory of the host country, with the degree of diversity in the people attracted to it, according to the difficulty of the means, according to economic conditions in both the host and sending countries, and with what he called the state of "progress" in both countries. Insofar as diversity has been concerned, the United States, with a wide variety of climates and environments, has attracted many more immigrants, and from more diverse sources, than has Canada. In Ravenstein's time the difficulty of the means chiefly involved the availability and affordability of transportation: Today, when most immigrants arrive by jet, it is more likely to involve legal difficulties—getting permission to leave, to enter, or, in some cases, both.

A striking example of how changes in transportation have affected immigration volume and settlement patterns may be found in the

Puerto Rican experience. Even though the island was annexed in 1898 and there were no significant legal difficulties to prevent its inhabitants from entering the United States, very few Puerto Ricans came to the mainland until after World War II because of the high cost and infrequency of transportation. Just after the war's end, when cheap air transportation to the mainland was inaugurated, Puerto Rican migration flourished. And since this transportation—originally set up to fly tourists to the sun—was based in New York, Puerto Rican barrios developed there first rather than in Florida, which was closer.

Since most migration has economic causes, economic conditions, in both the host country and the homeland, naturally have a great effect on the volume of migration. That immigrants went back to Europe during the Great Depression of the 1930s reflected not so much prosperity in Europe but the fact that it is easier to be poor in a poor country than in a rich one. Today the Polish government is doing a thriving business in attracting Polish-American social security recipients to Poland, where the zlotys their dollar retirement checks can buy translate into a higher standard of living than they could afford in the United States. Ravenstein's notion about "progress" affecting volume of migration is best shown by the case of Japan, which no longer sends working-class immigrants abroad: Those who come now are likely to be managers of American branches of Japanese companies.

The chief value of Ravenstein's laws is that they provide us with a way of thinking comparatively about the migrations of various peoples in different parts of the world and in various eras and, above all, of reminding us that migration is a universal process. At the same time we must remember that migration is carried on by individual immigrants who, although their actions may conform to larger patterns, are each acting on what, to them, is a unique combination of motives. In a great many cases the precise reasons for migrating, and even more for migrating to a particular place at a specific time, are so complex that not even the migrants themselves can properly assign the appropriate weight to the various factors that impelled them to move. And although we live in an age of increasingly sophisticated ways of counting and estimating, it is important to realize that there is no such thing as a calculus of motives, that there are no computer programs for the human spirit.

Emigration from Europe in Comparative Perspective

Of the roughly fifty-five million immigrants who have, in historic times, come to what is now the United States, nearly seven out of ten came from Europe, and without in any way minimizing the importance of the Africans, Asians, and Latin Americans who have come, we must never forget that, until the 1960s, Europeans predominated in migration, not only to the United States but to other nontropical settler societies all over the world. (In many tropical societies to which Europeans migrated—for example, the Caribbean—unfree and semifree migrant laborers, usually Africans and Asians, greatly outnumbered the Europeans.) The phenomenon of migration from Europe is, therefore, of paramount importance to any understanding of American immigration.

The classic essay on this topic, by the British historian Frank Thistlewaite, was presented to an international congress of historians in 1960.[17] Ironically, this was the point in history when the emigration of Europeans from Europe was becoming less numerically important than migration into Europe by persons from the New World, Africa, and Asia. Part of this latter flow represented, in the case of previously imperial countries like Britain, France, and the Netherlands, the coming of immigrants from their former colonies. (Some young left-wing scholars at the University of Birmingham waggishly titled an otherwise dreary book on the topic *The Empire Strikes Back.*) Other parts of the flow represent what the Germans call *Gastarbeiter,* or guest workers. These people, mainly from southern Europe and Turkey, were supposed to build Volkswagens, and so on, and then go back home. Many have stayed. The fact that the main Hamburg public library now has a large section of children's books in Turkish is but one piece of evidence showing that the German government's plans have been thwarted by ordinary people, who so often in history manage to do what they want in spite of the efforts of the authorities.

Thistlewaite's paper pointed out first of all that of the fifty-five million Europeans* who emigrated overseas between 1821 and 1924, some thirty-three million, or three-fifths, went to the United States. Nevertheless there was also a massive movement—twenty-two million persons—elsewhere, so that to represent migration from Europe as simply an aspect of United States history was to misunderstand it. He was

*That the rough totals for modern immigration to the United States and of Europeans overseas, 1821–1924, each work out to fifty-five million is a momentary statistical coincidence. At the current rate, the U.S. figure will be sixty million in a decade.

highly critical of previous scholarship, pointing on the one hand at the historians of immigration who tended to treat emigration from Europe as a simple case of "American fever" and on the other at those historians of Europe who had all but ignored this topic. Even today one can search the indexes of most books (including textbooks) on the general history of modern Europe in vain looking for references to emigration, or sometimes finding only citations of the important but ephemeral flights of the Huguenots or of the émigrés of the French Revolution.

Thistlewaite noted, as had historians of previous generations, the tremendous growth in European population during even the peak immigration years. Unlike his predecessors he had at his disposal relatively sophisticated demographic research, which enabled him to point out that even while Europe was sending so many of its people overseas, its share of the world's population was increasing from about a fifth to about a fourth. He noted that in no decade of the period had emigration drawn off as much as 40 percent of that decade's natural increase of population—that is, the excess of births over deaths.

One of the factors that distinguishes European emigration to the United States from that which went elsewhere is the sheer diversity of those who went to the United States as opposed to the relative homogeneity of the emigration to the other major receiving states. The 22 million European emigrants who did not go to the United States went chiefly to the British dominions—4.5 million to Canada and fewer than 1 million to Australia—and to Atlantic South America—5.4 million to Argentina and 3.8 million to Brazil. Immigration to Canada and Australia was dominated by persons from the British Isles; in Australia that domination continued until after World War II. As late as 1947, nearly three-quarters of its foreign born, 73.1 percent—were of British birth. Italians and Spaniards dominated immigration to Argentina, contributing, according to the 1914 census, more than a third each. Most of the former came from south of Rome, most of the latter from Galicia, so that Argentines still call all new Spanish arrivals "Gallegos." In Brazil, where 2.3 million immigrants were enumerated in São Paulo alone between 1882–1934, Italians at first predominated. About 73 percent of the more than nine hundred thousand immigrants who arrived between 1887 and 1900 were from Italy, mostly north of Rome. In the same period Spain and Portugal supplied about 10 percent each. In the twentieth century, Italians, mostly from south of Rome, were only 26 percent, followed closely by Portuguese and Spaniards at 23 percent and 22 percent, respectively. The largest other group was Japanese, who started migrating to Brazil's coffee fields after the United States was closed to them. Some one hundred thousand went to Brazil between 1908 and 1930.[18]

In the other major receiving countries, the relative incidence of immigrants in the population was higher than in the United States. In the decade 1901–10, for example, the rate of immigration into the United States was just over one thousand per one hundred thousand of population; it was about fifteen hundred for Canada and three thousand for Argentina. In some countries, the incidence of immigration rose and fell dramatically, most markedly in Australia. The 1861 census showed that three of five Australians were foreign born, 62.8 percent; by 1901 they were only one in five, 22.8 percent. In the United States the percentage of foreign born during the peak immigration years is one of the most amazingly consistent of our historical statistics: Between 1860 and 1920 it was never lower than 13.2 percent nor higher than 14.7 percent. Thus, during the sixty-year period in which the Civil War was fought, the nation became industrialized and urbanized, and other dizzying changes took place in every phase of American life, a constant one American in seven was of foreign birth.

Although the vast majority of these immigrants were Europeans—and by far the largest groups of non-European foreign born reported by the census were Francophone and Anglophone Canadians, the descendants of Europeans—the ethnic composition of the American foreign born changed over time. In 1860 at least 84 percent of the foreign born were either British (53.1 percent) or German (30.8 percent). Most of the British were, in fact, Irish, and the German figure understates the number of German ethnicity, as many of the Swiss and French reported by the census, which was concerned with nationality by place of birth, were German speakers. By 1890 the British/German majority had shrunk to 63.9 percent: 33.8 percent British and 30.1 percent German. In 1920 the two national groups were only a little over a quarter of the 13.7 million foreign born: 15.6 percent were British and 12.1 percent German. There were almost as many Italians as Germans and more than twice as many Eastern Europeans as Britons.

But if we restrict ourselves to examining gross numbers of immigrants by nationality or ethnicity we miss much of the drama and substance of the immigrant experience. First of all very large numbers of immigrants came and went. In 1904, for example, some 10 percent of the nearly two hundred thousand Italians who entered the United States told immigration officials that they had been here before. In South America the percentage of remigrants was much higher. For years thousands of Italians took ship every November after the European harvest to work in the wheat and flax fields of northern Argentina, only to return to Italy in time for the spring planting. Argentines called them *las golondrinas,* for the swallows whose annual migration was almost as long from southern California to the pampas and back. Thistlewaite compares this to the post-1920 seasonal migrations of Mexicans to American harvests as far north as Montana and Michigan and sees these movements as distinguished only by "the remarkable ocean ferry." Nor was it just the unskilled who were itinerants. English housepainters pursued their trade in the United States in the spring, traveled to Scotland for the summer, and then down to England for fall and winter. Sometimes there were intermediate stops for years or even generations. Many Irish left not for the ocean voyage to the New World but for the short eastward trip across the Irish Sea to Liverpool. Later some of them or their descendants migrated to the United States, Canada, or Australia, and—in the American statistics at least—the descendants would be recorded as English. Many Eastern European Jews first went to London or Manchester, often learning a trade like cigar making or improving their skills as tailors before moving on to New York, Chicago, or Philadelphia.

And, as some of the foregoing suggests, migratory patterns often, but not always, first involved journeys within the nation of origin, or in the Old World, before traveling to the New. Between 1876 and 1926, for example, 8.9 million Italians went to the Americas. During the same period at least 7.5 million went to other parts of Europe and North Africa, while uncounted millions migrated within Italy. How many participated in all three kinds of migrations we just do not know. If, in the early and mid-nineteenth century, Norwegian and Swedish immigrants tended to come to places like Wisconsin, Minnesota, and Nebraska directly from overpopulated rural districts, after industrialization set in many first migrated from those same rural districts to Scandinavian cities and then to American urban centers such as Chicago and Jamestown, New York.

We cannot quantify these intro-European movements to the degree that we can do so for transatlantic migration, partly because overseas migration has attracted so much more attention. It is clear, however, as Professor Friedrich Edding of Kiel University has noted, that the "sum total of all these movements was always greater than the volume of overseas migration." He suggests that in the decades before World War I seasonal shifts took between 1 and 2 million persons annually across national frontiers. For the interwar period the demographer Dudley Kirk has estimated that some 10 million Europeans west of the

Soviet Union were living outside the country of their birth. Even more important "as many as 75 million were living outside of their native province or department [and] at least one-third of all Europeans were living outside the commune of locality of their birth."[19] Put another way, before World War II, some 150 million Europeans were living in Europe, but not in the places of their nativity. This movement was, in general, from rural areas to urban ones, from less-developed countries to more-developed ones within Europe.

Although all the industrialized nations and regions in Europe received migrants, France was the European country of immigration par excellence, a fact not revealed in many histories of France. By 1896 more than 250,000 Italians were there and by 1913, 500,000, which was roughly a third of the number of Italian born then in the United States. After the American Congress restricted immigration severely in 1921 and 1924, France received more immigrants than any other nation. By 1932 there were in France more than 900,000 Italians, 500,000 Poles, 330,000 Spaniards, and 300,000 Belgians, not to mention small but growing contingents from France's overseas empire. Among the colonials who lived in France for a time was the Vietnamese who became known to history under his revolutionary name, Ho Chi Minh.

From the Age of Discovery until the 1920s, European overseas migration in general and migration to the United States in particular was heavily male, although both the iconography and the political rhetoric about it, whether from Ronald Reagan or Mario Cuomo, stress families. Cartoonists hostile to immigration liked to draw breaded, threatening-looking figures to represent immigrants, but photographers were usually sympathetic and focused on women, children, and families. (The covers of at least two recent books on immigration have been graced by Lewis Hine's appealing portrait of a doe-eyed young Jewish woman wearing a babushka.) A rough generalization can be made about gender ratios: Through the peak immigration years and into the 1920s, perhaps two immigrants in three coming to the United States were males, most of them youths and young men. However, an even greater proportion of those immigrants who returned were males. Students of immigration have long associated the sex ratios existing among immigrant groups with the rate of return: The more women in a given group, the more likely all members of that group were to stay. According to the calculations of Professor Thomas Archdeacon, the "maleness" and "remigration rates" of specific immigrant groups yield a "high Pearson product-moment correlation of .689," that being a standard statistical device for measuring the relationship between two phenomena. The three groups least likely to remigrate (Hebrews, 4.3 percent; Irish, 8.9 percent; and Germans, 13.7 percent) are among the six "least male" groups (Irish, 46.4 percent; Hebrews, 54.3 percent; and Germans, 57.5 percent). At the other end of the spectrum the correlations are similar. The three groups most likely to remigrate are among the four "most male groups.[20]*

This, then, is the general picture. The people of a whole continent, Europe, were on the move internally and expanding into other continents as well. Its millions who came here and stayed to become Americans and the ancestors of Americans were largely young adults, predominantly male, and drawn mostly from the lower but not the lowest ranks of European society. Although they came for a wide variety of social and personal reasons, the rubric "economic betterment" covers the vast majority of cases. This is the overall impersonal picture no serious student can ignore.

And yet there is something more to be said. Coming to America, whether as a sojourner or a settler, was in many cases to partake in an adventure, a drama, even a dream. For many the adventure became a disaster, the drama a tragedy, the dream a nightmare. Not all the individual stories have happy endings, and many who journeyed to America, Argentina, or Australia, would have been better off had they never left home. Too often in writing about immigrants we forget that there were immigrant losers as well as winners, and that sometimes "winning" took generations.

Despite these and other negatives, American immigration was and is overall a success story. Although I have no desire to contribute to what Richard Hofstadter called "the literature of national self-congratulation," any historian of immigration must report a positive balance, must show a "bottom line" written in black ink. The positive balance is not just for the immigrants personally, or for some of them: It also reflects the positive gains of the nation. While there have been those, like MIT's Francis Amasa Walker (1840–97), the first president of the American Economic Association, who argued that the presence of immigrants caused the fertility of so-called "Americans" to decline, few scholars believe such things today. The demographer Richard A.

*It would be a mistake, I believe, to take the precision of these calculations, which are based on data generated in the 1920s by Imre Ferenczi, too seriously. Neither ethnicity nor remigration can be calculated as precisely as these figures suggest. This kind of misplaced concreteness is fairly common among historians who have, only in recent years, become skilled in statistical manipulation as opposed to counting. The data do, however, give a good approximation of what happened.

Esterlin writes that this is not only highly unlikely but if there had been no immigration, "fertility among the native stock might have been even lower than it was."[21]

Walker's comments reflected not only the increasing anti-immigrant feeling in late-nineteenth-century America but also the enduring notion that persons who came early in our history should be called colonists or settlers, while those who came later should be called immigrants. Such a view is obviously nonsense. Although there is some utility in distinguishing between immigrants and native born, the second generation of any immigrant group—those born in the country of choice—are native stock whatever prejudice or law may say about them. (In the Netherlands today, for example, the children of foreigners born there are not usually citizens: That occurs only with the grandchildren.)

The rest of this book will focus on the immigrants themselves, their children, and sometimes their children's children, from the earliest European and African "settlers" to today's jet-age migrants. And while it will be, as near as I can make it, objective history, it will also be something of a celebration. A celebration not of national self-congratulation but rather of human daring (or foolishness) on the part of those who deliberately chose to become strangers in a strange land.

2

English Immigrants in America: Virginia, Maryland, and New England

The colonial period in American history—from 1607 to the adoption of the Constitution—is almost as long as the history of the United States. Only in 1968 did the span of our national existence exceed that of the colonial era. But there were not many people here then. Fewer than a million came—some six hundred thousand Europeans and perhaps three hundred thousand Africans. Since about half the Europeans and all the Africans were, to one degree or another, unfree, the free immigrant was in the minority during the colonial period. (If these numbers seem small, remember that total French immigration to Canada, 1608–1760, was fewer than 10,000 persons.) Although these colonial immigrants represent less than 2 percent of those who have come here, their influence on our history is incalculable. Since the vast majority of Europeans were British—and most of those English—our language and culture became first an extension and then a variant of English/British culture. The Africans, subordinate rather than dominant, not only contributed greatly to the wealth of colonial America but were also in large part responsible for the greatest regional difference in our history—the distinct culture of the Old South, whose premier unreconstructed historian, Ulrich Bonnell Phillips, insisted that the "central theme of Southern history" was the presence of large numbers of blacks.[1]

The vexed question of where to draw the line and begin talking about American rather than British and African culture—along with the related questions about when one can begin talking about an American nationality and how long ethnicity persists—is not really soluble. The practice here will be to consider all who came Americans as soon as they land. (In at least some parts of nineteenth-century Norway people called those who *intended* to emigrate "Americans" even before they left.) I will often preface *American* with an adjective of nationality, capitalized, such as *Irish American,* and will eschew the hyphen except in such compounds as *Anglo-American* or *Afro-American.* This chapter will treat those, mostly English, who came in the seventeenth century; the next will treat the African immigrants; and the fourth, eighteenth-century Europeans, most of whom were not English.

In 1790, just after the colonial period ended, the United States took the first modern census. The takers of the first American census, which was (and remains) a constitutional requirement, enumerated 3,929,214 persons—some 3.1 million whites and 750,000 blacks. No attempt was made to count Indians, who were excluded from the constitutional mandate. Thus the white population had grown by natural increase—excess of births over deaths—some five times larger than its immigrant base, and the black population some two and a half times.

This rapid growth should not come as a surprise to us; it was certainly not a surprise to astute contemporary observers. From London, on the other side of the Atlantic, Adam Smith could see, as he noted in *Wealth of Nations* (1776):

> But though North America is not yet so rich as England, it is much more thriving, and advancing with much greater rapidity to the greater acquisition of riches. The most decisive mark of the prosperity of any country is the increase of the number of its inhabitants. In Great Britain . . . they are not supposed to double in less than five hundred years. In the British colonies in North America, it has been found that they double in twenty or twenty-five years. [Smith understated British growth but is on the mark for the colonies.] Nor in the present times is this increase principally owing to the continued importation of new inhabitants, but to the great multiplication of the species. . . . Labour is there so well rewarded that a numerous family of children, rather than being a burthen, is a source of opulence and prosperity to the parents. The labour of each child, before it can leave their house, is computed to be worth a hundred pounds clear gain to them. A young widow, with four or five young children, who, among the middling or inferior ranks in Europe, would have so little chance for a second husband, is there frequently courted as a sort of fortune. The value of children is the greatest of all encouragements to marriage. We cannot, therefore, wonder that the people in North America should marry very young. Notwithstanding the great increase oc-

casioned by such early marriages, there is a continual complaint of the scarcity of hands in North America.[2]

Smith's analysis, if somewhat hyperbolic, is of a piece with the far-sighted notions of the geographer-propagandist Richard Hakluyt, who, almost two centuries earlier, expressed compassion for his unemployed countrymen. Writing in the 1590s, before England had a transatlantic empire, Hakluyt felt that his England was "swarminge at this day with valiant youths rusting and hurtfull by lacke of employment." His solution was imperial: Send them abroad to plant colonies on American soil under the English flag.[3]

Virginia's Immigrants

But initially the English population in America did not thrive. The first English colony in North America, Walter Raleigh's Roanoke, vanished without trace, leaving only the record of the first European birth, the child Virginia Dare. The first successful attempt—if one can properly call it successful—at Jamestown was a demographic disaster, happily on a relatively small scale. Located, unwisely, in a malarial swamp, Jamestown had a staggering death rate from disease. At the end of the first year, roughly two-thirds of the original 108 colonists were dead. Two winters later came the notorious "starving time" when more than 80 percent of the immigrants died in six months, reducing the population from some five hundred to about sixty. During 1619–22 the speculators who ran the Virginia colony—which was a joint stock company organized for profits which never came—sent about 3,750 persons to supplement the 700 already there. By 1625 some three thousand had died, leaving only about 1,200 people. Edmund S. Morgan has argued that only after 1700 did Virginia births regularly exceed Virginia deaths.[4]

Naturally, colonial morale was not high. The surviving letters and other documents make grim reading. Always in search of a hero, textbook writers and others have contrasted the laziness and lack of enterprise of most of the colonists with the vigor and energy of Captain John Smith (1580?–1631). Smith, like Leif Eriksson, was a European with characteristics Americans can appreciate: a man of action rather than of contemplation, a hustler with a never-say-die spirit. Supply-siders, past and present, use Smith to tout the superiority of free enterprise over a planned society. The historian Karen Kupperman has recently shown, however, that nutrition rather than "shiftlessness" was crucial at Jamestown.[5] "[I]t is now clear," she writes,

> that the lethargy and apathy that so many observers spoke of was largely due to a combination of physical and psychological factors, stemming from nutritional defects, a complex of symptoms modern science calls "anorexia."

Compared to the wastage caused by disease, the great Indian "massacre" of 1623, in which 347 settlers lost their lives, was a minor demographic glitch.

Who were the immigrants who came to Virginia, and why did they go to such a place? To answer the second question first, they were victims of unscrupulous propaganda paid for by the London proprietors, which painted Virginia as a land of milk and honey. One pamphlet of 1624, for example, Richard Eburne's *A Plaine Pathway to Plantations,* published as the number of Virginia deaths peaked, noted that since few gentry would come, "those of a degree next unto Gentlemen, that is Yeoman and Yeoman-like men . . . that have in them some good knowledge and courage" would not only get land for themselves and their children "in perpetium forever" but held out to them the likelihood that they would, in time, "be advanced to places of preferment and government there." The proprietors were the spiritual ancestors of those modern promoters who sell the gullible Florida real estate that is underwater at high tide or advertise land in an Arizona development called "Shady Acres" on which the only possible shade comes from an occasional saguaro cactus. The historian A. L. Rowse has estimated that, up to 1624, the proprietors had lost some 200,000 pounds trying to make a go of Virginia. In that year the Crown revoked their charter.[6]

Thanks to the work of recent economic and social historians—beginning with that of Mildred Campbell and Abbot Smith in the 1940s—we can now give fairly precise answers to the question of who seventeenth-century migrants were.[7] The old notions, which were so flattering to the descendants of the FFVs (first families of Virginia) that the South in general and Virginia in particular had been settled by Cavaliers or the younger sons of the gentry, have been relegated to the historical scrap heap—although, to be sure, old legends never really die. The fact of the matter is that the only Englishman who came to North America in the seventeenth century as a settler—as opposed to a soldier or official—who was entitled to a coat of arms was Sir Richard Saltonstall (1586–1658) of Massachusetts. As the most recent student of seventeenth-century migration, David Cressy, has emphasized, the majority of those who came were male, young, single, destitute, and ignorant. What is becoming clearer and clearer is that migration to America was exceptional in seventeenth-century England *only* for the distance of the migration. Vast numbers of English people were migrating to and between towns, and above all to London. London was a terribly unhealthy place in the seventeenth and early eighteenth centuries, although not as unhealthy as early Virginia. The demographer E. A. Wrigley has calculated that for the period 1650–1750, London deaths significantly exceeded London births, and only a net migration of some eight hundred thousand persons into London during that period explains the metropolis's sustained growth.[8] And added to these numbers should be the seventeenth-century migration of some two hundred thousand persons to Ireland, England's closest plantation. Those who came to America, then, were but a minor fraction of England's migrants. Even more important, the migration experience—which included sending village children out to service in neighboring communities—must have been familiar to almost every English person. This helps to explain why it was possible to recruit so many English for the colonies. Even if we had no other information about the English economy in this period—and of course we do—the levels of migration alone would indicate great amounts of economic dislocation. That England was undergoing considerable economic growth in the colonial era did not mean that there was not also widespread poverty and economic misery: The two often go hand in hand, as they did, for example, in both the late-nineteenth-century United States and the Soviet Union during the 1920s and 1930s.

Indentured Immigrants

Since the migration to America was expensive, many immigrants were forced—sometimes literally, sometimes by dint of economic circumstances—to indenture themselves to labor for some planter or company, usually for a term ranging from four to seven years. (Seven years had long been the traditional period for apprenticeships.) The person who bought the indentured servant paid the ship captain or his agent the cost of the passage plus a profit. Obviously, if a servant died or successfully ran away before his or her term was up, the owner lost. But it was not as speculative as it might seem: In Virginia, for example, the owner got a grant of crown land (headright) for each immigrant purchased. Those servants who completed their terms of service got freedom dues, theoretically, at least, enforceable at law. In some cases this included land, tools, apparel, and provisions. In later-seventeenth-century Virginia, for example, the custom of the country called for the master to provide newly freed servants with three barrels of grain and garments, usually a suit of clothes.

Other immigrants were convicts. Governor Thomas Dale of Virginia, citing the Spanish example, had asked as early as 1611 that "all offenders out of the common goals condemned to die" be sent to him instead. It has been estimated that Britain sent some fifty thousand convicts to America in the eighteenth century and a smaller but indeterminate number in the seventeenth. (After the American Revolution the practice continued, with Australia as the new dumping ground.) Not all were gallows birds nor were most what David Souden's racily titled essay, "Rogues, Whores and Vagabonds," would indicate, but more than a few of both convict and indentured immigrants clearly fitted that description.[9]

Perhaps more representative is the group that Mildred Campbell studied, listed in two large volumes discovered in the archives of Bristol, a chief western British port of embarkation in the Age of Sail. The volumes record the names of, and sometimes other information about, 4,136 men and women who embarked as "Servants to the Foreign Plantations" from the port in the years 1654 to 1661. Destinations were listed as Virginia, New England, and the West Indies. Of the 1,600 males whose status can be determined, about 63 percent were listed as being from the agricultural classes, and of these almost half were either yeomen—small landowners—or the sons of yeomen fallen on hard times. Campbell also examined, but did not quantify, the sessions records of Middlesex County (London), in which she discovered that justices, many of whom had colonial investments, were particularly likely to send to the colonies those who had skills, such as carpentry, that would be useful in a new land. The Bristol records do not usually include anything about the status of women apprentices, but both boys and girls were sometimes sent to the colonies against their will. As early as 1618 a man had been sentenced to death for counterfeiting a document with the royal seal and using it to kidnap "rich yeomen daughters . . . to serve his Majesty for breeders in Virginia." But most of the kidnappings, surely, were of homeless and neglected children who ran loose in most urban centers. There were, in the American South in general and the Chesapeake colonies in particular, never enough "breeders" either among the immigrants or, for a long time, in the population at large. Careful studies of the fragmentary population data that have survived suggest that throughout the seventeenth century in Virginia, males outnumbered females by something like three to four and a half to one.

Some indentured servants, after working out their terms, did manage to get land and begin the process of upward social mobility promised in the pamphlets of Eburne and his like. As early as 1629 seven members of the Virginia legislature were former indentured servants, as were Charles Thomson (1729–1824), secretary of the Continental Congress, and Matthew Lyon (1749–1822), a congressman from Vermont and Kentucky. Others, like Benjamin Franklin's maternal grandmother, eventually married the men who bought them. Despite much research we cannot even begin to generalize about what happened to indentured servants. Clearly, many died while still servants; given the sex ratio, many must have died later but without issue; and surely, most of those who did get and keep land became small landowners—the poor whites who were so numerous in southern American history.

In the first half of the seventeenth century, as we have seen, most immigrants—indentured or not—soon died. By the second half of the century, the majority seem to have survived their period of indenture. Although most who came from England surely had some expectation of what we would call upward social mobility, many were never able to rise out of the servant class. If an individual did get land, it was likely to be marginal, in an area where the danger from Indians was great and/or it was not possible to transport tobacco—the only sure cash crop—to the tidewater wharves for transshipment to England. Instead of becoming their own masters, many became tenant farmers on land owned by the person who had owned their indenture. For these reasons, Edmund S. Morgan has concluded, Virginia's freemen were "an unruly and discontented lot," and he suggests that the problems, actual and anticipated, with this growing and dangerous class, were one of the reasons that Virginians switched from using unfree European immigrants to using unfree Africans toward the end of the century. And, of course, Virginia's example was followed by the rest of the South.

The indenture system continued throughout the colonial period and, like any social institution, changed over time. Bernard Bailyn has subjected a register of 9,364 emigrants from Britain in the period from December 1773 to March 1776 to exhaustive and imaginative analysis. Nearly half—4,472 (47.8 percent)—were indentured servants, or redemptioners. (This somewhat overstates their incidence, since, as Bailyn notes, those who paid for "cabin" or better accommodations were often not enumerated. The same would be true during the peak immigration years—first-class passengers usually did not have to go through Ellis Island.) Four out of five of the immigrants in this group who went to Maryland, Pennsylvania, and Virginia, were indentured servants, showing the colonies where at that time white labor was wanted the most. By this time many of the indentured servants were highly skilled and could negotiate conditions. In what he calls an extreme example, Bailyn cites correspondence from a London merchant to a Baltimore firm that wanted a foreman for its rope walk—an establishment for making ship's rigging. The proposed foreman was forty years old and had twenty-four years experience in London: he would sign an indenture of two years if provided with passage, five hundred pounds of beef and pork, five hundred pounds of flour, a house for himself and his wife, and forty pounds a year.[10]

More commonly, the indentee had little or no bargaining power and was often placed on public sale and inspected much as Negro slaves had come to be examined. A contemporary and possibly spurious poem—spurious in that the writer may not have been a former servant—told in rhyme of the indignity of being made to:

> *. . . walk, to see we were compleat;*
> *Some view'd our teeth, to see if they were good,*
> *Or fit to chew our hard and homely food*
> *If any like our look, our limbs, our trade,*
> *The Captain then a good advantage made.*

Similarly, a German redemptioner in Pennsylvania (they will be discussed in chapter 4) remembered years later:[11]

> We had to strip naked, so that the prospective purchasers could see that we had perfectly developed and healthy bodies. After the purchaser had made a selection, he asked: "How much is this boy or this girl?"

Naturally some resented their lot and ran away, perhaps one in thirty. The pages of eighteenth-century colonial newspapers were dotted with ads for runaways, both white indentured servants and black slaves, who on occasion escaped together. From the detailed descriptions in the ads, it is clear that many owners were aware of what their servants and slaves looked like. (That always-methodical Virginia planter, George Washington, kept careful descriptions of his servants and slaves, and of what clothes they were wearing.) An ad in the *Pennsylvania Gazette* for September 28, 1752 read:

> Ran away on the 18th inst. at night from on board the ship Friendship, Hugh Wright, Commander, now lying at William Allen Esquire's wharf, James Dowdall, a servant man, a laborer, lately come in, but has been in many parts of this continent before; he is about 5 ft. 4 inches high, has short hair, but neither cap nor hat: Has on a blue frieze coat and jacket, a check shirt, leather breeches, and blue yarn hose: speaks as a native of this Province; he is at present greatly infected with the itch, and not able to travel far. Whoever secures the said James Dowdall so that he may be brought to the said Commander, or to Wallace and Bryan on Market Street Wharf, shall have 40s. reward and reasonable charges paid by
>
> WALLACE AND BRYAN

On the eve of the revolution, William Crean advertised in the *Pennsylvania Ledger* for April 22, 1775:

> RANAWAY from the subscriber . . . An English servant MAN, named *John Shaw,* of a swarthy complexion, has a scar on his right cheek: Had on an old felt hat, an old blue coat, an old light coloured jacket, red flannel double breasted jacket, old white breeches with holes in the knees, old blue rig and firr stockings, old shoes with shoe packs, a pair of odd buckles, and an old coarse shirt; he is about five feet six inches high, about twenty years of age. . . . Four dollars reward, and all reasonable charges paid. . . .

Similarly, William Bordley and Woolman Gibson "the 3d." ran an ad in Dunlap's *Maryland Gazette* on September 12 of the same year:

> Ran away last night, from the subscribers, an English servant man named JOHN SCOTT, a square well made fellow, about five feet four inches high, a good complexion, something sunburnt, wears his own brown straight hair; took with him a blue cloth coat with a red plush cape, a good scarlet knit jacket with callimanco back, old buckskin breeches, a pair of double channelled pumps capped at each toe, but may change his cloaths, as he stole a very good dark claret coloured coat with yellow buttons; he is a very artful scoun-

> drel. Also a white woman [who is nameless in the ad] somewhat taller than the fellow, of a very dark complex, a ring-worm on her upper lip, a small scar on one cheek; she has left behind a mulatto bastard, for having which she is bound to appear next court; it is probable they will pass for man and wife. TWENTY SHILLINGS Reward for each if taken in the county, and FORTY SHILLINGS each if taken fifty miles from home. . . .

Not surprisingly, runaways who were caught had additional time added to their indentures. Women servants who became pregnant often had as much as two years added to their term of service, which helps explain why so many, relatively, of the female runaways sought are described as either pregnant or as having borne a bastard child.

Richard Graham bought space in the *Virginia Gazette* of July 21, 1775 to offer "Twenty Dollars Reward" for:

> JOHN ECTON DUCRECT, a native of *Berne* in *Switzerland,* who speaks very good *French* and tolerable *English* and *Italian.* He is about 5 feet 9 inches high, pitted with the smallpox, and very swarthy, almost as dark as a mulatto; wears his own hair, with a false tail, and is generally powdered, being a barber by trade. I have heard, however, since he went off, that he intended to cut off his hair and wear a wig. He has been used to travel with gentlemen, and will probably try to get into employ that way, or with some of the barbers in *Williamsburg,* as he was seen at doctor Todd's tavern, on the way there, the 22d ult. He took with him a suit of brown and a suit of green clothes, the brown pretty much wore, and has some rents on the back of the coat, the green almost new, a pair of new buckskin breeches, trousers that button down the legs, some white thread stockings and white shirts, a powder bag and some shaving materials, a prayer-book in *French,* and some old commissions for officers in the *Swiss* militia, by which he will probably try to pass. One of his testicles was swelled, which he says was occasioned by a kick he got on board the *Justitia,* capt. *Kidd* [not *the* Captain Kidd; he was hanged in 1701], the ship he came in. Whoever secures the said convict so that I can get him again, shall be paid the above sum. . . .

How many servants ran away? How many were caught? No one can even begin to estimate the number but, obviously, advertising for a runaway servant was no more unusual in the mid-eighteenth century than advertising for a lost dog or trying to sell a used car is today. But clearly *one* of the reasons that the institution of indentured service was eventually replaced by that of African slavery was that, by their color, Africans were identifiable as being, presumably, someone's property.

Why, one might ask, given the terrible mortality rates in colonial Virginia, did the proprietors and the British government continue to allow people to come and, in fact, encourage immigration? The answer is quite simple: Because it was profitable. The production of tobacco became the *raison d'être* for the existence of Virginia (and Maryland and the Chesapeake generally) and the government quickly began to tax it quite heavily. Tobacco grown in the Spanish Caribbean was already being marketed in England in 1617, when the first shipment of tobacco was sent from Jamestown. It did not command the high prices of the Spanish stuff, which at eighteen shillings a pound was a real luxury item, but the three shillings realized by each pound were the light at the end of the tunnel as far as the long-term profitability of the Chesapeake region was concerned. (Although centuries later, tobacco sellers made Sir Walter Raleigh a noted brand name, it was John Rolfe, the husband of Pocahontas, who brought the successful West Indian seeds to Virginia.) And while tobacco brought wealth to some Virginians, it was the king who, as Edmund Morgan puts it, "stood foremost" of those who profited from the sale of the "noxious weed," as James I called it. And, with the same kind of hypocrisy that allows contemporary governments to require health warnings from a surgeon general or a health minister to appear on tobacco products while at the same time they subsidize and tax tobacco, James issued the first antismoking pamphlet while his government encouraged and taxed its growing in America and its import into England. During the six years between 1697 and 1702, for example, economic historians have estimated that nearly one hundred thousand tons of tobacco were imported into England from Virginia and Maryland—more than half of it was reexported—and that the Crown received nearly two million pounds in duties.

After about the middle of the seventeenth century, when mortality began to decline somewhat, a new kind of immigrant began to appear: Reasonably well-to-do persons and their sons came in small but significant numbers in order to profit from the tobacco trade. It is to this period that most of the great families of colonial and antebellum Virginia can be traced: the Byrds, the Carters, and others. It was they, rather than the survivors of the early decades, who became the planter aristocracy, although, to be sure, many of the new men on the make were shrewd enough to marry the widows of earlier colonists who had either large amounts of land or headrights to land. By the time of the American Revolution, Thomas Jefferson could complain that Virginia planters were a species of property attached (by their debts) to certain Scottish mercantile houses—and he could have pointed out (but didn't) that the Crown usually made, in taxes and duties, more from the labor of an indentured servant (or slave) than either the worker or his master, and perhaps more than both combined. Yet, when all is said and done, the Virginia planters, large and small, built a good life for their class largely on the undercompensated labor of unfree workers. That some of those workers and, more often, their descendants also eventually made it, is one of the glories of American history, but it should never be allowed to obscure the fact that for the Crown, the absentee investors in the American market, and the planter class itself, this result was only an incidental and largely unexpected by-product of a desperate search for profit.

Maryland's Immigrants

A somewhat different pattern of living developed in adjacent Maryland, where the first settlements had been overflow from Virginia.[12] The initial impetus for what became Maryland stemmed from the life and experience of George Calvert, first Lord Baltimore (1580?–1632), a onetime secretary of state to James I. Calvert was different from most of the nobles who had fingers in the American pie in at least two ways. First of all, after 1625, he was a public Roman Catholic who wanted his colonies to be a haven for well-to-do Catholic gentlemen. Second, he was no absentee promoter; he went to the New World as he had previously gone to Ireland to participate in English colonization there. His first attempt at a New World colony was in Newfoundland. His high hopes for a colony in that place did not survive one winter's residence. As he ruefully wrote to Charles I:

> [F]rom the middst of October to the middst of May there is a sad face of wynter upon all this land, both sea and land so frozen for the greatest part of the tyme as they are not penetrable, no plant or vegetable thing appearing until it be about the beginning of May, nor fish in the sea[;] besides the ayre so intolerable cold as it is hardly to be endured.

The point of Calvert's tale of woe was to plead for a grant of land in a more temperate climate, as he asked to serve his sovereign "by planting of Tobacco."

Calvert actually went to Virginia with some forty persons, mostly Catholics, survivors of his Avalon colony in Newfoundland. The staunchly Protestant Virginians, although they desperately needed settlers and capital, wanted no Catholics in their midst, and Calvert returned to England to seek a royal charter for a colony north (but not too far north!) of Virginia. Shortly after Calvert's death, his son Cecilius (1605?–75) received such a charter, giving the Calvert family broad proprietary powers over the proposed colony.

The Calvert colonization of Maryland began in 1634. It was a highly stratified migration. Of the first group of nearly 150 immigrants, a double handful were gentlemen, the rest their servants. To widen the gulf between them, the gentry were all English Catholics (including two Jesuit priests), while the servants were all English Protestants. Population grew very slowly in a heavily male population (only after about 1700 would there be natural increase), reaching four hundred persons about 1640 and then slumping to less than half of that when political troubles led to emigration to other colonies. A boom in tobacco prices then brought renewed immigration, some in family groups of "middling" status, although servants still predominated.

By the middle decades of the seventeenth century, Maryland, as opposed to Virginia, was what Russell Menard has called a "good poor man's country." He has painstakingly traced what happened to the 137 servants who came into the colony between 1648 and 1652. Just more than half—seventy-two, or 53 percent—can be found in the colony's records as freemen. Sixteen of these either died quickly after becoming freemen or disappeared from the records. But probably forty-two of the original group became landowners, about 30 percent. Another six or seven, some 5 percent, although they remained laborers, married, established households, and participated in local government. This was a greater degree of social mobility than existed in Virginia, but we should note that it was a minority phenomenon. Maryland was a good country for *some* poor men and women, but not for most. And, statistically speaking, those enjoying upward social mobility did not enjoy it long. The typical servant who survived to become a landowner married in his late twenties–early thirties and died in his early forties.

Schoolbooks, anxious to promote brotherhood, usually cite the so-called toleration of Catholicism in Maryland as if it had been a forerunner of contemporary ecumenicism and pluralism. Nothing could be further from the truth. Even the Catholic gentry was bitterly divided over questions of religious policy. The Calverts, wily political animals who managed to thrive in a country gone Protestant, wanted private or household Catholic worship; the Jesuits and others of the gentry wanted it made public and perhaps exclusive. Much political unrest—not all of it religiously motivated—culminated in an insurgency led by a Protestant association. The proprietary charter was revoked in 1691, although the Calverts kept their property rights; the Church of England was established in 1692; and in 1695 the capital of the colony was moved from Catholic St. Mary's to Protestant Annapolis. Early in the next century the third Lord Baltimore changed his family religion to Anglican and shortly thereafter (1715), the fourth Lord Baltimore, Charles Calvert (1699–1751) had the proprietorship restored to him. As would happen so many times in American history, the religious culture of the later immigrants triumphed over that of the first settlers.

New England's Immigrants

A very different pattern of immigration and settlement developed in New England—a pattern shaped by both the economic and geographic realities of northeastern American life and by the migration decisions of the people who came.[13] From the time of Henry Adams, who wrote about them just after the American Civil War, the archetypes of Cavalier and Puritan have been used to differentiate between those who settled in Virginia and those who came to New England—the first inspired by gain or politics, the second by religion. As we have seen, the Cavalier immigrants simply did not exist. And while there were, to be sure, plenty of Puritans who came to New England, and religious colonization was an important factor, they and it were not the typical mode. Despite what Americans are assured by schoolbooks, politicians, and other unreliable guides to history, the Pilgrims and the Puritans were a distinct minority of those who came to New England in the seventeenth century.

Scholars have been telling us for some time now that the image of the American Puritans being harried out of the land by wicked churchmen such as Archbishop William Laud (1573–1645) was a self-serving myth first popularized by third-generation American Puritans such as Cotton Mather (1663–1728). Even the Puritan minority among the seventeenth-century migrants to New England were not, generally, refugees. Most of those who came were actively recruited by promoters who promised them that they would have a better life in America.

What does seem to distinguish the New England migrants from others who came to America in the seventeenth century is that they were organized into family groups, were possessed of a great deal of agricultural and craft skills, and most seemed to have enjoyed some degree of economic security in England. In addition, there was an extraordinary number of educated men among the leadership of this migration, many of whose wives and daughters, although lacking in formal schooling were well educated too. Further, most of the organizers of the migration were not absentee investors but came as immigrants to the New World themselves. Partially because of the education and consciousness of history of so many early New England leaders, we have a very full and relatively detailed record of what happened there, which is one reason that the seventeenth-century immigrants to New England have been more closely examined than perhaps any other similarly sized community in human history.

The fabled Pilgrim fathers (and mothers) who arrived off Cape Cod in the *Mayflower* on November 9, 1620, were a minority even of the group of just over a hundred colonists on board. Several were indentured servants, while the immortal Miles Standish was a nonreligious hired gun. Even among seventeenth-century religious settlers to New England, the Pilgrims were a minority, as they were radical Protestants who wanted to separate from the Church of England and had first emigrated to Leiden, in Holland, to do just that. They represent a special kind of migration experience that would be repeated again and again in American history: the small religious group migrating as a body and bringing with it religious leaders and special notions about how it wanted to live in the New World. Like most utopian attempts, Plymouth did not last too long, being absorbed into Massachusetts before the end of the century, and its historical significance is much, much smaller than its place in the American imagination. The Pilgrims were the kind of people American mythmakers, beginning with Henry Wadsworth Longfellow's poem *The Courtship of Miles Standish* (1858), liked to imagine we were descended from. Plymouth also occasioned the writing of the first great book written in what is now the United States: William Bradford's *History of Plimmoth Plantation,* which was finished in 1651 but not published until 1856. Bradford, Plymouth's second governor, presided only over a ministate, but his memoir-history is a classic, especially when compared to the flatulent memoir-accounts ground out by today's superpoliticians. In any event, just before it was absorbed by Massachusetts in 1691, Plymouth had some seven thousand persons, while Massachusetts had about fifty thousand.

The quantum leap in Massachusetts' population was the so-called Great Migration of the 1630s when some twenty-one thousand persons migrated to New England. This group was less than a third of the approximately sixty-nine thousand Britons who crossed the Atlantic in that decade. History has called this migration Puritan, and certainly its leaders, such as John Winthrop (1588–1649) and his son, John (1604–76), who would become governors of Massachusetts and Connecticut, respectively, fitted that description. David Cressy argues that "ordinary workmen with moderate to low social status formed the core of the migration to New England." His analysis of passenger lists of eleven vessels of the 1630s, which carried 996 passengers to New England, bears this out. These included 286 adult men, 242 of whom were identified with occupations; 151 adult women; 176 boys under eighteen; 169 girls under eighteen; 86 males and 80 females whose ages are unspecified; and 18 persons unidentified as to either age or gender. For those identified by gender, the group is almost 58 percent male. The total is nearly 5 percent of the Great Migration, but there is no way of knowing whether the sample is representative. The occupations were grouped as follows:

Professionals and clergy	5	2%	(Only one minister.)
Yeomen	5	2%	(Underrepresented; in England they were about 10% of heads of households.)
Husbandmen	49	17%	(Less prosperous than yeomen but not poor.)
Laborers	10	3%	(Clearly underrepresented, with other agrarian occupations, yeomen and husbandmen, totaling only 22% here)
Artisans	123	43%	(Includes one merchant. While many, such as the 16 carpenters, were clearly needed, the 50 from the textile trades represented England's depressed industry. Most of them, like today's migrant auto workers, wound up doing something else.)
Servants	50	17%	(See below.)
Unknown	44	15%	(Almost certainly these are at the lower end of the social scale.)

The list clearly understates the number of servants, as many of them were boys and girls under eighteen, who were, Professor Cressy estimates, at least a quarter of the immigrants. The servants were young, like most migrants. For one group, Cressy has calculated the mean age of male servants at nineteen, with a range of twelve to thirty years, while female servants had a mean age of twenty-three, with a range of eleven to sixty-five years. Most served for three or four years, but some had terms more than twice that long. While most servants worked in households or for small enterprises, others worked at industrial employments. A 1653 inventory of the Lynn, Massachusetts, ironworks for example, shows thirty-five Scots servants valued at ten pounds each and two English boys with four and six years to serve valued at eight and thirteen pounds. Court records show many disputes between servants and masters over the terms of service, with most, not surprisingly, settled in favor of the latter.

Servants who completed their terms became regular if unusually humble members of the community. While there was not as much upward mobility as in the Chesapeake, it did exist. The Boston town records for 1638 relate one success story:

> Thomas Pettit, having served with our brother Oliver Mellows this three and a half [years], shall have a house plot granted unto him.

But, as Cressy tells us, most ex-servants lived obscure lives as wage earners or small landowners, and many probably migrated elsewhere. These modest competences were probably more than most would have had if they had stayed in Britain.

Why did all these people—masters, artisans, and servants—come to New England? Most historians insist, some quite stridently, that it was a Puritan migration, pure and simple. Samuel Eliot Morison, for example, who early in his career had pointed out that "some came to catch fish," had by the end of it hardened his position to:[14]

> Puritanism was responsible for the settlement of New England. . . . The New England people, almost to a man, were English and Puritan.

Clearly, religious motives were important, but so were others. David Cressy has listed a dozen factors, sacred and secular, that shaped migrants' decisions, drawn largely from his close readings of letters and autobiographies. They are, not necessarily in order of importance: "attractive information about New England; anxiety about the future in England; desire to ease spiritual malaise; desire to enjoy purity of religion . . .; escape from religious persecution; desire to advance the gospel among Indians; giddy humour, restlessness or adventure; desire to join earlier migrants; flight from personal adversity, career stagnation or family disruption; desire for economic betterment . . . land and opportunity in New England; influence of ministers, family and friends; influence of recruiters."

It would be foolish either to try to deny the influence of Puritans and Puritanism in the settlement of New England or to attempt to quantify that influence. Religion was clearly more important there than in either Virginia or on the Chesapeake, although it played a part in those colonies too. The seventeenth century was a religious age, and the hand of God was seen regularly where today we would assign other causes. Let us listen to one piece of testimony, that of Roger Clap, a twenty-one-year-old servant in Devonshire, who was the youngest of five sons and had few prospects in England. As he explained it in his autobiography, written half a century later for his children—he fathered ten sons and four daughters—he had never even heard of New England until he learned that a nearby clergyman and some followers planned to go.

> My master asked me whether I would go. I told him, were I not engaged unto him I would willingly go. He answered me, that should be no hindrance. I might go for him, or for myself, which I would. I then wrote to my father, who lived about 12 miles off, to entreat his leave to go to New England, who was so much displeased at first that he wrote me no answer, but told my brethren that I should not go. Having no answer, I went and made my request to him, and God so inclined his heart that he never said me nay.

Although the motives for migrating were mixed, the process of migration in the 1630s and after was becoming regularized, at least to a degree. It was vital to have the right kinds and quantities of supplies. As one of the first ministers in New England, Francis Higginson, wrote to prospective immigrants back home that they must bring everything with them, "for when you are once parted with England you shall meet neither with taverns nor alehouses, nor butchers, nor grocers, nor apothecaries' shops to help what things you need, in the midst of the great ocean nor when you come to land." In time, of course, such shops did exist in Boston and other New England towns, but provisioning for the journey across the "great ocean" was a continuing problem, especially so in the seventeenth century.

By the 1630s there were printed lists—one of them compiled by Higginson—of what prudent immigrants should bring with them. These early immigrants' guides, with titles like *Proportion of Provisions Needful for Such as Intend to Plant Themselves in New England for One Whole Yeare,* listed dozens of items:

> Food—Meal, one hogshead . . . Beef, one hundredweight . . . Cheese, half a hundred . . . Salt to save fish, half a hogshead; Clothing—Shoes, six pair . . . Leather to mend shoes, four pair . . . shirts, six; Tools [for a family of four or five]—Axes 3. One broad axe and two felling axes . . . Hoes 3. One broad of nine inches, and two narrow of five or six inches; Hardware—Nails of all sorts . . . Hooks and twists for doors; Household implements [for a family of six]—One iron pot, One great copper kettle . . . A spit . . . Two skillets, platters, dishes and spoons of wood.

Obviously, it was expensive to transport a family across the Atlantic, an expense beyond the means of most British families. The poor people who came to New England came as servants, and it is clear that many families went into debt to finance their migration. It is equally clear that many, if not most of those who did so, acted on the belief that they would be improving their standard of living: Those who had owned land in England could expect to own more in New England; husbandmen could become yeomen, and even servants, as we have seen, could at least hope for a significant increase in their status.

Even properly provisioned, however, it was no simple matter to leave England. There were layers of bureaucracy to surmount and yards of red tape to be cut. To leave Britain's western harbors, like Bristol, immigrants were supposed to have a license, or passport. (Originally this was a document allowing the holder to "pass the port," not one for identification and good treatment abroad.) After 1634 emigrants were also supposed to take a loyalty oath and have certificates of good character from a local minister. As Richard Mather, the founder of that clan in America, described his May 1635 leavetaking:

> This day there came aboard the ship two of the searchers, and viewed a list of our names, ministered the oath of Allegiance to all [of] full age, viewed our certificates from the ministers in the parishes from whence we came, approved well thereof, and gave us tickets, that is licenses under their hands and seals to pass the seas, and cleared the ship, and so we departed.

But, before one could depart, one had to find a ship, which was a matter of searching and negotiation. There were no regularly scheduled transatlantic sailings until after the Napoleonic Wars, and such crossings would not become common until the second half of the nineteenth century. Once emigrants had reached an agreement with a captain—and most arrangements were made by groups—and arrived at a port, a delay of weeks was common and one of months not unusual. Although fares varied, five pounds per adult was not uncommon. The Massachusetts Bay Company tried to standardize tariffs at that rate, with a two-thirds' fare for children aged eight to twelve, half for those four to eight, one-third for infants under four, with "sucking children not to be reckoned." Transporting a horse or a cow cost at least twice as much as an adult fare. To get a notion of relative cost, a husbandman might net fifteen or twenty pounds a year, so the fare for a family of six would be quite expensive. And the fare did not include provisions, which (as we have seen) had to be supplied by each passenger or family.

Although the thought of a passage overseas was frightening—and clearly dissuaded many would-be immigrants—seventeenth-century crossings were remarkably safe. Of one hundred and ninety-eight known voyages to New England in the 1630s, only one resulted in shipwreck, and even then not all were drowned. Propagandists like William Wood might assure those thinking about going to New England that "a ship at sea may well be compared to a cradle rocked by a careful mother's hand, which though it is moved up and down is not in danger of falling," but that probably did no more to ease the qualms of the nervous than do statistics about how few airlines crash today. In fact, there were many more shipwrecks in coastal waters than on the high seas, and the return trip to England was more dangerous than the westward voyage because of British weather, as stormy then as now.

Nevertheless, the ships themselves seem incredibly tiny. Some were only forty feet in length—shorter than a city bus!—although the average in the seventeenth century was probably twice that. The memoirist John Josselyn came over on a large ship that carried 164 passengers. They and the crew shared less than fifteen hundred square feet of deck space—less than ten square feet per person—and a hold barely high enough for a man to stand erect in.

Seasickness was, of course, endemic. The small vessels pitched and rolled alarmingly. More serious were scurvy and other forms of malnutrition, dysentery, and, of course, any contagious disease that broke out on the crowded little ships. Although precise mortality rates are not available for the seventeenth and eighteenth centuries, or for the first half of the nineteenth, most authorities agree that a smaller percentage of seventeenth-century immigrants died than did those of the next 150 years.

Mortality was, of course, related to the length of the voyage. Although some seventeenth-century immigrants made five-week passages, these were unusually rapid. Eight to twelve weeks–fifty-six to eighty-four days—was in the normal range, while slow or unlucky trips actually took as much as eighteen, twenty, or in one extreme case, twenty-six weeks—half a year!

Most trips were made in summer, and few were made between September and the end of February. There were both northern and southern routes—the former quicker but more hazardous; the latter, in hot weather, much more uncomfortable. Many vessels did not make direct crossings but made intermediate landfalls as far north as the Faeroes or as far south as the Canaries, both to pick up supplies and to do a little business.

Most seamen were at best indifferent to their passengers and at worst, downright hostile. William Bradford complained about "a proud and very profane" sailor on the *Mayflower* who delighted in cursing and tormenting seasick passengers; when, as it happened, he was the first on board to die, the future governor saw it as "a special work of God's providence." On two seventeenth-century voyages that we know about, superstitious sailors simply threw overboard women whom they considered to be witches.

For those who made a safe passage, claims for the special providence of God could be made. David Cressy has speculated that many immigrants on ships dominated by Puritan ministers may well have had their religious beliefs strengthened by their shipboard experiences, which included daily services and other religious exhortations. The Atlantic passage was part of the common experience of all immigrants to America in the Age of Sail, except for those few who crossed the wider but less stormy Pacific. The bad experiences endured by some were undoubtedly one more factor that persuaded them to stay rather than undergo them again. Others, to be sure, seemed to relish it: there are few things so exhilarating, in retrospect, as past dangers surmounted.

Some immigrants to New England, however, even in the seventeenth century, risked the more dangerous return trip and became returned immigrants. As early as 1630 some two hundred immigrants decided to return to England, and many if not most immigrant ships carried some returnees. John Winthrop's journal is dotted with references to return migrants, especially those who were shipwrecked or, in one instance, taken as slaves by Algerian pirates. In good seventeenth-century fashion, Winthrop saw this as a mark of God's displeasure with those "who have spoken ill of this country, and so discouraged the hearts of his people." The colony also, from time to time, deported dissenters such as Anne Hutchinson and Roger Williams, either to neighboring colonies or even to England, and some orphans and other dependent persons were also sent back across the Atlantic. Others returned to England planning to come back, and many actually did. And, finally, as the English Revolution progressed into civil war, perhaps a hundred prominent New England leaders, including college graduates, returned to their native land thinking it might be on the road to redemption. Since the records on return migration are less detailed than those on migration, it is difficult to be exact, but David Cressy believes that as many as one in six of the seventeenth-century migrants may have returned to England, either permanently or temporarily. If this is so, it involved some 3,500 persons who might be styled sojourners rather than settlers.

Thus was the initial settlement of New England accomplished, by the combined efforts of literally hundreds of small enterprises or expeditions. By 1700 there were perhaps ninety thousand Europeans and descendants of Europeans living there, more than four times the twenty-two thousand in 1650. This increase was not chiefly due to immigration: David Galenson has calculated that there was a negative net migration of some six thousand persons in the second half of the seventeenth century, so the gain was due to natural increase—the excess of births over deaths.[15] And the rate—doubling in a little less than twenty-five years—seems to have been just what Adam Smith said it was. Until the coming of the Irish in the nineteenth century, New England's population growth—it reached one and a quarter million in 1800—would be due to such increase. Many, if not most, of those who left went not back to England but to other parts of the New World, and, after the Revolution, they migrated almost exclusively within the United States. Although Benjamin Franklin was hardly typical, his migration to Philadelphia in 1723 is symbolic both of the Yankee desire to succeed and the narrowing of opportunity within New England.

By 1700 the overwhelming majority of New Englanders were at least second-generation Americans, and New England had become—and would remain for more than a century—the most homogeneous of Britain's New World colonies and, later, the United States. This was not just because many of New England's leaders wanted it that way, although the increasingly hostile attitudes of New Englanders to outsiders were quite pronounced. As early as 1718 fears had been expressed in Boston that "these confounded Irish will eat us all up," and at the end of the next decade a Boston mob "arose to prevent the landing of the Irish."[16] And these Irish, to be sure, were not Catholics but the Protestant Scotch Irish, whose coming will be discussed in chapter 4. Immigrants, in whatever century, tend to go where economic opportunity seems to be, and even if New England's welcome had been more friendly, most eighteenth-century immigrants would have gone elsewhere.

New England's ninety thousand in 1700 was a little less than a third of the nearly three hundred thousand Europeans and their descendants in Britain's New World colonies. Some thirty-five thousand lived in the Caribbean and are beyond our purview, as are the three thousand in Newfoundland. Nearly another third lived in the southern colonies, most of them on the Chesapeake, and in Virginia, while about a hundred thousand lived in the middle colonies.

The eighteenth century would see much heavier immigration, most of it non-English, chiefly Scotch Irish and Germans. Most of this immigration went where economic opportunity was perceived—the middle colonies—and will be discussed, as noted, in chapter 4. But by the beginning of the eighteenth century, if not before, the character of what was to become the United States was pretty well set. It would be English and Protestant. And it would rapidly adopt migration myths that had little relation to what had been seventeenth-century realities. In 1775, for example, the Second Continental Congress, as part of its justification for taking up arms, insisted that Americans were all the descendants of persons who had left the Old World "to seek on these shores a residence for civil and religious freedom." That could have described only a minor fraction of the Europeans who had come and, of course, not one of the three hundred thousand immigrants from Africa.

This 1591 engraving shows a French explorer in Florida receiving homage from Indians. The coat of arms shown on the pillar is that of the Bourbons, then kings of France. (*Florida State Archives*)

One of the earliest depictions of Manhattan and New York harbor by a European artist from *Beschrijvinghe Van Virginia, Nieuw Nederlandt*, Amsterdam, 1651. (*I. N. Phelps Stokes Collection, Miriam and Ira D. Wallach Division of Art, Prints and Photographs, New York Public Library, Astor, Lenox and Tilden Foundations*)

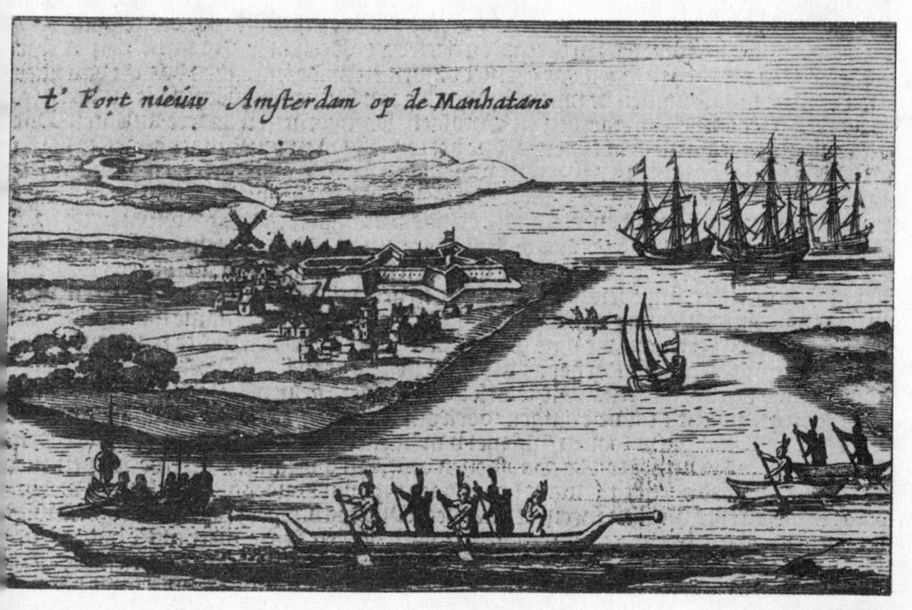

René Robert Cavelier, Sieur de la Salle (1643-87), a major architect of French settlement in North America. (*Library of Congress*)

Father Junipero Serra (1713-84), the Spanish Franciscan who founded the chain of missions that were the nucleu of European settlement in California. (*United States Postal Service*)

A romantic nineteenth-century depiction of the first landing of the Pilgrims. (*Library of Congress*)

John Winthrop (1588–1649), governor of Massachusetts Bay Colony. (*Library of Congress*)

George Calvert, Lord Baltimore (1580?–1632), the founder of English settlement in Maryland, from a nineteenth-century engraving based on a family portrait. (*Library of Congress*)

This magnificent Benin bronze, *Oba with Two Attendants*, probably dates from about 1590. (*University Museum, University of Pennsylvania*)

This illustration first appeared in the London printing of Phillis Wheatley's *Poems on Various Subjects, Religious and Moral*, published in 1773. It may or may not be an accurate likeness. (*Rare Book Collection, UNC Library, Chapel Hill*)

"A Slave Auction at the South," *Harper's Weekly*, July 13, 1861. This engraving, like the one following, was intended to be unsympathetic to slavery. (*Library of Congress*)

"The Slave Deck of the Bark *Wildfire*, brought into Key West on April 30, 1860," *Harper's Weekly*, May 20, 1860. The *Wildfire* had been captured by U.S. naval vessels. (*Library of Congress*)

Although all American immigrant groups have formed ethnic institutions, none did so more assiduously than the very large German American communities in the Midwest between the end of the Civil War and World War I. These men worked at the Jackson Brewery, one of dozens of local breweries in Cincinnati, "the Queen City of the West," in the 1870s. (*Cincinnati Historical Society*)

Cincinnati's Schmittie Band, in its very European uniforms, was but one of the many German contributions to American music. (*Cincinnati Historical Society*)

In Cincinnati and some other Midwestern cities, there were public schools that did much of their teaching in German. These children attended the German Fulton School in Cincinnati's East End in 1908. The two black pupils are exceptions; most black youngsters in Cincinnati went to segregated schools. (*Cincinnati Historical Society*)

These women, deaconesses of the German Methodist Motherhouse, comprised the nursing staff of Cincinnati's Bethesda Hospital in 1913. Other German ethnic institutions included a diverse vernacular press, a theater, and an old people's home, the Deutsches Altenheim. (*Cincinnati Historical Society*)

This unfamiliar likeness of Benjamin Franklin at age fifty-five is from a print by the artist Chapman, done in the early 1820s. (*The Bostonian Society/Old State House*)

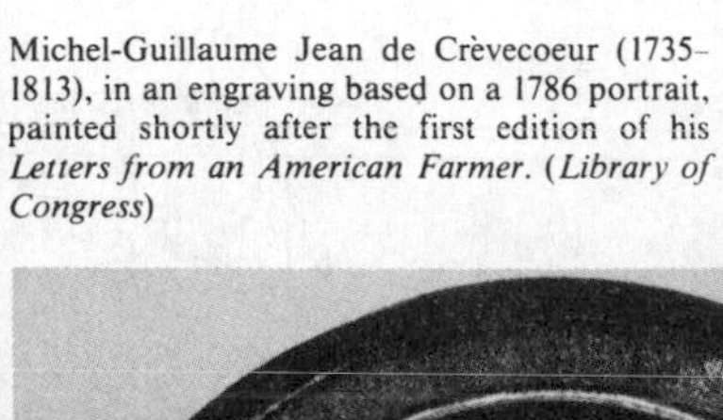

Michel-Guillaume Jean de Crèvecoeur (1735–1813), in an engraving based on a 1786 portrait, painted shortly after the first edition of his *Letters from an American Farmer*. (*Library of Congress*)

An idealized picture of the immigrant deck of a mid-nineteenth-century vessel from a German publication of 1849. The original caption reads, "The Inside of the emigrant ship, Samuel Hop." (*National Park Service: Statue of Liberty National Monument*)

Irish emigrants ready to board a vessel for New York from Queenstown (now Cobh), Ireland, 1874. The inscription on the trunk in the right-hand corner reads, "M. Fitzgerald, Passenger, New York, Steerage." From *Harper's Weekly*, September 26, 1974. (*Courtesy of the New-York Historical Society, New York City*)

Castle Garden was the immigrant depot on the Battery in New York. Here recruiters are signing up Irish and German immigrants for the Union Army in 1864. Note that one of the bounty signs on the wall at the left is in German. (*Museum of the City of New York*)

Castle Garden interior, full of immigrants, 1871. (*Museum of the City of New York*)

Immigrants having their names entered into the Castle Garden depot's register. (*Museum of the City of New York*)

WANTED!

3,000 LABORERS

On the 12th Division of the

ILLINOIS CENTRAL RAILROAD

Wages, $1.25 per Day.

Fare, from New-York, only - - $4 75

By Railroad and Steamboat, to the work in the State of Illinois.

Constant employment for two years or more given. Good board can be obtained at two dollars per week.

This is a rare chance for persons to go West, being sure of permanent employment in a healthy climate, where land can be bought cheap, and for fertility is not surpassed in any part of the Union.

Men with families preferred.

For further information in regard to it, call at the Central Railroad Office,

173 BROADWAY,

CORNER OF COURTLANDT ST.

NEW-YORK.

R. B. MASON, Chief Engineer.

H. PHELPS, Agent,

July, 1853.

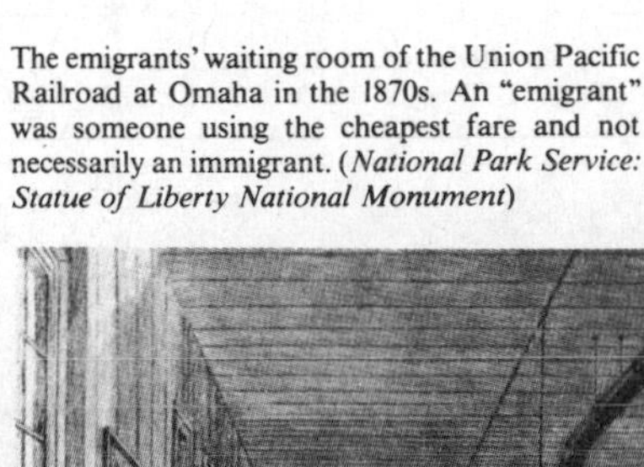

This July 1853 handbill represents a standard way that immigrants were recruited by employers. Note that the fare from New York to the job site in Illinois was a little less than four days' wages. (*National Park Service: Statue of Liberty National Monument*)

The emigrants' waiting room of the Union Pacific Railroad at Omaha in the 1870s. An "emigrant" was someone using the cheapest fare and not necessarily an immigrant. (*National Park Service: Statue of Liberty National Monument*)

This 1948 postage stamp commemorated the putative beginnings of nineteenth-century Swedish immigration. Relatively few would have traveled by ox-drawn Conestoga wagon. (*United States Postal Service*)

A typically anti-immigrant cartoon by the nativist artist Thomas Nast (1840–1902), who was of German birth. The subcaption is a play on the by-then-infamous decision of Chief Justice Roger B. Taney in the Dred Scott case of 1854, which proclaimed that blacks had "no rights that the white man was bound to respect." From *Harper's Weekly*, April 22, 1871. (*Dover Publications, Inc.*)

The Hilton family of Custer County, Nebraska, British immigrants, posed on their farm with a prized possession, about 1889. The instrument probably just arrived on the mule-drawn wagon. (*Solomon D. Butcher Collection, Nebraska State Historical Society*)

A group of German American musicians at a picnic in a public park, Madison, Wisconsin, 1897. (*State Historical Society of Wisconsin*)

A group of Norwegian immigrants being rowed out to their ship to begin their journey to America, 1906. (*Norwegian Folk Museum, Anders B. Wilse*)

3

Slavery and Immigrants from Africa

The African slave trade was a great international crime. Spanning the years from the middle of the fifteenth century to the middle of the nineteenth, involving all the nations of western Europe and every colony and nation in the New World, it was a major factor—whose significance is largely ignored by historians of Europe and the United States—in world history and in the development of capitalism. It involved the transportation of more than nine million persons to the New World. If we take the New World as a whole during the colonial period, four or five Africans immigrated for every European who came. Thus the impact of Africa on the New World is incalculable, and it is impossible to imagine what the post-Columbian history of the Western Hemisphere would have been had no Africans come.

The slave trade was a business entered into for profit. It was an integral part of western European imperialism. The slave trade stimulated investment, abetted the growth of European merchant marines, and was crucial in the development of ports such as Liverpool, which later became important places of embarkation for immigrants. Much of the profits accumulated in the slave trade eventually flowed into other branches of capitalist enterprise, particularly in England, France, and the northeastern United States, and accelerated the industrialization of those places. Similarly, most of the profits realized in the slave-dependent plantation economies of the colonial world were eventually expatriated to the various mother countries. Sidney Mintz has recently shown how one slave-based industry—the growing, refining, and distribution of cane sugar—contributed to the development of capitalist economies all over the world. Although sugar was not grown in significant quantities in the American colonies, its influence nevertheless affected the development of merchant capitalism there. One of the famous triangular trades that were the initial foundations of their economy involved New England merchants, mariners, slaves, molasses, and rum. Ships from New England, many of them based in Newport, Rhode Island, took slaves from Africa to the West Indies. There they acquired cargoes of molasses—a by-product of sugar refining—which, when brought to New England, were the basis for the manufacture of rum, some of which was exported to Africa and used as part of the price of a new cargo of slaves. These and other triangular trades were not necessarily completed by one ship sailing all three legs—slave ships, for example, became highly specialized—and many of the northern merchants who profited from them eventually put some of their profits into manufacturing.[1]

Although slavery existed in every North American colony including Canada, it eventually became predominant only in the southern colonies and states. But it was not just southerners who profited from what they came to call their "peculiar institution." Long after the United States outlawed the slave trade, northern merchants and industrialists profited from slavery. A symbiotic relationship developed between southern cotton planters and northern cotton textile manufacturers, all resting on the unfree labor of African immigrants and their descendants. Not only did cotton provide the raw material for New England's first great industry, but profits accumulated in that industry were eventually reinvested elsewhere. In addition some textile manufacturers even developed cheap and durable fabrics that were exported to the South and the West Indies for slave clothing, such as it was. In hundreds of ways no historian or economist has fully traced, African slavery is inextricably linked with the development of western—and American—capitalism.

But for our purposes, the slave trade was one of the major means of bringing immigrants to the New World in general and the United States in particular. Many historians of immigration simply ignore the slave trade, arguing, implicitly or explicitly, that it was not immigration. Others, such as Maldwyn A. Jones, author of a widely used text on immigration, although he questioned "whether or not the Negroes were immigrants in the strict sense," did acknowledge that the slave trade made "one of the most significant contributions to the peopling of the colonies" and devoted just over a page to it. More recently, Thomas Archdeacon, in the best existing survey, raises no doubts about the appropriateness of treating slaves as immigrants, devotes, all told, half a dozen pages to slavery and the slave trade, and quickly examines the African cultural baggage the slaves brought with them.[2]

Even these expanded treatments, it seems to me, do not go far enough. The evolution of American historical writing has been such that historians of black Americans and historians of immigrant Americans are organized into different subfields. Black history, long ignored by all but a handful of black scholars, became, in the decades after World War II, perhaps the most exciting and innovative genre practiced by historians of the United States, as August Meier and Elliott Rudwick have recently demonstrated. Nowhere is this more apparent than in the flowering of studies of slavery. But one unfortunate result of this separate flowering has been that historians of black Americans have all but ignored free immigrants who were black, and that historians of immigration have all but ignored not only immigrants who were slaves but also have not paid much attention, until very recently, to black immigrants from the Caribbean, even though in the twentieth century those immigrants and their descendants have exercised an influence in American life out of all proportion to their numbers. This influence has been particularly notable in literature, politics, sports, and, most recently, in the American military. (For example, General Colin Powell, who in 1989 became the first black chairman of the Joint Chiefs of Staff, is the son of Jamaican parents.) But leaving aside the question of the impact of individuals, mere numbers alone suggest a more detailed treatment than is customary.[3]

At the end of the colonial period, roughly every fifth American was either an African immigrant or the descendant of one. African immigrants are, ultimately, the ancestors of more than 10 percent of the American people, so a case could be made for devoting about 10 percent of a book on the peopling of America to them. That will not be the case here because the kinds of materials needed for that kind of a history simply do not exist. Most of the truly important questions a social historian wants answered involve knowing something about how people reacted to their historical situation. Letters, diaries, reminiscences—the kinds of sources used in the last chapter for seventeenth-century European immigrants—do not exist for Africans. They were a nonliterate people. The contemporary documents generated about African slaves are all written by slave traders, masters, and other white observers. The great popularity of Alex Haley's fraudulent creation of an African past for his family—*Roots* may have been poetically true, but it was historically a fabrication largely borrowed from a novel—indicates the hunger for such documentation. It is a hunger that can never be satisfied.[4]

What can be done, then, is limited by the nature of the historical record. We can say something about the nature of the Africa from which the slaves came, about the slave trade itself, and about the American aspects of it; and, finally, we can describe, in some detail, how the system of slavery evolved in the colonial United States. Were I writing a history of slavery in the United States or of black Americans, there are a whole variety of materials on which to draw which give the point of view of at least some blacks. The poems of Phillis Wheatley, the letters and calculations of Benjamin Banneker, the vast writings of Frederick Douglass are all primary documents of great significance for the later eighteenth and nineteenth centuries. But for the first generations of Africans in America, there is nothing. Think, for example, of writing a history of seventeenth century New England without Bradford's *History* or Winthrop's letters, or the letters and sermons of Roger Williams. Or of Virginia without the writings of John Smith. They would be histories without people, without anecdotes. We do not even know the names of the Africans who came here, only what their masters decided to call them.

The Myth of the Negro Past

In 1941 the American anthropologist Melville J. Herskovits published *The Myth of the Negro Past,* which revolutionized the way educated persons regarded African history. The myth that Herskovits demolished was a dual one: First of all, it had been believed that black Africa, the homeland of American Negroes, was a cultural desert that had contributed nothing to the rest of the world and, therefore, that the slaves who came here were "primitive savages without even the vestiges of a viable culture"; second, that whatever culture Africans might have had in the Old World was totally lost, except, perhaps for some "savage" survivals in music and dance.[5]

Africa is, of course, a large continent, stretching from the Mediterranean littoral to the Cape of Good Hope, from the Red Sea to the Atlantic, more than eleven million square miles as opposed to slightly more than three million for the United States. Within that vastness a multiplicity of cultural levels existed, from Egypt, one of the first highly developed civilizations, to groups that were still at the hunter-gatherer stage. But the African slave trade to the New World was almost completely restricted to a coastal strip of tropical Africa between the Senegal and Congo rivers. Few slaves seem to have been taken from more than three hundred miles into the interior.

The area just north of the locus of the slave trade, between the Sahara and the rain forest, was one of the major centers where human culture developed, along with the better-known Fertile Crescent of the Near East, the Indus and Yellow River valleys and pre-Columbian America. In each of these places—none of which is in Europe—a very high degree of agricultural productivity took place during Neolithic times, which, in turn, allowed for a rather dense population, specialization of labor, a complex social organization, a steeply stratified society, urbanization, and some kind of theocratic-monarchal government. Whether the agriculture that developed in the western Sudan was an indigenous development or a transfer from the Fertile Crescent is a matter of debate among scholars. The basic grain crops cultivated there were pearl millet and sorghum.

We still know very little about the archaeology of West Africa, and most of what we do know is of relatively recent origin. (Since, before Herskovits, everyone "knew" that there was nothing there, no one looked; there was no Trojan myth to attract a Schliemann to West Africa). Nor were such discoveries consistent with colonial rule in Africa. What the British called Rhodesia, after the imperialist Cecil Rhodes (1853–1902), Africans now call Zimbabwe, after an ancient southern African civilization. Since Herskovits, and particularly since African independence began in the 1950s, archaeologists and historians have been recreating the African past. The earliest of the West African states south of the Sahara about which we now have knowledge is Ghana, which was much farther north than contemporary Ghana. The first Ghana probably arose in the fourth century and reached its peak in the tenth. Weakened by wars with the Muslims to the north, Ghana succumbed to the rising empire of Mali in 1240. Mali and its rival kingdom of Songhai were codominant in the western Sudan from the thirteenth to the fifteenth century. Both states eventually became Muslim, that having replaced indigenous religious origins in perhaps the seventh century. For a time Mali absorbed Songhai; at its peak, under Mansa Musa (1307–32), it stretched from the Sahara to the rain forest, from close to the Atlantic Coast to beyond the Niger. Even larger was Songhai, which regained its independence in the later fourteenth century and became the largest of the West African kingdoms, reaching almost to Lake Chad. Its greatest leader, Askia Muhammad I (1493–1528) established the University of Sankore at Timbuktu, which became a major center of learning in the Muslim world.

South of the Sudan is the rain forest, the region from which, as we have seen, most slaves were taken. Developments there were similar but considerably later. The grains on which the civilizations of the Sudan were based were not suitable for the rain forest. Other crops and other techniques were developed; the staples were not grains but root crops, the chief of which was yams. Agriculture seems to have been established there by the beginning of the Christian Era. The Bantu-speaking peoples of the Lower Niger Valley, having a developed agriculture, began to migrate into central, eastern, and southern Africa, largely displacing gatherers such as the Pygmies and the Bushmen.

By perhaps A.D. 1000, population growth and the division of labor enabled the creation of despotic states, probably modeled in part on those of the neighboring Sudan. By the time the slave trade began, about the middle of the fifteenth century, the socioeconomic structure of the societies of the area that became the "slave coast" varied widely, from small enclaves in which kinship ties were the font of all authority to relatively large states with perhaps as many as a million inhabitants and complex political institutions, including kingship. Although some generalizations fit all of these societies—none had developed a written language—a wide range of cultures and practices existed. Descent, for example was traced matrilineally in some societies, patrilineally in others. Even if one confined oneself to discussing the more complex and larger of these West African societies—the Ashanti, the Dahomeans, the Mossi, and the Yoruba—the differences are such that an intelligible rendering in the space available here is impossible. Instead I will describe some of the more striking sociocultural characteristics that are clearly part of what Herskovitz called the "Negro past."

Instead of being populated by the howling savages that inhabit the Tarzan movies and such epics as *King Solomon's Mines,* these societies had complex economic, political, and social organizations. The societies were, of course, basically agricultural, with some fishing and much raising of poultry. Among the Ashanti, men and women worked in the fields; the Dahomeans and the Yoruba allocated these tasks to one gender—female in the former, male in the latter. Beyond that there was much craft specialization, including ironworking, bronze casting, wood carving, pottery making, basketmaking, and weaving. The aesthetic beauty of much of these West African works of art has been increasingly appreciated during the twentieth century, and African forms have been an acknowledged and important influence on such modern artists as Pablo Picasso. Benin bronzes are particularly valued by contemporary museums, although an increasingly wide range of West African art is now being exhibited—as a visit, for example, to the Michael Rockefeller wing of the Metropolitan Museum in New York will quickly demonstrate.

Political organization was uneven. In some instances there seem to have been relatively autonomous villages; in others, particularly Dahomey, there were kings whose power approached that of European absolute monarchs. But there and elsewhere, women had significantly more political power: In Dahomey, major officials had female counterparts, known as "mothers," who supervised their work. In Ashanti the queen sister or queen mother actually nominated the king, while in many of the other societies such figures exercised real economic and political power. Slavery was common if not universal in West Africa, although we know very little about how the institution operated there.

West African religions were generally animistic; some gods represented the sky and the elements, others represented animals. A serpent god, often a python, was quite important in many if not most societies. Most religions involved offerings and sacrifices, sometimes including human sacrifices.

How much, if any, of West African culture survived the middle passage and the first slave generation? Although such black American scholars as W. E. B. DuBois (1868–1963) and Carter G. Woodson had correctly asserted that Africanisms were present in Afro-American culture early in the twentieth century, these were more like inspired guesses and acts of faith than research-based hypotheses. Herskovits's work, based on ethnographic field research on both sides of the Atlantic, clearly established the existence of African cultural transfers in the New World. Herskovitz and his successors now generally agree that these transfers were particularly significant in some parts of the Caribbean and in Brazil, and that what now exists in those regions is a synthesis of African and European cultural traditions. Religion, some family institutions, linguistic elements and folk tales, mutual aid societies, and music and dance forms now existing in the New World south of the United States have unmistakable African content. Afro-Cuban and Afro-Brazilian music and dance and Haitian religion are good examples. The latter, Haitian *vodun,* usually rendered *voodoo* in English, comes from a Dahomean word and is now a mixture of African and Christian beliefs and practices.

For the United States such survivals are less common. This is chiefly due to the lower proportion of blacks to the total population and the absence of large runaway slave or maroon enclaves where African institutions could evolve. There is also evidence that many American slave owners deliberately chose a work force from assorted African peoples just as, in the late nineteenth and early twentieth centuries, American steelmakers deliberately recruited polyethnic labor forces. In each instance the purpose was not to create a "melting pot" but to inhibit cooperation among members of the work force. Some American slaveholders had ethnic preferences among Africans. South Carolinian slave buyers seemed to prefer Senegambians and Angolans and did not want Ibos (from today's Nigeria), believing, probably falsely, that the latter were more prone to suicide. Most Virginia masters seem to have had no such prejudices, so that Ibos represented perhaps 2 percent of the Africans brought to the Palmetto State but almost 40 percent of those brought to the Old Dominion.

The clearest evidences of African survivals in the United States are linguistic. Since Lorenzo Turner published *Africanisms in the Gullah Dialect* (1949), scholars have recognized that certain oddities of vocabulary and syntax among the black inhabitants of the Sea Islands off the coast of South Carolina and Georgia are not, as was once believed, variants of sixteenth-century English but are clearly of African origin. The Sea Islands, relatively isolated and with a very high proportion of blacks to whites, are just the kind of locale in which the likelihood of linguistic survivals would be the strongest. Other survivals, hard to isolate, include music and dance, but just how much of jazz, for example, is African and how much Afro-American, is—and probably always will be—a matter of heated debate among experts. What cannot be doubted is that African influence is there, and that what was once Afro-American has become characteristically American. Similarly, it is clear that much of the content of Joel Chandler Harris's (1848–1908) Uncle Remus stories comes from folk material with African origin. Particularly striking are the themes revolving around the rabbit as trickster, a motif found in much West African folklore. Other survivals can be associated with certain craft techniques.[6]

Not all of Herskovits's claims are easily documented, and some have been disputed. This is particularly true about his claims for African origins of certain patterns in Afro-American family life and religion. E. Franklin Frazier, for example, in his *Negro Family in the United States* (1939), disagreed vehemently with Herskovits's notions about the origins of black family structure in the United States. Both agreed—as do most other scholars—that a matriarchal-pattern was particularly noticeable among lower-class Afro-American families. Herskovits related this, as he was wont to, to matriarchal patterns found in Africa; Frazier, on the other hand, held that these patterns evolved from the historical experiences, during slavery and after, of black Americans. Such arguments cannot be resolved here and may never be fully resolved. What is important for our purposes is to note that African immigrants—like others—brought cultural baggage with them, baggage that, under the most inhospitable circumstances, some of them managed to keep and adapt to their New World experience.[7]

How Many Africans Came?

The African slave trade existed for over four centuries, from the mid-fifteenth century to about 1870. Although it began before Europe knew of the New World, it was, as we have seen in the case of Virginia, the New World plantation economy with its insatiable demand for labor to grow cash export crops, which shaped it. Nearly 10 million persons were kidnapped out of Africa, all but about 350,000 of them for sale in the Americas. The following table, from Philip Curtin's *The African Slave Trade: A Census* (1969), is about as accurate as the evidence will now allow: Many of the numbers may well be, as Curtin himself says, "plus or minus 20 percent of actuality."[8]

The first thing to note in the table is how relatively few slaves were brought to what is now the United States: 427,300 persons (British North America plus Louisiana), or not quite 4.5 percent of Curtin's total. When one compares this with the almost incredible numbers that were brought to tiny Caribbean islands—say the 387,000 to Barbados's 166 square miles—it becomes immediately clear that, from a demographic point of view, slavery in the North American colonies and the United States was quite different from that in the Caribbean. Most of those brought to Caribbean islands like Barbados—current population some 250,000—clearly died without issue; those brought to what became the United States were the ancestors of more than 20 million persons. Put another way, as of 1950, before heavy emigration of blacks from the Caribbean had begun, the descendants of the 4.5 percent of the African emigrants to the New World who came to the United States represented almost a third—31.1 percent—of all the black people then living in the Western Hemisphere. This is not to argue that slavery was, somehow, "better" in the United States, but it does call into question the notion put forth by some nineteenth-century abolitionists and twentieth-century scholars like Frank Tannenbaum and Gilberto Freyre that North American slavery was harsher than the more "lyrical" Latin American varieties. Slavery was, above all, a business, and most of the major decisions concerning it were based on the slaveholders' notions about what constituted the most profitable course of action. If slave owners in the American Upper South decided that it was profitable to breed slaves for sale, and thus perhaps provided somewhat better care,

Table 3.1

Estimated Slave Imports into the Americas, by Importing Region, 1451–1870 (in thousands)

REGION & COUNTRY	1451–1600	1601–1700	1701–1810	1811–1870	TOTAL
British North America	—	—	348.0	51.0	399.0
Spanish America	75.0	292.5	578.6	606.0	1,552.1
British Caribbean	—	263.7	1,401.3	—	1,665.0
Jamaica	—	85.1	662.4	—	747.5
Barbados	—	134.5	252.5	—	387.0
Leeward Islands	—	44.1	301.9	—	346.0
St. Vincent, St. Lucia, Tobago, & Dominica	—	—	70.1	—	70.1
Trinidad	—	—	22.4	—	22.4
Grenada	—	—	67.0	—	67.0
Other BWI	—	—	25.0	—	25.0
French Caribbean	—	155.8	1,348.4	96.0	1,600.2
Saint Domingue	—	74.6	789.7	—	864.3
Martinique	—	66.5	258.3	41.0	365.8
Guadeloupe	—	12.7	237.1	41.0	290.8
Louisiana	—	—	28.3	—	28.3
French Guiana	—	2.0	35.0	14.0	51.0
Dutch Caribbean	—	40.0	460.0	—	500.0
Danish Caribbean	—	4.0	24.0	—	28.0
Brazil	50.0	560.0	1,891.4	1,145.4	3,646.8
Old World	149.9	25.1	—	—	175.0
Europe	48.8	1.2	—	—	50.0
São Thomé	76.1	23.9	—	—	100.0
Atlantic Islands	25.0	—	—	—	25.0
Total	274.9	1,341.1	6,051.7	1,898.4	9,566.1
Annual average	1.8	13.4	55.0	31.6	22.8
Mean annual rate of increase[a]	—	1.7%	1.8%	−0.1%	

SOURCE: Philip D. Curtin, *The African Slave Trade: A Census* (Madison, Wis., 1969), p. 288.

[a]These figures represent the mean annual rates of increase from 1451–75 to 1601–25, from 1601–25 to 1701–20, and from 1701–20 to 1811–20.

it should not be argued that they were therefore morally superior to the West Indian planters who, on the same basis, decided that it was more expedient to work their slaves to death and buy more.

The slave trade itself, although it lasted for more than four centuries, peaked during the "enlightened" century, 1721–1820, when 60 percent of all slaves brought to the New World were transported. (The century and a half after 1700 accounts for 80 percent.) The total loss of life involved in the slave trade is incalculable. Anywhere from 13 to 33 percent more slaves were taken out of Africa on ships than were ever delivered to the New World, a loss of anywhere from 1.25 to 3.15 million African lives. To this must be added unknowable numbers of lives lost in warfare related to the slave trade in Africa, in transit from the interior to the coast, and in holding pens and barracoons waiting for shipment.

If we look at the data just from British North America, we find that 87 percent of all the slaves brought there or to the successor United States came between 1701 and 1810, and that nearly 13 percent of all slaves imported came *after* the slave trade—but not slavery itself—had been outlawed by the American Congress. Those 50,000 Africans were the first illegal immigrants in our history. (For reasons that I do not understand, Curtin gives no figures for slaves brought to North America before 1701, although clearly there were some. Edmund S. Morgan has estimated that Virginia, which would have had the bulk of seventeenth-century North American slaves, had a black population of between six and ten thousand in 1699, not all of whom would have been of African birth.)

By 1750 there were perhaps 250,000 blacks in North America, while at the first census in 1790 more than 750,000 Afro-Americans were enumerated. Probably no more than a fifth of these latter, 150,000 persons, were African born; the rest were at least second-generation Americans. Some 690,000 were enslaved; the others existed in a condition that John Hope Franklin has properly described as "quasi-free." By that time their color had become, by a process that cannot be fully documented, a badge of servitude and inferiority. And it was this racial characteristic—color—that made African slavery in the New World different from slavery in Africa, and that made the black experience—and later the Asian experience—in the United States different from that of white ethnic groups. Members of the latter, could, sometimes even in the immigrant generation, shed most if not all of the visible ethnic characteristics and "melt" into the general population. For blacks (and

Asians), except for those with sufficient white ancestry to "pass," no such disappearance was possible.

Although the slave trade was demographically unbalanced—much more than a majority of all slaves landed were young males—natural increase began to balance the sex ratio in the United States even before the official end of the slave trade. By 1820 there were 899,000 black males and 873,000 black females, a 50.7 percent–49.3 percent split, and by 1840 enumerated black females slightly outnumbered black males. By 1850, when we have reasonably reliable statistics about comparative mortality, the death rate of slave infants under one year of age was at least 25 percent higher than for comparable white babies, and the life expectancy of slaves was some 12 percent lower than that of whites.

Slaves and their descendants lived primarily in the South; until after World War II more than 90 percent of black Americans lived in the region once dominated by slavery. Although from the first some slaves were involved in craftsmanship and later in industrial processes, and, as Richard Wade has shown us, significant numbers of slaves lived and worked in cities, the plantation was, par excellence, the *locus classis* of slavery. While, of course, conditions differed from plantation to plantation, as they would differ from job site to job site for later immigrants, we can make some rough generalizations about what the material conditions of slave life were like.

First of all, there was a great deal of hard physical labor, the tasks of the agricultural year with few labor saving devices and crude tools. A typical weekly food ration would include 3.5 to 4 pounds of salt pork and a peck of corn for an adult male—"no more peck of corn for me" is a line from one well-known freedom song—with smaller rations for women, children, and the aged. In rice-growing areas, rice was substituted for corn; and molasses, salt fish, and vegetables, sometimes from slave gardens, were often added. Clothing was crude and uncomfortable: In the heyday of the cotton kingdom northern and European clothing factories produced special cheap lines "for slaves only." Although one would not know it from the reconstructed "slave quarters" on "historic site" antebellum plantations, most slaves lived in windowless (and stoveless) one-room huts ten to fifteen feet square with dirt floors.

As was and would be true with other immigrant groups, the emergence of a native-born Afro-American majority created a new kind of person. By the time of the American Revolution most blacks were more American than African, but society used their color to keep them separate, apart, and segregated from as much of American life as it could. Black Americans, like all of us save Amerindians, are descended from relatively recent immigrants; they have shared most of the experiences of American life, but those experiences have been, at least in part, modified by their special position in U.S. society. No one ever expressed this unique position better than W. E. B. DuBois who wrote, early in this century:

> One ever feels his twoness—an American, a Negro, two souls, two thoughts, two unreconciled strivings; two warring ideals in one dark body, whose dogged strength alone keeps it from being torn asunder. The history of the American Negro is the history of this strife—this longing to attain self-conscious manhood, to merge his double self into a truer and better self.[9]

4

Other Europeans in Colonial America

Although when compared to the New World colonies of France and Spain—in which almost all Europeans were of the nationality of the mother country—the colonies that became the United States seem polyglot, the fact of the matter is that by the time of the first census in 1790, just over three-fifths of the white population was calculated to have been of English stock, and more than two-fifths of the rest came from the British Isles. Put another way, continental European stock accounted for a little more than one white person in seven; one white in fourteen could not be assigned an ethnicity by the scholars who produced the table below more than half a century ago.

In this chapter we shall examine the immigration to the American colonies, of non-English European groups, most of whom came in the late seventeenth or eighteenth century. More people came to British North America in the eighteenth century than in the seventeenth, and we must remember that the total immigration for the period before 1790 was in the neighborhood of a million persons, only about 2 percent of all those who have ever come. Yet the importance of these early ethnic groups, and the precedents provided by their presence, are of incalculable significance for American history. Already there were distinct ethnic settlement patterns. Afro-Americans, some 20 percent of the population as we have seen, were concentrated in the South; and in two states, South Carolina and Virginia, they constituted more than two-fifths of the population, 43.7 percent and 40.9 percent respectively. Germans, just 8.6 percent of the white population, were a third of the population of Pennsylvania. Dutch, just 3 percent of the white population, were about a sixth of New York and New Jersey. French, 2.3 percent of the whites, were more evenly distributed than any of the other continental groups; while Swedes, 0.7 percent of the whites, were

Table 4.1

National or Linguistic Stocks in the United States, 1790 (whites only)

STATE	ENGLISH	SCOTCH	IRISH		GERMAN	DUTCH	FRENCH	SWEDISH	SPANISH	UNASSIGNED	TOTAL
			ULSTER	FREE STATE							
Maine	60.0	4.5	8.0	3.7	1.3	0.1	1.3	—	—	21.1	100.0
New Hampshire	61.0	6.2	4.6	2.9	.4	.1	.7	—	—	24.1	100.0
Vermont	76.0	5.1	3.2	1.9	.2	.6	.4	—	—	12.6	100.0
Massachusetts	82.0	4.4	2.6	1.3	.3	.2	.8	—	—	8.4	100.0
Rhode Island	71.0	5.8	2.0	.8	.5	.4	.8	0.1	—	18.6	100.0
Connecticut	67.0	2.2	1.8	1.1	.3	.3	.9	—	—	26.4	100.0
New York	52.0	7.0	5.1	3.0	8.2	17.5	3.8	.5	—	2.9	100.0
New Jersey	47.0	7.7	6.3	3.2	9.2	16.6	2.4	3.9	—	3.7	100.0
Pennsylvania	35.3	8.6	11.0	3.5	33.3	1.8	1.8	.8	—	3.9	100.0
Delaware	60.0	8.0	6.3	5.4	1.1	4.3	1.6	8.9	—	4.1	100.0
Maryland and District of Columbia	64.5	7.6	5.8	6.5	11.7	.5	1.2	.5	—	1.7	100.0
Virginia and West Virginia	68.5	10.2	6.2	5.5	6.3	.3	1.5	.6	—	.9	100.0

Table 4.1 (continued)

National or Linguistic Stocks in the United States, 1790 (whites only)

STATE	ENGLISH	SCOTCH	IRISH		GERMAN	DUTCH	FRENCH	SWEDISH	SPANISH	UNASSIGNED	TOTAL
			ULSTER	FREE STATE							
North Carolina	66.0	14.8	5.7	5.4	4.7	.3	1.7	.2	—	1.2	100.0
South Carolina	60.2	15.1	9.4	4.4	5.0	.4	3.9	.2	—	1.4	100.0
Georgia	57.4	15.5	11.5	3.8	7.6	.2	2.3	.6	—	1.1	100.0
Kentucky and Tennessee	57.9	10.0	7.0	5.2	14.0	1.3	2.2	.5	—	1.9	100.0
Area enumerated	60.9	8.3	6.0	3.7	8.7	3.4	1.7	.7	—	6.6	100.0
Northwest Territory	29.8	4.1	2.9	1.8	4.3	—	57.1	—	—	—	100.0
Spanish, United States	2.5	.3	.2	.1	.4	—	—	—	96.5	—	100.0
French, United States	11.2	1.6	1.1	.7	8.7	—	64.2	—	12.5	—	100.0
Continental United States	60.1	8.1	5.9	3.6	8.6	3.1	2.3	0.7	0.8	6.8	100.0

SOURCE: *Annual Report of the American Historical Association for the Year 1931* (Washington, D.C.: U.S. Government Printing Office, 1932), p. 124.

NOTE: Typical of its time, the table ignores Afro-Americans, who comprised, as we have seen, almost a fifth of the whole population, so that, for example, the 60% figure for English is really 60% of 80%, or 48%. Thus, even in 1790, there was no English ethnic majority. Considered as an ethnic group, Afro-Americans rank second in number.

8.9 percent of the white population of Delaware and 3.9 percent of white New Jersey.

Clearly, the Southern and Middle Atlantic states were the major loci of non-British migrants. Conversely, both blacks and continental Europeans were underrepresented in New England. The thorniest problem in enumeration, as we shall see, is the vexed question of the Scotch-Irish, who are ignored in the table above, and have to be teased out of the Scotch and Irish/Ulster categories. The anachronistic Irish/Free State heading, betrays one of the purposes for which this table was originally constructed: to buttress and make more scientific the National Origins Act of 1924, as will be shown in chapter 10. In any event, almost all the Irish in the chart were Protestants, as were all but a very few European immigrants in the Colonial Era.

Germans

Although a few Germans came early in the seventeenth century—a handful of skilled miners from the south and scattered individuals who arrived under Dutch auspices, most of whom were probably part of traditional seasonal migrations between Germany and the Netherlands—the start of organized migration of Germans can be located as precisely as can that of the Jamestown colonists or the Pilgrims. Before discussing these community founders, who were from Krefeld on the Lower Rhine and established what became Germantown, Pennsylvania, in 1683, it is appropriate to examine the question of what we mean when we say "Germans." There was, of course, no Germany as a national state until 1871. Ethnic Germans—persons speaking German as a mother tongue—came to the United States from a number of places. The *Harvard Encyclopedia of American Ethnic Groups (HEAEG)* has seven separate essays on ethnic Germans: Germans, Alsatians, Austrians, Germans from Russia, Hutterites, Luxembourgers, and Swiss. The *HEAEG* adopts an essentially political definition of Germans; mine be largely ethnic. Germans of Austrian nationality, for example, will not be treated separately. Two further difficulties must be pointed out. Although it will be noted later that in the nineteenth and twentieth centuries German immigrants were of all three major confessions—Protestant, Catholic, and Jewish—most of the treatment of German Jews will occur in discussions of Jewish immigration rather than that of German. (Jews are clearly an American ethnic group; Protestants are clearly not; for Roman Catholics a case for ethnicity can at least be made, but the treatment here will be traditional. Catholics will be treated largely by nationality, although ethnic problems within American Catholicism will be analyzed.)[1]

And, finally, there is the confused question of the so-called Pennsylvania Dutch. The "Dutch" tag was clearly a misunderstanding in colonial times, one that probably stemmed from both a traditional English usage that used the term *Deutsch* to describe anyone in the area from the mouth of the Rhine to its source in Switzerland and from hearing immigrants describe themselves as "Deutsch or "Deitsch" (that is, German) and transmuting that into "Dutch." For centuries in America, any ethnic German was likely to be nicknamed "Dutch."

The Krefelders—thirteen Quaker and Mennonite families who had been proselytized to settle in this new (est. 1681) colony by William Penn—came under the leadership of their minister, Francis Daniel Pastorius (1651–1720?). They were the forerunners of perhaps a hundred thousand ethnic Germans who came in the colonial period, and, as such, they were properly celebrated on both sides of the Atlantic during the tricentennial of German immigration in 1983. But their radical religious posture—and the great publicity given to Amish and other German pietistic sects—has created the false impression in the popular mind that most German immigrants in the colonial period were religious dissenters. Such was not the case. Most were members of the two chief German Protestant churches: the Lutheran and the Reformed. Dissenters did continue to come: Particularly noteworthy were Swiss Mennonites, who got a grant of some ten thousand acres in what is now Lancaster County (Amish are a subsect of Mennonites); in 1719 there began the migration of Westphalian Dunkards, another Anabaptist group; Schwenkfelders began to come in 1734, while in 1741 more than seven hundred Moravians founded the Pennsylvania towns of Bethlehem and Nazareth. All these groups have increased and multiplied and expanded settlement into the neighboring states of New York and Ohio and into the farther Midwest, most notably in Iowa.

Large-scale migration of Germans to America did not begin until a quarter of a century after the founding of Germantown. Politics and economics, rather than religion, were the major push factors. As a part of the War of Spanish Succession (called Queen Anne's War in America), which went on intermittently between Britain and France between 1702 and 1713, Louis XIV attacked some of Britain's German allies, devastating much of the Rhineland in the process. The region was already beset with overpopulation, heavy taxation, and, in 1708–19, suffered an exceptionally hard winter. Some 13,000 German Rhinelanders, or Palatines as they were called, emigrated to England by sailing down the Rhine and across the North Sea. The British responded by setting up what would be called today refugee camps* around London for the Protestants while sending some 3,500 Catholics back to Germany. The camps, as they usually do, created problems of sanitation and public order. Some of the London poor, resentful that foreigners should get aid denied to them, attacked the camps. The government quickly shipped most of the Palatines elsewhere: About three thousand eight hundred were sent to Ireland, perhaps six hundred to North Carolina, and, most disastrously, nearly three thousand to New York as part of a plan to have them manufacture naval stores along the lower Hudson River around Newburgh. Anywhere from a seventh to a quarter of them either died en route or shortly after their arrival (sources differ), and by 1712 the whole project was abandoned. Some Germans stayed on the Hudson, others migrated to the Mohawk Valley and other frontier areas in New York, but most found their way to Pennsylvania, which would become the mecca of the Germans.

Despite the trials and tribulations of these Palatine pioneers, the success of some of them—especially those who got to Pennsylvania—spread both by "America letters" and by proprietors and their European agents who were anxious for settlers, eventually resulted in a growing migration. Most came from Southwest Germany, but all tended to be called Palatines in America. Until 1727, when Pennsylvania began keeping what seem to be good records of those who landed by sea, we are not too sure of numbers, but historians believe that the arrival of at least three ships in 1717 carrying 363 Germans marks the onset of relatively heavy migration. Between 1727 and 1740 about six immigrant ships arrived each year; this quickened to more than ten a year until the Seven Years War (1756–63), which virtually stopped the trade. Between the end of that war and the outbreak of the American Revolution, about six immigrant ships a year arrived.

Most of the ships sailed from Rotterdam, which the emigrants reached only after a four-to-six week journey down the Rhine. Large numbers of German immigrants, perhaps half to two-thirds in the period up to 1817, arrived in Rotterdam so impoverished that they had to indenture themselves to pay their passage. Unlike indentured servants from the British Isles, many or most of whom were forced into indenture, most Germans chose it deliberately and thus were called

*The word *refugee* did not come into use until after the French Revolution, which also caused the coinage of the word *émigré,* from which the terms *emigrant* and *immigrant* derive.

contemporaneously free willers or redemptioners. Redemptioners actually executed two contracts. The European contract with the ship captain bound the immigrant either to pay his fare on arrival—some had their fares paid by friends or relatives already here—or to be sold to some purchaser. The American contract was with the purchaser and set forth the terms and conditions of the indenture. One such American contract from 1800, found by Günter Moltmann in the Hamburg archives, reads, in part:

> [Martin Ernst Scheffler] of [his] own free will has bound [himself] servant to [Jonathan Bradfield], for the consideration of [Eighty] Dollars paid to [Jacob Sperry] for his passage from [Hamburgh], as also for other good causes [he] the said [Martin Ernst Scheffler] . . . by these Presents doth bind and put [himself] Servant to the said [Jonathan Bradfield] to serve him, his Executors and Assigns . . . for . . . [Three] Years [and] faithfully shall serve, and that honestly and obediently in all Things, as a good and dutiful servant ought to do. And the said [Jonathan Bradfield, his] Executors and assigns . . . shall find and provide for the said Servant sufficient Meat, Drink, Apparrell, Washing and Lodging and at the Expiration of [his] Term the said Servant shall have two complete Suits of cloths one whereof to be new and [Twenty] Dollars in money.

Half a century before, a German critic of the system, Gottlieb Mittelberger, wrote one of the most graphic descriptions we have of the last step in the process, the redeeming of the contract in America:

> Every day Englishmen, Dutchmen and High Germans come from Philadelphia and other places, some of them very far away . . . and go on board the newly arrived vessel [and] from among the healthy they pick out those suitable. . . . They negotiate with them as to the length of the period for which they will go into service in order to pay their passage. . . . When an agreement has been reached, adult persons by written contract bind themselves to serve for [from three to six years]. . . . The very young, between the age of ten and fifteen, have to serve until they are twenty-one. . . . It often happens that whole families—husband, wife and children—being sold to different purchasers, become separated, especially when they cannot pay any part of the passage money. When either the husband or the wife has died at sea, having come more than halfway then the surviving spouse . . . must also pay for or serve

> out the fare of the deceased. When both parents have died at sea, having come more than halfway, then their children . . . must be responsible for their own fares as well as those of their parents.

Why did redemptioners agree to the bargain? Obviously because many of them found the conditions of life in Europe intolerable. When the economist Friedrich List questioned some emigrants from Württemberg about it he was told that "they preferred being slaves in America to being free townsmen" in Germany. Others told him that "their misery could not weigh on them more heavily that what they had to face at home." (The analogy to slavery, although made by some historians, is overdrawn.) It is fascinating to note that in isolated instances redemptioners were bought by free blacks. In New Orleans it was made illegal for blacks to "own" whites; in Baltimore, in 1817, two German families were purchased by blacks, but the local German American community was so horrified at this that it purchased the contracts. Nor was life for the German indentured servants generally as hard as that of the English and Irish who toiled in the tobacco fields of seventeenth-century Virginia and Maryland.

The German migration, in addition to experiencing generally better working and living conditions, was largely a family migration. It was not at all unusual for some members of a family to start free, while others had to be sold to make good on their European contracts. In such cases it was often possible for the free family members to buy up the contracts of the unfree members after the former had earned some money. It was also possible to pay part of the "freight" and thus shorten the term of service. To be sure, some redemptioners tried to break their contracts by running away, and one can find advertisements seeking the return of Germans. But most redemptioners seem to have enjoyed upward social mobility as a result of their bargains.

Moltmann, the leading student of the phenomenon, argues that "on the whole it was not a bad system." He quotes with approval a contemporary observer, Moritz von Fürstenwarther, who wrote from Philadelphia in 1819:

> I must agree with the general opinion of judicious people, that even those who have paid their freight and come here freely, farmers as well as craftsmen, even if they have some money, do better to bind themselves in the beginning. . . . During the time of servitude they learn the language, the customs, the differences in performing crafts, they learn about local conditions [and after they gain their

> freedom] they are able to work independently as craftsmen or, if they are farmers, to buy a few acres of land. Almost all who came here as redemptionists ten or twelve years ago and began in servitude are now well-off. I personally know of several people who settled here twenty or thirty years ago and who are now capitalists.

There is an obvious urban bias in von Fürstenwarther's emphasis; the vast majority of redemptioners as well as other German immigrants of the period became farmers.

The success story should not blind us to the fact that there was exploitation involved in the trade. Using figures for the end of the redemptionist era, in early the nineteenth century, a freight could range from sixty-five to eighty dollars, as we have seen in Martin Scheffler's contract. An unskilled laborer might make as much as a dollar a day at that time, so, conservatively, let us say that Martin could have earned eight hundred dollars during his indenture. Jonathan Bradfield got four years of labor for a tenth of that. To be sure, Bradfield had expenses—food, clothing, and freedom dues—but it was clearly a good bargain for both sides.

If it was a good bargain, why didn't the system continue? For one thing, the whole system of indentured servitude, with its feudal overtones, was simply no longer consonant with the ideology of American life, although we must note that it lasted, on an ever-diminishing scale, until at least 1830, more than fifty years after the Declaration of Independence. For another, the increasingly capitalist nature of production in the United States made such a system, which tied the employer to providing year-round maintenance for the worker, an anachronism. In addition, there was a breakdown in the system in 1816–17, when a combination of terrible weather conditions (contemporaries called 1816 "the year without a summer," probably caused, at least in part, by the global spread of sunlight-obscuring ash from the eruption of the Indonesian volcano Sumbawa) and the resulting poor agricultural yields produced an oversupply of prospective redemptioners. So many wanted to come that Dutch shipping firms both raised the price of a freight and accepted only those who seemed most likely to be salable in America. The result was chaos on both sides of the Atlantic. One estimate speaks of thirty thousand prospective emigrants stranded in Amsterdam alone. Eventually a royal decree—a forerunner of the infamous LPC or "likely to become a public charge" clause of American immigration law—barred the entry into Holland of those who could not get some resident of the kingdom to assume responsibility for them, so that other thousands were forced to make the ignominious trip home, if they could get there.

In the United States, the arrival of overcrowded ships with many ill and abused redemptioners not only created a scandal but also unbalanced the labor market so that many redemptioners had to accept most unfavorable contracts. The growing German American populations in the port cities were stirred into action. In Baltimore, for example, the arrival of the *Jufvrouw Johanna* with hundreds of distressed passengers led to the rechartering of the German Society of Maryland in 1818. The beginning of its charter speaks for itself:

> Whereas, the arrival of Germans and Switzers from Europe . . . have induced a number of persons in this State to associate themselves for the purpose of removing or lessening their distress in a strange land . . .

The Maryland German Americans successfully lobbied for protective legislation. An 1818 Maryland act "Relative to German and Swiss Redemptioners" provided for the appointment of a "register" who had to approve apprenticeship contracts for the immigrants. Contracts were restricted: All under twenty-one had to be provided with at least two months' schooling a year; terms were limited to four years; no one could be detained on board an arriving vessel for more than thirty days; while so detained they had to receive "good and sufficient provisions," and their terms of service could not be lengthened because of this extra expense; children could not be required to pay the passage of deceased parents, nor surviving spouses those of dead partners; the register could remove to shore any sick emigrant or one who had been mistreated; and, finally, if a redemptioner was not sold within sixty days, the master would have no further lien on him. Similar laws were enacted in Pennsylvania and Delaware. This legislation clearly added much risk for the shipowners and presaged the end of the system.

There is one special group of German immigrants who must be mentioned: the so-called Hessians, German mercenary soldiers who fought for the British in America during the American Revolution. Despite the name, not all were from Hesse. Authorities estimate that perhaps as many as five thousand remained in America, some of whom were deserters, others prisoners of war who chose not to return. Nearly a thousand had been taken in Washington's Christmas Eve attack on Trenton in 1776. There have been no modern studies showing what became of these men: The assumption is that most gravitated to German American settlements in Pennsylvania and elsewhere and were absorbed into them.

Whether redemptioners or free settlers, most Germans, like most colonial Americans, became farmers and farm workers. The majority had been peasants in the Old World, although craftsmen were well represented among them: The first group from Krefeld, for example, had been weavers. The Germans were not only farmers but very good farmers, especially by American standards. Unlike most American farmers, who essentially mined the soil and then moved on, the Germans paid more attention to manuring and other improved farming methods. To be sure, Pennsylvania had some of the best soil. Frederick Jackson Turner, always happy to have a mechanistic answer, argued that German farmers could be associated with limestone soils, but the social geographer James Lemon has argued that for the wealthiest Pennsylvania farmers, Quakers and Mennonites, their "social philosophy of discipline and mutual aid" rather than good soil should be seen as the key to success.

Although German Americans from Pennsylvania participated in the westward movement along with everyone else, there has been a high degree of persistence of self-conscious German Americans, particularly in southeastern Pennsylvania. The culture they created was not German, but German American. Pennsylvania Germans, called a separate ethnic group by the *HEAEG,* created a culture that has continued into the twentieth century. Its artifacts have ranged from elite to peasant forms, from sophisticated portraiture to primitive styles in art, for example. Folk arts abounded, most notably the *Fraktur* style of manuscript decoration that flourished from about 1750 to 1850. While Pennsylvania German religious music is widely known, secular songs were produced in great profusion for a long time. Some were political, such as the song written for the 1890 gubernatorial race—"Der Delamater Kummt Net Nei," "Delamater Won't Get In" (he didn't)—or paeans to modern technology as "Die Ford Maschin," which appeared as the Model T, became ubiquitous. An 1880 translation of *H.M.S. Pinafore* proved highly popular.

Religious services in German continued to be common as late as 1930 in mainstream—Lutheran and Reformed—Pennsylvania German churches and are still used by many of the sectarian groups. Some of the mainstream churches still hold an annual "Commemorative Service in the Language of the Fathers," and there are a number of organizations, most notably the Pennsylvania German Society, devoted to the recording and preservation of Pennsylvania German culture. The long persistence of a special variant of European culture in the midst of a modernized culture over some two centuries is unique in our history, although, as we shall see, there are other examples of cultural persistence. What must be remembered is that Pennsylvania German is an American culture and not merely a transplant from the Old World.

Scotch Irish

The term *Scotch Irish* is not used outside the United States. It refers to Presbyterians from the Scottish lowlands who were settled in Ulster (now Northern Ireland) in the seventeenth century. Some two hundred thousand went to England's first plantation, and it has been estimated that eventually perhaps two million of their descendants migrated to the United States and Canada; probably about a hundred thousand had come to the American colonies before the American Revolution, almost all of them after 1717, some forty thousand in the 1770s. Both push and pull factors, plus the fact that migration was a part of group experience, help to explain why so many of a relatively small population chose to cross the Atlantic.[2]

In Ulster, the Scotch Irish occupied a marginal, middle position. As Protestants who were generally loyal to the British crown, they were anathema to the largely Catholic native Irish; but as dissenters they were forced to pay tithes in support of the Anglican church and suffered many civil disabilities. Even more bothersome were economic conditions, and it is economic conditions, both general and specific, which most clearly triggered the Ulster emigration. Like their American cousins, the Scotch Irish were part of the British Empire and had to subsist within a mercantile system that placed the interests of the metropolis above those of the colony. Thus, for example, the Wool Act of 1699 severely damaged local manufacturing by forbidding export of finished products—but not raw wool—that might compete with goods made in England. And, even more galling to more people, they suffered from the chronic problem of all Irish tenants, regardless of ethnicity—rack renting (rents demanded at the highest possible rates) by English landlords and their agents. The fact that America was advertised to be a country without landlords, where there was land for those who wanted it, was surely the major pull, although for some, especially Presbyterian divines who were often prominent in plans for emigration, the chance to enjoy greater religious freedom was also significant. One Scotch-Irish minister, Francis Makemie (1658?–1708), came to Virginia in 1698 and organized the first presbytery in the New World in Philadelphia in 1706.

The Scotch-Irish immigration was not a steady flow, but was concentrated in five pulses—1717–18, 1725–29, 1740–41, 1754–55, and 1771–75—totaling fourteen years in an almost-sixty-year period. These pulses provide, as James Leyburn has put it "a chart of the economic health of northern Ireland," although, to be sure, the terminal date of the last surge was dictated by American rather than Irish events. Hugh Boulter, Anglican archbishop of Ireland, in a group of letters written during the second pulse described some of the conditions that pushed Scotch Irish out.

> The scarcity and dearness of provisions still increases in the North. Many have eaten the oats they should have sowed their lands with. . . . The humor of going to America still continues . . . if we knew how to stop them, as most of them can get neither victuals nor work, it would be cruel to do it. . . . We have had three bad harvests together. . . . Above 4,200 men, women and children have been shipped off from home. . . . The humor has spread like a contagious distemper, and the people will hardly hear anybody that tries to cure them of their madness. The worst is, that it affects only Protestants, and reigns chiefly in the North, which is the seat of our linen manufacture.

Although the eighteenth-century Scotch Irish settled in every part of the colonies, like other immigrants of the period they tended to shun New England. They concentrated on the middle colonies, particularly Pennsylvania, which had generous land policies and whose chief city, Philadelphia, was the major port of debarkation of the emigrant trade. Large numbers of Scotch Irish came as indentured servants, and the emigrant trade from the Ulster ports—Belfast, Londonderry, Newry, Larne, and Portrush—seems to have been particularly venal. Many ships were overcrowded and underprovisioned, which sometimes resulted in tragedies. One of the most horrible involved the *Seaflower* out of Belfast for Philadelphia in 1746. Forty-six of her passengers starved to death, and the sixty who survived did so only by resorting to cannibalism.

Large numbers of the Scotch-Irish servants were skilled craftsmen and, after the collapse of the Ulster linen industry in 1771–72, many involved in that trade emigrated. Scotch-Irish servants seem not to have had the social mobility that Germans did—perhaps because they did not have the support of a compact ethnic community. However, some of them prospered, like Charles Thompson (1729–1824), who after emigrating from Londonderry as a youth made his mark in business and politics, serving as the secretary of the Continental Congress from 1774 to 1789.

It was from Pennsylvania that the settlement of the interior of the southern colonies proceeded, a settlement that was chiefly Scotch Irish but that also contained large numbers of Germans and lesser numbers of Swiss and Welsh. The route was the Great Philadelphia Road, which proceeded west from the port city across Chester and Lancaster counties, turned southwest at Harris' Ferry (today's Harrisburg), where it crossed the Susquehanna, passed through York and Adams counties, and after traversing the western neck of Maryland, headed down Virginia's Shenandoah Valley and then across North Carolina's central plateau on into mid-South Carolina, terminating at Camden. Migration along this route occurred in stages. The Germans pioneered the settlement of the upper Shenandoah Valley in the late 1720s–early 1730s, but in the later thirties and the forties, Scotch Irish predominated, particularly in the central and southern valley. Two counties there, Augusta, birthplace of Woodrow Wilson, and adjoining Rockingham, claim to be the most heavily Scotch-Irish counties in the United States. In the later 1740s Scotch Irish began to move further south, past Roanoke into the North Carolina Piedmont, where they had been preceded by groups of Scots who had reached it via an overland trek from the coast through the Cape Fear Valley. By the 1760s the Scotch-Irish migration had reached South Carolina. It continued to include immigrants who were still coming in at Philadelphia, but it now encompassed second-generation younger sons and their families whose parents had settled along the route. By the time of the Revolution, the North Carolina backcountry had some sixty thousand settlers and South Carolina's eighty thousand, with Scotch Irish the largest single element and perhaps a majority. While much has been written—and properly so—about the American-born pioneers after the Revolution who made their way west through the Cumberland Gap to South Pass and beyond to the Pacific, not nearly enough has been written about the immigrant's way south along the Great Philadelphia Road. It was *the* major internal migration route of the Colonial Era.

A number of myths have arisen about the Scotch Irish. The term itself did not come into general use until the 1830s and 1840s, when it seemed necessary for the Protestant Irish to differentiate themselves from the incoming Catholic Irish. In the Colonial Era and during the early republic they were most often referred to as Irish. The myths began to arise at about the same time as the new identification took hold. They were chiefly that: (1) the Scotch Irish had been especially desired as hard fighters on the frontier; and (2) that the Scotch Irish, to a man, had been patriots rather than Tories during the American Revolution. These myths can also be associated with the first of a number of Scotch-Irish Americans to become president, Andrew Jackson. (The latest was Ronald W. Reagan.) Jackson (1767–1845), the son of immigrant parents, was, as John William Ward has taught us, a mythic figure, a "symbol for an age." And in his person as well as in his persona he embodied both of the ethnic myths: the frontier Indian fighter and the anti-British patriot.

While many Scotch Irish did settle on the frontier, this was more due to the time of their arrival than to any propensity for being pioneers. Some American officials, it should also be noted, such as Ulster-born James Logan (1674–1751), the provincial secretary of Pennsylvania, did deliberately settle Scotch Irish as a kind of frontier shield. In any event, from Pennsylvania south to Georgia, large numbers of Scotch Irish settled on the North American frontier, and those who survived quickly learned the kinds of tactics appropriate there. The tactics of Europe were of no more use to them than they were to General Braddock. The frontier settlements established by the Scotch Irish developed a rough-and-tumble style of life, like frontier settlements anywhere. And the Scotch Irish quickly developed a reputation for being hard to get along with. The same James Logan who welcomed them to Pennsylvania wrote in 1729 that "a settlement of five families from the north of Ireland gives me more trouble than fifty of any other people." What Logan may have been saying here is that the Scotch Irish were less docile than the Germans, and why not? They came knowing the language, and many, because of their experiences in Ulster, had reason not to be complacent about British authority, whether at first or second hand.

This anti-British antagonism did *not* translate into 100 percent support for the patriot cause. First of all, we must remember John Adams's seat-of-the-pants notion that about a third of the American people supported the Revolution, another third opposed it, and the final third tried to be neutral. Adams, of course, was talking about 1775–76, not the 1780s, when victory seemed surer. The same kinds of divisions existed among immigrant groups, except, as we shall see, for the Scots. Germans in the Mohawk Valley on New York's Indian frontier, for example, strongly supported the Revolution and opposed their Tory landlords; Germans in and around British-occupied Philadelphia, on the other hand, collaborated with the British. They sent their food to Philadelphia for cash, rather than to Valley Forge for patriotism. The same kinds of loyalty decisions, based on their particular circumstances, were made by the Scotch Irish. In the Carolinas, for example, where the prerevolutionary conflicts between the colonial elites of the coastal zone and the more plebian backcountry were particularly pronounced, the Scotch Irish of the backcountry tended to oppose the Revolution largely because it was led by the elite who were their enemies. Some authorities suggest that lateness of arrival of so many Scotch Irish—40 percent of the total in the 1770s—meant that many of them had not yet put down colonial roots and that others feared to lose newly gained Crown land grants if they backed the rebels.

Because they and the Germans were the largest non-English immigrant groups and large concentrations of both existed in Pennsylvania, many commentators, then and now, have compared the two. (There was also significant ethnic conflict between the two, but that will be discussed in the next chapter.) Commentators on agriculture, for example, have contrasted the orderly and efficient farms of the Pennsylvania Germans with the "indolence and carelessness" of the Scotch Irish. Some modern authorities have attributed this difference to the peasant mind-set of the Germans as opposed to the tenant attitude of the Scotch Irish, but the differences are also due to the nature of the land they farmed and the marketing possibilities that were open to them. Wheat grown in German-dominated Southeast Pennsylvania, for example, could be milled into flour and marketed in Philadelphia, either for consumption there or export elsewhere. Apart from local use, there were no such markets for frontier farmers—overland shipment charges tended to double the cost of bulk commodities for every hundred miles of shipment—so surplus grain was distilled into whiskey, a product with higher bulk value, but the market, even for whiskey, was limited.

Unlike the Germans, the Scotch Irish quickly acculturated to American life. They already spoke English, so language was neither a bar to acculturation nor an adhesive to keep the community together. Nevertheless, James Leyburn's insistence that, somehow, settlers from Ulster never underwent "minority" status, never felt like "marginal men," but were "full Americans almost from the moment they took up farms in the backcountry" seriously overstates the matter and seems like a modern addition—in social science terms—to the Scotch-Irish myth. For the eighteenth century, however, they were highly literate. A petition from one Ulster community for permission to settle in America contains signatures for 306 of 319 petitioners. In America a continuing cultural problem was the absence of an educated clergy—which Presbyterianism demanded—and a place to train them. After relatively short-lived stopgap measures—such as William Tennent's "Log College" outside of Philadelphia, (1726/27–1742)—Presbyterians founded the College of New Jersey (later Princeton) in 1736, Hampden-Sydney in Virginia in 1776, and Dickinson College in Pennsylvania in 1783. The New Jersey school, it must be noted, was founded by Scots rather than Scotch Irish, but Scotch Irish participated from the beginning and supplied one of its colonial presidents.

Scots

A very different history and profile are presented by the Scots, those who came to the colonies direct from Scotland without the searing experience of the Ulster migration. Almost all the migrants to Ulster were Lowland Scots—a people of partly Teutonic origin and speaking a basically Teutonic language, the form of English that became Lowland Scots. The Highlands, on the other hand, were the home of a Celtic population that had come from Ireland in perhaps the sixth century and still spoke the Celtic tongue they had brought with them. (In the Colonial Era, Lowlanders still called Highlanders "Irish.") Migrants from Scotland consisted of both groups and several religions, while almost all the Scotch-Irish Ulsterites were, as we have seen, Presbyterians. One of the great problems of the chart with which this chapter began is that since it was based on surnames as reported in the 1790 census, there is no way properly to separate persons from Ulster and their descendants from persons from either southern Ireland or Scotland and their descendants. The assumption of most contemporary scholars is that a considerable proportion of the 8.1 percent of the white population in 1790 assigned to "Scotch" should really be "Scotch Irish." Conversely, contemporary Americans searching for Scottish roots tend to adopt what are now regarded as quintessential Scottish symbols that were strictly Highland in origin and use, such as the tartan, whether their ancestors were Highlanders, Lowlanders, or Ulsterites. Professor Gordon Donaldson of Edinburgh has commented dourly that this cult is "deplorably unhistorical, especially when adopted by descendants of Lowlanders who thought the only good Highlander was a dead Highlander."[3]

Although at least one Scot was an original colonist at Jamestown, in the first century of American life many more Scots migrated elsewhere—to Ulster; into England; to the Hebrides, Orkney, and Shetland; and to the European continent—than came to the New World. While Ulsterites were colonists, Scotland was an independent nation, although it shared a monarch with England from 1603 to 1707, when the realms were finally merged into the United Kingdom. During the former period Scottish colonies were actually established in the New World—most notably and disastrously in the settlement at Darien (New Caledonia) in present-day Panama (1686), but also short-lived plantings in Nova Scotia (1629) and in what are now New Jersey (1683) and South Carolina (1684). The latter two were largely refuges for dissenters, the first for Quakers, the second for Presbyterians. Many of the Scots who were involved in these New World ventures—and in subsequent voluntary emigration there—were what Professor Donaldson has called "somewhat unrepresentative Scots," those who had been at least partially Anglicized in culture and were in close touch with English colonial investors and interests.

Involuntary migration of a very special kind was practiced by Scottish and British governments: Religious and political dissenters as well as criminals were deported to America. Before the 1730s more Scots may have come to American colonies as punishment than came voluntarily. All three times that Oliver Cromwell's forces defeated Scottish armies, hundreds of prisoners were sent to America, as were followers of the earl of Argyll's rebellion in 1685 and the better-known Jacobites taken—and not hanged—after the doomed pro-Stuart risings of 1715 and 1745. Among the religious dissenters sent across the Atlantic were Quakers considered troublemakers in England, Ireland, or Scotland, as were Scots Presbyterians and Anglicans at one time or another as the established church of Scotland changed. Some Highland clans maintained their Catholicism and sometimes migrated as groups, bringing their priests with them.

After the 1730s, as noted, voluntary migration predominated, although many Scots came as indentured servants, with the combination of voluntarism and coercion that indenture suggests. This is especially true with regard to the period from 1763 until the outbreak of the Revolution, when some twenty-five thousand Scots, largely from the Highlands, went to America. The Highlands were probably never fully self-sufficient, and raiding the more fertile Lowlands had been a form of economic activity for the armed Highland clans for centuries. The establishment of law and order and the demilitarization of the clans meant that the Highlands had a surplus population and that some of the most substantial figures in the Scottish elite, the tacksmen—large leaseholders who had been the clan chiefs' organizers of armed forces—became redundant as well. In some cases the tacksmen led mass migrations. In perhaps the largest of these, tacksmen brought some five thousand persons, in families, from the island of Skye in the Hebrides—more than a fifth of its population—to the Cape Fear Valley in North Carolina. And much smaller group migrations established settlements in New York's Mohawk Valley and on Prince Edward Island (Canada). Organized group migrations of Lowlanders were sometimes abetted by associations formed for that purpose in Scotland. The earliest of these was late in the colonial period—in 1773—and they continued to be formed and function well into the nineteenth century. Some of these societies not only planned and financed emigration but also bought land in America cooperatively.

Despite the multiplicity of religions among Scots of the period, the overwhelming majority of Scots—stay-at-homes and immigrants alike—was Presbyterian. For Highland Presbyterian migrants the Gaelic language was for a time a cohesive force, particularly when it was used in church services, but that practice had all but died out by the middle of the nineteenth century. Most Scots seem to have acculturated rather quickly; many of them, however, retained a kind of loyalty to their motherland, and it was not unusual for Scots who had "made it" here to leave some of their estate to benefactions in Scotland. Andrew Carnegie (1835–1919) is, of course, the archetype, but he was following a tradition that had been established by colonial migrants. Alexander Milne (1742–1838), who immigrated in 1776 and made his pile in New Orleans, left $100,000 to found a school in his hometown, while James Dick (1743–1828) bequeathed part of his estate for the benefit of schoolteachers in three Scottish towns. The concern of Scots for education was a reflection both of the values of Scots life and of the experience of those persons who came. A relatively high percentage of Scots were merchants and traders in American ports and numerous others were professionals of one kind or another. Scots were more prominent in American revolutionary politics than their incidence in the population would suggest. Of the fifty-six members of the Continental Congress that adopted the Declaration of Independence, no fewer than eleven were either of Scots birth or ancestry.

Yet despite this prominence in revolutionary politics, many Scots—perhaps even a majority—favored the Crown, and Scots and their descendants were certainly overrepresented among the 80,000 Loyalists who emigrated to Canada and Britain after independence was established. Lowland merchants and traders had continuing ties to the Scottish economy—remember Jefferson's remark about Edinburgh mercantile houses—and Highlanders, who might have been expected to oppose the Crown, which they had battled so often and so recently, often followed the course of their chiefs and tacksmen, who, as holders of Crown land in America, maintained that loyalty. In addition, many Scots were relatively recent arrivals who had not had time to develop local political loyalties. In any event, Scots are the only ethnic group likely to have contained a Tory majority, although, had they had the opportunity, many more than the five thousand Afro Americans who, promised freedom, left with the British, would have done so.

Irish

The history of the Irish in colonial times is highly complex and, as we have seen, it is often impossible to tell whether a given entry in a historical document refers to a person of Irish, Scotch Irish, or even Scots ancestry. But it is clear that Irish came and were sent to the New World in large numbers from the earliest times. In the seventeenth and eighteenth centuries English and British governments enacted a harsh series of penal laws against Catholics and Protestant dissenters, that is, those not members of the Anglican Church of Ireland. Catholics were forbidden to acquire land from Protestants unless they converted, could not lease land for longer than thirty-one years and, when a family head died, all land had to be divided among all descendants, although Protestants could and did practice primogeniture. By the time of the American Revolution, only about 5 percent of Ireland's land remained in Catholic hands, though Catholics represented about 75 percent of Ireland's roughly two million people. Dissenting Protestants were forbidden to hold government posts or become lawyers, but these laws were gradually abandoned by the early eighteenth century and were not much enforced against Presbyterians. Irish Quakers, however, were still subject to severe disabilities. In the Cromwellian period in the 1640s, thousands of Irish Catholics—military and political prisoners and their families—were shipped overseas, and some wound up in British North America.[4]

Irish already had a migratory tradition and had been crossing to England for seasonal work and occasional settlement since the middle ages. After the Cromwellian persecutions, more than thirty thousand, many of them members of elite or formerly elite families, left for the European continent, where Irish contingents of soldiers and others served Catholic monarchs well into the nineteenth century. However welcome Irish might have been on the Continent, they were generally unwelcome in the North American colonies, as we have seen from Lord Calvert's experience in Virginia. An undeterminable number of Catholic Irish came as indentured servants, but there were enough so that at least three colonies—South Carolina, Virginia, and Maryland—passed laws, probably ineffective, aimed at stopping or reducing the importation of Irish papists at the end of the seventeenth century.

And not all Irish Catholics were poor. One Charles Carroll moved to Maryland in 1681 after his estates in Ireland were confiscated. Almost a century later a descendant would add his name to the Declaration of Independence with a flourish—Charles Carroll of Carrollton. He was the only Roman Catholic signer. At times during the Stuart restoration Catholics were appointed to high positions in the colonies: Sir Thomas Dongan, who served as governor of New York (1682–88), was the most prominent of these. There were well-to-do Irish Catholics in New York City and Philadelphia by the mid-eighteenth century, and even in pope-hating Boston the Charitable Irish Society, which had been formed by Protestants, was willing to admit Catholic donors about the same time.

During the Revolution, most of the Irish seem to have been supporters of the patriot cause: Few can have had any serious affection for the British crown. Not only were there specific Irish units—such as the Volunteers of Ireland, raised in Philadelphia and containing both Catholics and Protestants—but John Barry (1745–1803), born in County Wexford, became the "father" of the American Navy.

If we know relatively little about the Irish Catholics in colonial America—the not-particularly-reliable 1931 survey at the beginning of this chapter (see table 4.1) indicates that there were more than one hundred thousand of them—it is because the overwhelming majority of them were poor and scattered: They were clearly overrepresented, not only in the port cities but also in their penal and charitable institutions. Large numbers of these were what John Carroll, the first American Catholic bishop, called "name only Catholics," since, in 1783 there were only 25,000 Catholic communicants in the whole country; many others had drifted into Protestant denominations because there were relatively few Catholic churches and priests anywhere in America. In the next century Irish missionary priests reported that the southern states were full of Protestant descendants of Irish Catholic settlers who had never seen a priest or been able to attend a Catholic service. The prejudice that existed against them—however abated by the Revolutionary Era, in which ideology was clearly more important than ethnicity—would be greatly heightened when very large numbers of their compatriots, many of these destitute, began to arrive in the next century.

Welsh

So few Welsh came to America in the colonial period that they are not even included in the table that began this chapter. Had they been scattered, as the Irish were, it would be all but impossible to say anything significant about them. But because their settlement was funded, well organized, geographically discrete, and persisted, it is not only possible but appropriate to say something about them here. Wales, which had about four hundred thousand people in the eighteenth century, was part of the so-called Celtic fringe of Britain, and the one in which the Celtic language—Cymraeg, or Welsh—tended to persist longer and more generally than those spoken by other Celts in the British Isles—Highland Scots, Irish, Manx, and Cornish. But the Welsh who emigrated in the Colonial Era were largely acculturated to English, as were most of the Welsh gentry by that time.[5]

The earliest group to come was a Baptist congregation that founded a settlement it called Swansea (1677) in Massachusetts near the Rhode Island border. The major migration of this era began in 1681 when Welsh Quaker gentry obtained a tract of perhaps 40,000 acres in Pennsylvania, but any hopes they may have had of establishing an exclusively Welsh Quaker settlement were short-lived, as within a few years not only Welsh of other confessions but English and Germans were settling in the same area. By 1790, the geographer James Lemon has estimated, there were perhaps ten thousand persons of Welsh birth or ancestry in southeastern Pennsylvania. In these and smaller Welsh settlements in other parts of Pennsylvania, Delaware, and the Carolinas, whatever Welsh language there was did not outlast the first generation, but Welsh place-names—Bryn Mawr, Radnor—still grace the Pennsylvania landscape. The degree of Welsh concentration in the colonial period can be seen in the fact that of twenty-one churches founded by the Welsh—nine Baptist, six Anglican, two Presbyterian, and four Quaker meetings—all but four were in Pennsylvania, and two of those were in neighboring Delaware. The Welsh in Pennsylvania were generally well-to-do, reflecting their origins. Lemon has calculated from 1782 Lancaster County tax rolls, that the Welsh, although only one percent of the population, were 7 percent of those paying more than forty pounds in annual taxes.

Dutch

The next three groups to be treated, the Dutch, French, and Spanish, were each at one time colonial powers, so that to a much greater degree than with, say, the Scotch Irish or the Irish, a fuller spectrum of persons came to the New World, including elite rulers, merchants, and soldiers, as well as ordinary settlers. Although this pattern is evident for all three groups, nowhere is it clearer than for the Dutch. The Dutch colonial empire in North America—as opposed to the Caribbean—lasted for just half a century, from the establishment of Fort Nassau (later Fort Orange and eventually Albany, New York) in 1614 to the conquest of New Netherland—which became New York—by the British in 1664. During that fifty-year period, according to the careful analysis of Ernst van den Boogaart, fewer than six thousand Dutch persons migrated to New Netherland and the former Swedish colonies on the Delaware, which the Dutch conquered in 1655. The estimated one hundred thousand Dutch persons at the first census were mostly descendants of those six thousand, illustrating very nicely Adam Smith's remarks about colonial American natural increase. There was only scattered Dutch migration during the remainder of the Colonial Era, most of that going to the New York area: In 1790 an estimated 80 percent of all Dutch immigrants and their descendants lived within fifty miles of New York City. The major exception to this clustering were small groups of Dutch Quakers and Mennonites who settled in Pennsylvania.[6]

The seventeenth century was, of course, the Dutch Golden Age. It was a period of great Dutch self-confidence, prosperity, and overseas expansion in which the small Protestant republic founded empires in Asia and the New World that endured for more than three centuries. The nation of the Dutch—the Netherlands—was forged in the epic struggle between the Protestant Hollanders and the Catholic Hapsburg Empire in the late sixteenth century and the first half of the seventeenth, with seven provinces north of the Scheldt River—the United Provinces—establishing their independence. Most of the Dutch who came to America—and almost all of those in the Colonial era—were Hollanders who spoke Dutch, a Low German linguistic form. Of the three other groups, the largest were the Frisians, a minority group in both the contemporary Netherlands and West Germany, whose language is separate although related to Dutch. Two smaller groups among immigrants to the United States have been Flemings—Flemish-speaking Belgians whose language is to all intents and purposes Dutch—and the Dutch-speaking people from East Friesland and other German areas bordering on Holland. United States Census reports follow national boundaries, not language or ethnicity, so that, for example, Frisians born in the Netherlands are listed as Dutch, while Dutch-speaking Flemings are listed as Belgians, along with the French-speaking Walloons.

New Netherland was a commercial enterprise, originally a spin-off from the fur-trading activities of the Dutch West India Company. One of its most distinctive characteristics was the establishment of feudal baronies, called patroonships, along the Hudson River. Life under such conditions did not seem attractive to relatively prosperous Dutch peasants, so that most of those who came, including some fifteen hundred indentured servants, were from the lower social orders. Robert P. Swierenga, with perhaps a little hyperbole, has written that

> the colony primarily attracted religious refugees and the poverty-stricken—Belgian Walloons, French Huguenots, Italian Waldensians, Dutch soldiers and Jewish settlers from an aborted Brazilian colony, teenagers from Dutch poorhouses and orphan asylums, families of unemployed day laborers from Amsterdam, agricultural tenants from landed estates, and a few political visionaries. [Scandinavians could be added to this list as perhaps three hundred Finns and Swedes came during Dutch rule.]

In addition, the colony attracted a largish migration of English and English American settlers, most of whom crossed to Long Island from southern New England. Perhaps 30 percent of the colony's population of ten thousand at the time of the English conquest in 1664 was non-Dutch. And, already in the Dutch years, New York City, more than any other city in North America, was gaining its reputation as a polyglot port. In 1643, when the population of all New Netherland was perhaps four thousand persons, a French Jesuit, Isaac Jouges, who had been captured by and escaped from the Mohawk Indians, visited New Amsterdam and made the most quoted remark in the history of colonial New York: that eighteen languages could be heard on its streets. Although it is difficult if not impossible to imagine what all eighteen might have been, the remark does illustrate the ethnic complexity that American cities—and particularly port cities, even provincial Boston—achieved very early in their history.

But, just as interesting as its diversity, is the example that New York gives us of ethnic persistence. Although the small Dutch population—which was not significantly augmented by immigration in the Colonial

Era after the English conquest—lost political power, it retained and in some cases augmented its economic power. In addition the Dutch had put down deep cultural roots: the Dutch Reformed churches continued to use the Dutch language and accept the discipline of the Classis of Amsterdam until more than a century after the conquest, and in the later eighteenth century they established two institutions of higher learning to train American clergy: Rutgers College (1766; originally Queens College) and New Brunswick Seminary (1784), both in New Jersey. Although the Dutch Americans did not enjoy much political power in either eighteenth-century colonial America or in the early republic—the last patroon, Stephen Van Rensselaer, did cast the determining vote in the House of Representatives during the disputed presidential election of 1800. Since that time Dutch stock has furnished three presidents of the United States, all elected from New York—Martin Van Buren and the two Roosevelts.

French

Although there were fewer than one hundred thousand persons of French origin in the United States (or what would soon become the United States) in 1790, they came from three different sources and had strikingly different histories. The first group, widely dispersed in scattered settlements in the long arc that runs from Detroit to New Orleans via Saint Louis, were original European settlers, all that remained of the French dream of colonial domain in North America. The second and third were ethnic and religious refugees, the Acadians and the Huguenots, whose almost polar American experiences can be made to serve as exemplars of the vexed question of assimilation/acculturation that is so central to American life. (The fourth and fifth streams of French-speaking migrants—the French Canadians of the nineteenth century and the Haitians of the twentieth—will be discussed later.)[7]

The first group was very small, and although French-speaking individuals played important roles in the American Revolution, in the West, and later in the development of the fur trade, they were soon swamped by emigrants from the states and other parts of Europe: the French of the Midwest are largely icons to be remembered on monuments and postage stamps and, of course, in place-names like Prairie du Chien, Wisconsin; Louisville, Kentucky; and Vincennes, Indiana. Only at the southern tip of the arc, in Louisiana, did French culture and institutions develop lasting roots.

Louisiana, once the name of a vast, empty domain several times larger than France, survives as the name of one southern state, and it is in this latter sense that the term will be used here. New Orleans, founded in 1717, and much of its hinterland remain at least psychologically French even today; Catholicism remains strong there; in Louisiana, alone of American states, French law—to be precise, the Code Napoleon—was the foundation of state law; and the primary geographical subdivisions are called parishes rather than counties. The initial French settlement was reinforced during the late eighteenth and nineteenth centuries by a small stream of French and French-speaking immigrants from Europe and the Caribbean; as late as 1860 there were fifteen thousand French-born residents in New Orleans.

But the crucial migration to Louisiana, in cultural terms, was the migration there of French-speaking refugees from what is now called Nova Scotia but which they called Acadia. These are the Acadians—or, as they came to call themselves, Cajuns—who have become a distinct ethnic group and whose entire history will be summarized here. In 1980 some eight hundred thousand Cajuns were living in south and south-central Louisiana, and another twenty thousand in the Saint John River Valley in Maine. (The Maine Acadians will not be further noted here.) They are descended from French immigrants to Canada, largely from Normandy and Brittany, who settled Nova Scotia in the seventeenth century. As a result of the Treaty of Utrecht in 1713, which ended the War of the Spanish Succession (Queen Anne's War), Nova Scotia and its inhabitants were ceded to England, as all French Canada would be fifty years later. After more than four decades under British rule, the six to eight thousand Acadians were driven from their farms by British soldiers and shipped into exile to various parts of British America. This exile was, of course, immortalized by Henry Wadsworth Longfellow in *Evangeline* (1847), but a century earlier few New Englanders had any sympathy for French Catholics.

No Acadians were originally sent to Louisiana—which was French until 1762, Spanish from then until 1800, and American after 1803—but some probably reached there as early as 1756. Relatively large-scale immigration began in 1765–66; Spanish officials, glad to receive industrious farmers, granted them parcels of land and supplied seed and other necessities for those who were destitute or close to it. For the next twenty years several thousand Acadians migrated to Louisiana from the American colonies, from the West Indies, and even from France, to which many had returned. Cajun settlement in south and southwest Louisiana was intensive and exclusive. Few others lived there until well into the twentieth century. Most Cajuns became and remained small farmers, as they had been in Nova Scotia, but a few became wealthy. By 1810 the census in two Cajun parishes—white population just over six thousand—showed the presence of more than twenty-five hundred slaves held by a tiny minority of whites. The agriculture they practiced was, of course, quite different from that of Nova Scotia: some, learning the ropes from Spanish neighbors, became cattle raisers on *vacheries* (ranches) in the prairie country of southwestern Louisiana. Most, however, lived in the bayou country.

Isolated and set off from most of their neighbors by their language and Roman Catholic religion, the Cajuns became a distinct ethnic group with their own cultural patterns. As a group they had little regard for education, and many Cajuns, until very recently, never learned to read or write. Very large families—ten or twelve children were not unusual—resulted in steady population growth from perhaps 35,000 in 1815 to some 270,000 in 1880. There was no significant migration from overseas during this period, but often when Cajun women married non-Cajun neighbors, the culture simply absorbed the outsiders, so that by the twentieth century there were Cajun-speaking families with such names as Schneider, Higginbotham, and Hernandez.

In the twentieth century the process of modernization and the intrusion of the state began to chip away at Cajun exclusiveness. In the outpouring of nativism after World War I, the use of Cajun French in schools was prohibited by the Louisiana constitution, and in some instances pupils were punished for using the language on playgrounds during recess. Compulsory school attendance to age fifteen, legally required after 1916 but not seriously enforced until 1944, resulted in more and more Cajuns being unable to speak French. Roads and other improved infrastructure, accelerated during the administrations of Huey P. Long and his successors in the years before World War II, made it easier for Cajuns to leave and outsiders to come in. The oil and gas boom of the 1950s and radio, movies, and television have all intruded on, modernized, and at least partially diluted Cajun culture.

Only after the culture was endangered were serious efforts made to preserve it. There are now several organizations in Louisiana that are attempting, in various ways, to preserve—and also to exploit—Cajun culture. The University of Southwestern Louisiana (Université des Acadiens) has a thriving Center for Acadian Folklore and Culture. A statewide Council for the Development of French in Louisiana actively fosters study of French culture, history, and language in the elementary and secondary schools. Louisiana law now mandates that school boards provide programs of instruction in French at all grade levels once a quarter of the district's parents petition for it, but these programs, paradoxically, further threaten Cajun French because what is taught is the French of contemporary France. And, in a move designed more to attract tourists than preserve a culture, the state has designated twenty-two parishes as "Acadiana," and what were once truly local folk festivals to celebrate Crawfish, Rice, and Yams, are now actively promoted to attract outsiders, both non-Cajun Louisianans and out-of-staters.

The Huguenot story, while also one of exile, is quite another matter. In 1685 Louis XIV, in his ongoing crusade to extirpate Protestantism in France, revoked the Edict of Nantes (1598), which had guaranteed Protestants freedom of worship. In the decade of the 1680s, in the face of severe and vicious persecution, Huguenots in France declined in number from about one million to some 75,000 in a population of perhaps twenty million. Most simply abjured Protestantism; fewer than 250,000 emigrated. It is fascinating—but fruitless—to speculate about what would have happened had Louis and his ministers sought to encourage Huguenot settlement in France's empty North American domain, but such was not the case. Between 1550 and 1620 France's Protestants had tried to found New World colonies from Rio de Janeiro to South Carolina, but all failed. Between then and the 1680s only scattered individual Huguenots came to the New World, largely to British North America and French Canada, but no Huguenot churches were established. These came only in the refugee migration of the 1680s and after. Jon Butler, in his meticulous study *The Huguenots in America* (1983), shows that fewer than two thousand came by 1700 and that only a few hundred came after that. These numbers are much smaller than historians have generally believed, and a paragraph or so on the Huguenot migration is *de rigueur* in general American histories that ignore more numerous groups, such as the Acadians. There are two basic reasons for this: First of all, the Huguenot, the Protestant fleeing Catholic oppression, fits very nicely into one of the major myths of American migration. Second, during the period in which they migrated and after, New England preachers used the real persecution that the Huguenots suffered as an analogy—which they believed but we now know to be largely false—with what their own forebears had endured in England. The redoubtable Cotton Mather (1663–1728), for example, was preaching about the Huguenots' misfortunes as early as 1681, before there were even a dozen of the refugees in Boston.

The Huguenots settled and established Huguenot churches in three very different American colonies: Massachusetts, New York, and

South Carolina. The first was the most homogeneous American region, the second most polyglot, and the third a center of slavery. In each a few Huguenots quickly rose to the very top level of society. André Faneuil came to Boston with some capital, and the family achieved spectacular success as merchants and ship owners. In 1742 it showed its civic spirit by building Faneuil Hall and presenting it to the city. A second-generation member of another rich merchant family, James Bowdoin (1726–90) became a revolutionary governor of Massachusetts. (The person most often identified as a Boston Huguenot, the silversmith and patriot Paul Revere [1735–1818], was the son of a Huguenot who arrived only in 1715.) In New York the De Lanceys became one of the richest and most prominent merchant families. So did the Manigaults in South Carolina; when Gabriel Manigault (1704–81) died he left an estate that included almost fifty thousand acres of land, much other property, and 490 slaves. While not all Huguenots, of course, were or became rich—some were actually poor—most seemed to have done very well, very fast. In South Carolina, where the records are best, Huguenots as a community achieved extraordinary success. Despite a brief period of anti-Huguenot nativism there around 1700, Huguenots were soon holding seats in the colonial legislature at a rate of more than twice their incidence in the white male population.

The Huguenots, then, most of whom came with skills and some of whom came with capital, achieved significant economic, political, and social success. Yet, as an ethnic group, they are a prime example of rapid disappearance. What Butler says about South Carolina Huguenots holds true, with certain variations, for all three major loci of colonial Huguenot refugee settlement:

> South Carolina Huguenots began very quickly to assimilate within this newly evolving society. . . . By mid-century these Huguenots simply were South Carolinians; aside from their surnames they were indistinguishable from all other settlers in the colony except, of course, slaves.

We thus see a paradox that will be a continuing one within American polyethnic society: Some ethnic groups survive, others do not. In terms of ethnic persistence, the Cajuns of Louisiana and Maine have been a success story; the Huguenots of New England, New York, and South Carolina, a failure. Yet in terms of the general standards of American society, it is the Huguenots, not the Cajuns, who are the successes. For the former, again quoting Butler, "everywhere they fled, everywhere they vanished"; for the latter, as we have seen, ethnic persistence has continued to the present. And while there are no formulas to explain persistence or its lack, these two cases can show us some of the factors involved in each.

The settlement patterns of the Acadians, concentrated and isolated, worked for persistence; those of the Huguenots, a very small group widely dispersed and intermixed with largely English-speaking settlers, worked for disappearance. The Huguenots also suffered because their religious institutions were so similar to those of their Protestant neighbors that Huguenot churches disappeared. Martin Marty has argued that ethnicity is "the skeleton of American religion." Perhaps. But in the case of the Huguenots it might be argued that religion was the skeleton of ethnicity, and that the dissolution of distinct Huguenot churches heralded the disappearance of the ethnic group. Similarly, as we have seen, the absence of Roman Catholic churches in the South meant that many Catholic indentured servants and their descendants were unchurched. The Cajuns, on the other hand, developed and maintained Catholic churches so that a discrete religion nourished a discrete ethnicity and vice versa. But in the final analysis, it was the successful acculturation to the standards of colonial American society, whether in Puritan Boston or secular Charleston, plus a steady assimilation in biological terms as more and more Huguenots, male and female, married non-Huguenots, that quickly promoted their disappearance as an ethnic group. Conversely, for the Cajuns, it was their successful resistance to some aspects of American culture, their lack of exogamous marriage and the absorption into Cajun culture of many of the few outsiders who married Cajuns, that enabled their long and largely unchallenged survival.

The question of survival—and of who or what survives—is a matter to which we shall return more than once. Although the notion of a melting pot is a myth, some groups did in fact, melt, and all groups, even those who persisted the longest, changed over time through contact with the new environment and new peoples. The ancestors of the Cajuns, seventeenth-century immigrants from Normandy and Brittany, spoke the dialects of their regions: Twentieth-century Cajun French would not be understood in Saint Lo or Brest. These changes sometimes bewilder the immigrants and their descendants. An older second-generation Japanese American couple, after a recent visit to contemporary Japan, expressed that confusion nicely:

"Our parents," one of them said while the other nodded in agreement, "were much more Japanese than the Japanese who are in Japan now."

Spanish

Although Spanish immigration is properly associated chiefly with Latin America, it is too often forgotten that Spanish explorers not only were the first Europeans to traverse much of the United States, but that they also founded the first successful European colony there. Saint Augustine in northern Florida was founded in 1565, forty-two years before Jamestown, fifty-five years before Plymouth Rock. And the proximate cause of its founding, more than half a century after Ponce de Leon's unsuccessful attempt in 1514, was to provide a base from which to destroy the French Huguenot colony in South Carolina, mentioned earlier. Most of the Spaniards who came to Florida in the sixteenth and seventeenth centuries were soldiers and priests, as first Jesuits and then Franciscans tried to convert and "civilize" the Indians. The former were unsuccessful and withdrew in 1572; the latter, supported by punitive military expeditions, managed to establish thirty-eight missions by 1655 and claimed some twenty-six thousand Indians as at least "partially converted." Florida remained Spanish until 1819; eleven years later the first U.S. Census taken there showed only thirty-four thousand persons, many of whom were Americans or Britons. Spanish population in Louisiana, which was Spanish only between 1762 and 1800, was minimal, consisting chiefly of officials, many of whom returned to Spain. And not all of these were ethnically Spanish: The most noted governor was named O'Reilly.[8]

West of the Mississippi the Spaniards and Spanish Mexicans were also the pioneer "European" settlers. They settled first in New Mexico in 1598, nine years before Jamestown. At the end of the next century, Spanish Mexicans made the first settlements in Texas—reacting, as in Florida, to a French presence. Similarly, there was no Spanish Mexican settlement in California until 1769, after the Russians had established a tiny post at Fort Ross, about one hundred miles north of San Francisco. In that year Fray Junipero Serra founded the first of his twenty-one missions in San Diego and the accompanying soldiery set up a *presidio* (military headquarters) there.

These settlements had very different histories. Those in Texas were sparse and largely unprofitable. By 1821, more than a century after its first settlement, there were perhaps four thousand Tejanos, non-

Indian residents of Texas. Shortly thereafter they were swamped by immigrant Americans so that, by 1835, when the Texas Revolution occurred, Tejanos were a distinct minority in what had been their own country. In California after the often brutal clerical pacification of the Indians in the choicest parts of the state, a thriving pastoral culture developed among the Californios. The population was about thirteen thousand—excluding Indians—at the time of American annexation (1848). The gold rush of the next year raised the population to perhaps one hundred thousand and the Californios, like the Tejanos, were a minority in what had been their land. In New Mexico, which was annexed at the same time as California, an entirely different situation prevailed. The Spanish-Mexican culture was not only firmly rooted, but there was also no immediate inundation by outsiders. The first U.S. census showed a population of sixty thousand, the vast majority of whom were of Spanish-Mexican origin. Such persons, who in New Mexico traditionally have called themselves Spanish Americans, remained a majority of the state's population until about 1940. Today, reflecting that history, New Mexico is our only officially bilingual state, Spanish having—in law—equal status with English.

For decades, except in New Mexico, which was a backwater, the question of Spanish-Mexican or Hispanic immigration and culture was almost totally ignored, but in the twentieth century the migration and settlement of such persons would become, as will be shown, one of the most important demographic developments.

Swedes

Sweden, briefly a major force in European power politics as a result of the Thirty Years War (1618–48), made a belated and short-lived attempt at founding an American empire. In 1638 a tiny settlement called Fort Christina was established on the Delaware by the Swedish West India Company on the site of present-day Wilmington. The Dutch from New Amsterdam had established a trading post—called Fort Nassau and the "city colony"—in the region two years earlier. The Swedes, who began with traders and soldiers, sent some farmers in 1639, and settlement slowly spread along the river valley. Even after the Dutch conquest of New Sweden, in 1655, some belated immigrants from Sweden continued to arrive. By 1664, when the Dutch, in turn, were conquered by the English, the total population along the Delaware was around eight hundred persons, mostly Swedish. Sporadic Swedish migration continued, much of it through Amsterdam.

The scholars who tried to assess the ethnic origins of the American people in 1790 put the number of persons of Swedish ethnicity at around twenty thousand, but, as Ulf Beijbom has noted, the figure seems clearly inflated. And a goodly number were not ethnic Swedes at all, but Finns. In the days when the concept of "immigrant gifts" was dominant in the literature of immigration, it was usually pointed out that a major contribution of the Swedes and Finns was the log cabin, more precisely the dovetailing technique by which it was put together. Many students of vernacular architecture, however, now doubt that the nearly universal log cabin was diffused solely from the Delaware region, although it seems to have been first used there, following practices already established in the forested areas of northern Europe. (The schoolroom pictures that show log cabins along with Pilgrims, Indians, and turkeys are anachronistic.)[9]

A few Swedes made their mark in revolutionary politics: John Morton, né Mortenson, was a Pennsylvania signer of the Declaration of Independence, and during 1781–82, John Hanson of Maryland was a presiding officer of the Continental Congress. But the Swedes of the Colonial Era established no lasting ethnic institutions, and while certain families may have remembered their heritage, there was no effective Swedish American ethnic group per se. That would have to wait for the renewal of Swedish migration, on a much larger scale, in the nineteenth century.

Jews

The Jewish experience in colonial America demonstrates that neither size nor geographic concentration was necessary for ethnic survival. Numbering perhaps two thousand as late as 1790, spread from New England to Georgia, Jews not only survived but thrived, establishing synagogues in five port cities: New York (1656), Newport (1677), Savannah (1733), Philadelphia (1745), and Charleston (1750). The problem of an educated clergy, so vexing to many colonial faiths, was not a problem for colonial Jews. They didn't have any. The first ordained rabbi did not come to America until 1840. The synagogues were run by their members, although, to be sure, they often looked for both financial support and authoritative interpretations of religious law from abroad.[10]

We think that we know precisely who the first Jews in North America were and when and why they came. In 1654, twenty-three Jewish refugees arrived in New York from Recife, having been expelled with other settlers after the Dutch colonies in Brazil were conquered by the Portuguese. Although the New York authorities did not want them either, pressure from the Dutch West India Company in Amsterdam, which had Jewish shareholders, enabled them to stay. They were Sephardic Jews, descended from the tens of thousands who had been expelled from Spain (and later Portugal) in 1492 and subsequently. In the seventeenth and eighteenth centuries, Jews were more statistically significant in some Caribbean colonies—Dutch Surinam and Curaçao and British Jamaica—than they were on the mainland, and many of the Jews who immigrated during that period came from or via the Caribbean rather than Europe. The Ladino-speaking Sephardi—Ladino is a variety of Spanish written in the Hebrew alphabet—were joined in the eighteenth century by Ashkenazi immigrants who used German or sometimes Yiddish, a Low German dialect written in the Hebrew alphabet. These latter came largely via London or the British Caribbean colonies, although some came directly from the Continent.

More than any other group that we know of in the colonial period, the Jewish community was continental as well as local. Jewish merchants in the various ports communicated and did business together, and—not surprisingly due to the small number of available marriage partners—interurban marriages were not uncommon. Nor were marriages outside the faith. Sometimes, as when Phila Franks, daughter of the prominent New York merchant Jacob Franks (1688–1769), the first Ashkenazic *parnas* (lay head of a synagogue) there, married the Huguenot Oliver DeLancey, she was simply lost to the faith. But soon, against all Jewish tradition, some congregations, including New York's Shearith Israel and Philadelphia's Mikveh Israel, allowed Jewish spouses in mixed marriages to be seatholders in the synagogue. A Philadelphia Jewish physician, Dr. David Nassy, who came from the Caribbean in 1792 and returned to Surinam in 1796, described the practice:

> There are [many families in which men are] lawfully married to Christian women who go to their own churches, the men to their synagogues, and who, when together, frequent the best society.

Jews, and to a lesser degree other dissenters, suffered legal disabilities in both Britain and the colonies. In 1740 Parliament permitted the naturalization of Jews, and in many colonies laws that discriminated were either repealed or ignored. Despite the religious neutrality of the American Constitution—and particularly the "no establishment" clause of the First Amendment—some state laws continued to prescribe religious tests for officeholding into the second half of the nineteenth century. Social and personal anti-Semitism existed, but the official tone was secular and tolerant, as the famous exchange between President George Washington and the Newport synagogue indicates. This was best symbolized, perhaps, by the parade held in Philadelphia to celebrate Pennsylvania's ratification of the federal Constitution. As Benjamin Rush (1745–1813) described it:

> Pains were taken to connect ministers of the most dissimilar religious principles together, thereby to show the influence of a free government in promoting Christian charity. The Rabbi of the Jews, locked in the arms of two ministers of the gospel, was a most delightful sight.

In this chapter and the previous one we have examined the non-English groups in the American colonies, generally using the first census of 1790 as a cutoff point. The next will show how these groups interacted with the largest ethnic group, the English—whom John Higham calls the charter group—and with each other, and how and to what degree they formed a new people called Americans.

5

Ethnicity and Race in American Life

Perhaps the most persistent rhetorical question in our history is, What is an American? The most famous—and misleading—answer to that question was published in the year before the American Revolution ended. Its French author, a Norman petty nobleman, Michael-Guillaume-Jean de Crèvecoeur (1735–1813), is a most curious person for Americans to accept as a definer of their nationality, but it has become almost obligatory to do so. Crèvecoeur came to America with the French army and, after Montcalm's defeat, toured extensively in the British colonies and settled and was naturalized in Orange County in upstate New York. A Loyalist during the Revolution, he abandoned the United States in 1780, although his book, published two years later under his American pseudonym, J. Hector St. John, was titled *Letters from an American Farmer.* In 1783 he returned to the United States as French consul in New York, where he stayed for seven years before returning permanently to France. In the most quoted passage in his book he asked:

> What then is the American, this new man? He is either an European or the descendant of an European, hence that strange mixture of blood, which you will find in no other country. I could point out to you a family whose grandfather was an Englishman, whose wife was Dutch, whose son married a French woman, and whose present four sons have now four wives of different nations. *He* is an American, who, leaving behind him all his ancient prejudices and manners, receives new ones from the new mode of life he has embraced, the new government he obeys, and the new rank he holds. He becomes an American by being received in the broad lap of our *Alma Mater.* Here individuals of all races are melted into

> a new race of men, whose labours and posterity will one day cause great changes in the world.

Although even casual students of intellectual history will recognize that effusion as a product of French romanticism rather than American experience, it is easy to see why it became and has remained so popular. In the first place it was flattering—"will one day cause great changes in the world"—and, to later generations, the almost uncanny prefiguration of the myth of the melting pot makes Crèvecoeur seem prophetic. The polyethnic family he describes, while certainly possible, especially in New York, was clearly not representative. And the notion that an immigrant could shed his culture the way a snake sheds his skin is nonsense. More than a century later a lesser known immigrant from Romania, Marcus Eli Ravage (1884–1965) wrote a much more realistic analysis of the problem of acculturation.[1]

> The alien who comes here from Europe is not the raw material that Americans suppose him to be. He is not a blank sheet to be written on as you see fit. He has not sprung out of nowhere. Quite the contrary. He brings with him a deep-rooted tradition, a system of culture and tastes and habits—a point of view which is as ancient as his national experience and which has been engendered in him by his race and his environment. And it is this thing—this entire Old World soul of his—that comes into conflict with America as soon as he has landed.

Neither of these nearly polar views is, of course, a substitute for modern social analysis. During the eighteenth century and before, there was an intermediate step between being a European and an American: It was being a colonial. Nicholas Canny and Anthony Pagden have analyzed what they call the development of "colonial identity in the Atlantic world" in the three centuries before 1800. They describe a process in which colonizing individuals and groups come to see themselves as distinct and different *both* from people back home and from indigenous persons in the place of settlement. Thus Lowland Scots transplanted to England's first plantation in Ireland quickly differentiated themselves from stay-at-home Scots on the one hand and the native Irish on the other.[2] Yet if we imagine two immigrant Ulstermen, one Presbyterian Scotch Irish and the other Catholic Irish, on the Carolina frontier in the 1760s, both would be likely to make primary distinctions, not between themselves, but between settlers and Indians, between settlers and slaves, and perhaps between backcountry people and coastal gentry.

This is not to say that there was no ethnic conflict in the colonies or that it was not important. Maldwyn Jones was correct in devoting an entire chapter of his survey, *American Immigration* (1960), to what he called "Ethnic Discord, 1685–1790."[3] But before we discuss and detail that discord, we must spend some time looking more closely at the attitudes of the most numerous ethnic group in the American colonies, the English.

English Attitudes

Toward the end of the seventeenth century, probably about 90 percent of all the settlers in British North America were of English birth or descent. The development of African slavery and the increased immigration of non-English groups caused this percentage to drop steadily, so that, as we have seen, by the first census a century later, that percentage was down to 48 percent of the total, 60 percent of the white population. In two of the original states, New Jersey and Pennsylvania, English were actually a minority of the white population, an estimated 47 percent in New Jersey and barely more than a third—35 percent—in Pennsylvania. There the minority English founded immigrant protective societies: In Philadelphia in 1772 self-conscious English residents founded the Society of Sons of St. George for the "Advice and Assistance of Englishmen in Distress," emulating similar organizations founded there by Scots (1749), Irish (1759), and Germans (1766). Majorities, or those who psychologically feel themselves to be majorities, don't really need to organize, and nowhere else in colonial America were such groups established by English people: Even in the nineteenth and twentieth centuries, new English immigrants organized few such groups.

The dominance of English culture and institutions was even more overwhelming than that of English and English-descended persons: language, law, the predominance of English and later British economic power, were overwhelming, and have left indelible marks on American society. But very soon after settlement began native-born Americans—second-generation English Americans—came to outnumber the immigrant generation, the founding fathers and mothers. In New England, for example, the most English part of the American colonies, there was relatively little immigration from anywhere after the English Civil War of the mid-seventeenth century, and the tremendous population increase in New England from less than twenty-five thousand in 1650 to just over a million at the first census is almost entirely due to natural increase. All things considered there may have been as much or more emigration from New England as immigration to it in that period. Whole groups, as we have seen, went to Long Island (and to other parts of New York), some individuals, the most famous of whom was Benjamin Franklin, went to other colonies or to Great Britain and, in the aftermath of the Revolution, many Loyalist exiles found havens in Canada and England.

It is quite clear that, even before the Revolution produced a crisis in loyalty, there were conflicting claims of loyalty for most English Americans and a colonial identity was evolving. On the one hand, persons who had never seen it referred to England as "home," and among the elite, education in England—at the Inns of Court, for example—and travel there, were common. On the other hand, the very different conditions of American life, differing views of what today we call national security, and, eventually, economic conflict all helped to produce an American, as opposed to an English American, outlook. These differences, as we shall see, came to a head during and after the American Revolution, which has been called the crucible of American nationalism. But less than twenty years before the Revolution a fourth-generation English American like John Adams (1735–1826) could refer to himself enthusiastically as a "British American," a description he would avoid once the Revolution began.

But long before the Revolution began—in fact, from the earliest days of English colonization—there were conflicts between English and other groups and between non-English groups in the colonies. For convenience we can divide these into "race relations" and "ethnic relations." The term *race relations* sounds modern and is in fact of twentieth-century coinage, but the fact of race relations, between English and Indians, between English and blacks, and, eventually, between all whites and Indians and blacks, was a fundamental if largely ignored aspect of colonial life.

Race Relations

The relationships between European settlers and Native Americans can be described in a very narrow spectrum. Early on, in Jamestown, Plymouth, and elsewhere, the settlers learned essential agricultural techniques from Indians, techniques that were probably crucial to their survival. Soon after the initial peaceful coexistence, conflict broke out, usually stemming from Indian resistance to white aggrandizement. The conflicts usually resulted either in the extermination of the Indians of a given locality or their expulsion to the west. In Massachusetts what we can call proto-Indian reservations—areas in which the remaining Indians were to be segregated—were developed as early as the mid-seventeenth century. The colonists called such places "Indian plantations," "Indian villages," or "praying towns." As the last designation suggests, Christianization was often part of the "taming" process.

Until very recently, most of what writing there has been about Indian-white relations in the Colonial Era—and after—had assumed, implicitly or explicitly, that however brutal and rapacious most whites in direct contact with Indians had been, the purposes of the missionaries who went among them were benign and benevolent. The path-breaking work of Robert Berkhofer a quarter century ago has destroyed this notion: he, and a number of other scholars since, have shown clearly that Christian missionaries had little regard—to put it mildly—for the Indian's culture and that they consciously advanced the political and economic goals of settler society.[4] From John Eliot (1604–90), the Puritan "apostle to the Indians," through the recently canonized Junipero Serra (1713–84), who established missions in California, the martyred Marcus and Narcissa Whitman (1847), missionary-settlers in Oregon, and the clerical reformers who helped the federal government run its reservations in the late nineteenth and early twentieth centuries, there persisted one constant demand: that Indians cease being Indians, that they commit cultural suicide, or what has been termed ethnocide. This racial policy established a fateful pattern for American ethnic relations. In years to come this same kind of demand—what has been called the demand for Anglo-conformity—would be made of European ethnic groups: assimilate on the terms of the prevailing norms.

Not surprisingly, some Indians were enslaved, and although the subject has been little studied, Indian slavery in North America seems to have been practiced most extensively in South Carolina. Even there, however, it did not last long. Since Indian wars, smallpox, and other epidemics greatly reduced the available Indian population, the number of Indian slaves fell drastically after 1719, and African slavery came to dominate. Reliable data are hard to come by. One report of the governor and council in 1708 lists 4,080 whites—3,960 free and 120 servants; 4,100 Negro slaves—1,800 men, 1,100 women, and 1,200 children; and 1,400 Indian slaves—500 men, 600 women, and 300 children. Thus slaves were 57 percent of the population, and a quarter of the slaves were Indians. The leading student of the subject, John Donald Duncan, has counted the advertisements about slaves in the *South Carolina Gazette* for a period beginning in 1732. In the next ten years some 3,343 Negroes, 15 white servants, and only 4 Indians were advertised for sale; during 1732–52 the paper ran ads seeking the return of 678 runaway blacks, 191 whites, and 14 Indians. Clearly Indian slavery was no longer a statistically significant factor; nonetheless it had existed and is but another indication of the state of race relations in the American colonies.[5]

White-black race relations, of course, are even less ambiguous; Christian missionaries were not much interested in black souls, although most masters did see that their slaves became Christianized. Unlike Indians, who were largely sloughed off—the Constitution's dismissive "excluding Indians not taxed" is typical—Negro slavery and thus white-black race relations were very much a matter of law. Although the English common law did not recognize slavery and in the celebrated Somerset case, Lord Mansfield, a British judge, ruled in 1772 that all slaves held in England must be set free, with the double standard that came to characterize British imperialism on racial matters, slavery in British colonies took on the full color of law.[6] Slavery existed in the North and South from the seventeenth century, but only in the states from Delaware south were elaborate slave codes developed. While in the southern states slaves, whether red or black, were primarily rural and, as we have seen, of increasing and eventually paramount importance in southern agriculture, in the North slaves and free blacks were almost all urban residents. Only in very recent years have historians, most significantly Gary Nash, begun to explore the dimensions of black life in northern cities before the Civil War. In only one northern city, New York, was slavery well established. In 1740, when the city's population was some 12,000 there were about 2,000 slaves, one-sixth of the population. New York's brief slave code was nearly as strict as those of the Deep South, and it was in New York that the most appalling atrocity of northern urban race relations of the colonial period occurred: the so-called New York slave revolt of 1741. As a result of judicial proceedings that lasted more than a year, 150 slaves were imprisoned, 18 hanged, 13 burned, and more than 70 shipped to the West Indies; since the "conspiracy" supposedly involved a Negro-Catholic plot to take over the city, 25 whites were also imprisoned and four of them hanged. Was there a revolt? Probably not. The pioneer Negro historian George Washington Williams wrote in 1883 that it was "one of the most tragic events in all the history of New York or the civilized world."[7] Contemporary historical opinion is more likely to doubt the conspiracy while recognizing that there were criminal elements at work. Much of the evidence involved a series of arsons that were used as a cover for burglaries. As in most conspiracies, real and imagined, the testimony of coconspirators was vital. New York had had a real slave rebellion in 1712, in which nine whites were killed, which probably helped account for the degree of hysteria.[8] But perhaps the most instructive thing about the 1741 slave revolt is the way in which this example of mass hysteria has been almost totally ignored, except by specialists, as opposed to the continuous examination and reexamination of the other great example of colonial hysteria—the Salem witchcraft trials of 1692—the details of which are known, as Macaulay would say, by every schoolboy. The American mythos does not include the notion of race relations in the colonial North, although religious hysteria is admissible.

No full account of the majority of black Americans in colonial cities who were not slaves can be given here, but their status can be summarized: They were neither slave nor free. In Massachusetts, where there may have been 1,300 blacks in a population of 62,000 in 1710—about 2 percent—a number of laws regulating black behavior had been enacted by then. Black slaves and servants—along with similarly enslaved Indians and mulattoes—were subject to curfew; interracial marriage and sexual intercourse between blacks and whites were forbidden, and free Negroes were forbidden to entertain or harbor nonwhite servants in their homes without the consent of the latter's masters. Yet free blacks could testify in courts, own and transfer property, and were otherwise given status by the law. Since assimilationist pressures upon blacks were largely absent in both North and South, the direct analogy with treatment of European immigrants does not exist. But when, in the mid-nineteenth century, immigrants began to come from Asia in significant numbers, the same kinds of patterns of race relations developed in the Far West as had existed on the East Coast. It also seems clear that having long-established discriminatory legal codes of behavior to govern race relations made discriminatory ethnic relations seem more "natural."

Ethnic Relations

During the era of the American Revolution—as in other great crises in our history—ethnic differences were played down. In the *Federalist Papers,* for example, John Jay was thankful that

> Providence [had] been pleased to give this one connected country to one united people—a people descended from the same ancestors, speaking the same language, professing the same religion, attached to the same principles of government, very similar in their manners and customs.

As we have seen, this was nonsense, and New Yorker John Jay (1745–1829) knew it better than most: He not only lived in the polyglot city but was himself of Huguenot ancestry. In the nearly two centuries before the Revolution, the attitudes of the predominant English Americans blew hot and cold about "foreign immigrants," prefiguring an ambivalence that has continued to prevail in American society up to the present day.

We must not imagine that there was a uniform ethnic attitude among the English-descended colonists: Differences of geography and class were also important, as were differences of religion. Attitudes in Massachusetts were not the same as those in South Carolina; an employer of indentured labor would not have the same views as a wage earner. A member of a religious minority would often be more tolerant of "strangers" than would a member of whichever group happened to constitute the local majority religion.

But there were similarities, the most fundamental of which was an emergent and growing English nationalism, which, especially when mixed with militant Protestantism, produced an extreme cultural arrogance. While English Americans in the colonial period at one time or another expressed hostility toward every other European ethnic group present in North America, the greatest and most long-lasting animus was directed, not surprisingly, against the two largest such groups: the Irish and the Germans.

Prejudice against the Irish was more general, if for no other reason than the fact that Irish were distributed throughout the colonies and the Germans more concentrated. Even in Massachusetts, where 90 percent of the ethnically identifiable white people were believed to be of English descent, some 4 percent of the same population—fifteen thousand persons—were believed to be of Irish heritage. There were at most, twelve hundred persons of German origin or heritage in a state that had nearly three hundred seventy-five thousand inhabitants. The widespread prejudice against the Irish was, as we have seen, unmixed with distinctions about where in Ireland they hailed from: Persons who today would be classified as Scotch (or Ulster) Irish, southern Irish, or English Irish were all just Irish to eighteenth-century Americans. Yet we must understand that this prejudice was not reflected in statute law. Except for the anti-Catholic legislation, which at one time or another existed on the statute books of every colony—even, as we have seen, in Maryland—no legal distinctions were made among free white persons. We must also remember that much of the anti-Catholicism of colonial America was, by definition, rhetorical. Since there were by the end of the era only about twenty-five thousand practicing Catholics in the colonies—and almost all of those in Pennsylvania and Maryland—most Americans had never seen a Catholic, which perhaps made them all the more frightening when they did begin to come a couple of generations later.

Discrimination against Germans was another matter: Its focus was, naturally enough, in Pennsylvania, where Germans were a third or more of the population from the middle of the eighteenth century. John Higham has argued that "the first major ethnic crisis in American history" involved the Pennsylvania Germans whose settlements seemed to be an unassimilable alien bloc to many of Pennsylvania's non-German majority. Included among the latter was Benjamin Franklin, who, it could be argued, was a founding father of non-religious American nativism. His complaints about "German boors" are notorious and his only-half-humorous suggestion that his fellow English-speaking Pennsylvanians learn German so that they would not become strangers in their native land sounds very much like the anti-Spanish arguments of the zealots of the contemporary organization called US English (discussed in chapter 16).

Franklin, who tried to boss colonial Pennsylvania politics, was angry with the Germans because, in one of the earliest examples of ethnic bloc voting in our history, they had voted the wrong way and supported his enemies in a series of elections. The Philadelphia German Lutheran leader, Henry M. Mühlenberg, described the marshaling of ethnic voters in one colonial election: Some six hundred German voters assembled at the German schoolhouse and then "marched in procession to the *courthouse* to cast their votes."[9] Franklin's pamphlet, *Observations Concerning the Increase of Mankind* (1751), shows him not only anti-German but, like so many of his fellow Americans, imbued with a broad racism.[10]

> Why should the Palatine Boors be suffered to swarm into our Settlements, and by herding together establish their Language and Manners to the Exclusion of ours? Why should Pennsylvania, founded by the English, become a Colony of *Aliens,* who will

> shortly be so numerous as to Germanize us instead of our Anglifying them, and will never adopt our Language or Customs, any more than they can acquire our Complexion.
>
> Which leads me to add one Remark: That the Number of purely white People in the World is proportionally very small. All Africa is black or tawney. Asia chiefly tawney. America (exclusive of the new Comers) wholly so. And in Europe, the Spaniards, Italians, French, Russians and Swedes, are generally of what we call a swarthy Complexion; as are Germans also, the Saxons only accepted, who with the English, make the principal Body of White People on the Face of the Earth. I could wish their numbers were increased. And while we are, as I may call it, *Scouring* our Planet, by clearing America of Woods, and so making this side of our Globe reflect a brighter Light to the eyes of Inhabitants of Mars or Venus, why should we in the Sight of Superior Beings, darken its people? Why increase the Sons of Africa, by Planting them in America, where we have so fair an Opportunity, by excluding all Blacks and Tawneys, of increasing the lovely White and Red? But perhaps I am partial to the Complexion of my Country, for such Kind of Partiality is Natural to Mankind.

Nor were Germans the only ethnic target group in Pennsylvania. The Quaker establishment, Gary Nash tells us, was particularly concerned, not about the Germans who were often its allies, but about the Scotch Irish, who they feared would take over the colony. Quaker pamphleteers attacked Presbyterians complaining of an almost global conspiracy, "a perpetual Presbyterian holy war against the mild and beneficent government of the Kings of England."[11] The Scotch Irish, in turn, attacked Germans and Quakers, as, among other things, being soft on Indians.

Other kinds of ethnic resentments found their way into the language: The phrase "Dutch treat"—meaning no treat at all—for example, describes the alleged stinginess of the Dutch and probably expressed resentment at their generally superior status—they had, after all, got in on the ground floor—in New York and New Jersey. That there was not more ethnic conflict than there was may well be attributable, as George Frederickson and Dale T. Knobel have suggested, to the fact that so many of the non-English were on the frontier: "Physical space effectively inhibited interethnic tension or at least impeded active discrimination against disapproved minorities."[12]

In any event, John Jay's propaganda about ethnic unity to the contrary, there was significant ethnic discord before the American Revolution. And although the Revolution is, quite properly, viewed as the crucible of American nationality, it also created important fracture lines not only in the American people but in every ethnic group in the new nation.

The most obvious evidence of these fractures is the incidence of emigration from the new United States, emigration being the most profoundly un-American act that one can imagine. Although American historians have enjoyed dwelling on the consensual nature of American society in general and the American Revolution in particular, proportionally speaking, the eighty to one hundred thousand persons who fled the country—mostly to Canada or Britain—were a larger segment of the American people than the perhaps half a million who left France after 1789 were of the French people. The American loyalists, in addition, helped to ensure that Canada would not be drawn into the political orbit of the new nation. As we have seen in the last chapter, loyalists came from many ethnic groups, and non-English or English-descended persons may well have been overrepresented among them. Very recent—post-1763—immigrants certainly were. Some twenty-eight thousand of various ethnicities went to Nova Scotia, most of them taken to Halifax by British ships. Scots Highlanders, from the Mohawk Valley and elsewhere, who had fought on the losing side, were settled *en bloc* in what they named Glengarry County in Ontario; smaller ethnic enclaves were created by Scots Presbyterians, Dutch, and Germans, who each settled along the Saint Lawrence or on the shores of Lake Ontario. There were Huguenot and Jewish loyalists as well; particularly notable were the De Lanceys of New York who fled to Canada and whose vast estates were seized.[13]

The revolutionaries not only tried to play down ethnic discord, but sometimes wrote as if no ethnic differences existed. Jefferson, in the Declaration of Independence, spoke blithely of "one people," prefiguring Jay's line in *The Federalist,* and in the Declaration of the Causes and Necessity of Taking Up Arms of a year earlier, which he coauthored with John Dickinson, a bald reference was made to "our forefathers, inhabitants of the island of Great-Britain." (George Washington, in his 1796 Farewell Address, partially written by the West Indian–born Alexander Hamilton, was more circumspect: He used the "one people" line but carefully noted that Americans were either "citizens by birth or choice of a common country" and that, "with slight shades of difference," they had "the same religion, manners, habits, and political principles.")

Jefferson did make reference to immigration in the Declaration. Part of his bill of particulars against "the present King of Great Britain" was that "[h]e has endeavored to prevent the population of these States; for that purpose obstructing the Laws of Naturalization of Foreigners; refusing to pass others to encourage their migration hither." He also commented on race relations, making it clear that his United States was a white man's country, further charging that the king had excited "domestic insurrections amongst us, and has endeavoured to bring on the inhabitants of our frontiers, the merciless Indian Savages, whose known rule of warfare, is an undistinguished destruction of all ages, sexes and conditions."

Immigrants and the Constitution

The revolutionary consensus continued into the era of the Constitution. The Articles of Confederation, agreed to in 1777 and adopted in 1781, did not deal with the crucial questions of immigration, naturalization, and citizenship, but provided only in Article 4, that "the free inhabitants" of each state "shall be entitled to all privileges and immunities" in other states, "paupers, vagabonds and fugitives from Justice excepted." (The right of poor migrants to cross state lines would not be fully established until the Supreme Court's ruling in *Edwards* v. *California* in 1941.)

The Constitution, ratified in 1788, was more specific. Its only direct mention of immigration—and the only direct mention to date, as immigration is not discussed in any of the amendments—referred to a very special kind of migration, the slave trade. In Article 1, Section 9, dealing with the powers of Congress, it provided:

> The Migration or Importation of such Persons as any of the States now existing shall think proper to admit, shall not be prohibited by the Congress prior to the Year one thousand eight hundred and eight, but a Tax or duty may be imposed on such Importation not exceeding ten dollars for each person.

The words *slave* and *slavery* do not appear in the original Constitution, although the subject was a bone of contention at the Constitutional Convention. In the famous three-fifths compromise, slaves are "other persons." Gouverneur Morris (1752–1816), who did the final polishing of the language, did not approve of slavery and kept that word out. He did permit the word "Indian" to be used. "Slavery" only appears in the

Thirteenth Amendment (1865), which abolished the institution. Congress did abolish the slave trade as quickly as it legally could, in March 1807. Jefferson had invited it to do so in his message of December 1806:

> I congratulate you, fellow-citizens, on the approach of the period at which you may interpose your authority constitutionally to withdraw the citizens of the United States from all further participation in those violations of human rights which have been so long continued on the unoffending inhabitants of Africa!

As we have seen in chapter 3, despite the prohibition, perhaps fifty thousand slaves were illegally imported from Africa and elsewhere in the New World after the 1807 prohibition.

The Constitution also empowered Congress, in Article 1, Section 7, "to establish an uniform Rule of Naturalization" but provided no particulars, so that naturalization has been a kind of political litmus test of the national climate of opinion about immigrants and immigration. The Constitution made certain distinctions between native-born and naturalized citizens. In Article 1, Section 2, it ordained that "no Person shall be a Representative [in Congress] who shall not have . . . been seven Years a Citizen" and set the requirement higher for senators who had to have been "nine Years a citizen." And Article 2, Section 1, barred any future immigrant from the presidency, restricting that office to "natural born citizens" and persons who were citizens at the time of the Constitution's adoption. This provision, of course, applies to the vice-presidency as well. But rather than emphasizing the negative nature of these provisions, what should be noted is that the founding fathers not only expected immigration to continue but also enabled immigrants to hold all but the very highest office in the land. Immigrants have, in fact, held every office in the American government to which they are eligible, save only Chief Justice of the United States.

Changing Standards of Naturalization

The first Congress in 1790 took advantage of the specific constitutional empowerment and passed a naturalization statute. In keeping with the revolutionary consensus, which was still in effect, it was the most generous in our history, at least for white people who were not indentured servants. It provided that "free white persons" who had been in the United States as little as two years could be naturalized in any American court. However this meant that immigrant blacks—and later immigrant Asians—were not supposed to be naturalized; it said nothing about the citizenship status of persons who were not white and born here: their status, until 1870, would depend on the law in the individual states, as no national citizenship was created until that time. Whether free blacks were citizens or not depended on which state they lived in.[14]

Even more perplexing was the status of American Indians. From the time of Chief Justice John Marshall's decision in the case of *Cherokee Nation* v. *Georgia* (1831), it was established that Indian tribes were, in Marshall's words, "domestic dependent nations," and that Indians therefore were aliens, and not citizens. Marshall based his ruling on the brief but potent commerce clause, part of Article 1, Section 8, which empowered Congress "to regulate Commerce with foreign Nations, and among the several States, and with the Indian Tribes." And since Indians were not "free white persons," they were aliens who could not be naturalized. In practice many states and localities treated acculturated Indians as if they were citizens—allowed them to vote, hold office, and so on—and, as early as the Grant administration, Eli Samuel Parker (1828–95), an American Indian and a former general in the Union Army, held office as commissioner of Indian affairs. In 1887 Congress, in the Dawes Act, conferred citizenship on acculturated Indians not living on reservations, citizenship that many western states and localities refused, in practice, to recognize. Finally, in 1924, Congress made all native-born American Indians citizens.

But the revolutionary consensus soon began to break down as American politics was polarized around those two archetypical figures, Alexander Hamilton and Thomas Jefferson, and our first political parties, the Federalists and what became known as the Democratic-Republicans. The Federalists perceived that many of the staunchest supporters of their opponents were recent immigrants from Ireland and France, immigrants who were essentially political exiles in the era of the French Revolution. It was not ordinary immigrants who most concerned the Federalists but, rather, distinguished ones who became anti-Federalist editors and pamphleteers. These included such men as the English scientist Joseph Priestley (1733–1804) and Jefferson's friend Victor Marie Du Pont (1767–1827). The anti-immigrant feeling, which had arisen toward the end of Washington's second term, came to a head during the administration of John Adams (1797–1801) and was exacerbated by the first undeclared war in our history, the so-called quasi war with France of 1798–1800—which was fought entirely at sea. The second Naturalization Act, in 1795, only five years after the first, increased the necessary period of residence to five years. Three years after that, as part of the package of bills known to history as the Alien and Sedition Acts, the now-desperate Federalists, realizing that they were about to lose political power, required an alien to file a declaration of intention to become a citizen at least five years before becoming a citizen and increased the period of residence to fourteen years. In addition, all aliens were required to register with federal officials. Other parts of the Alien and Sedition package, all of which were enacted in less than a month in the summer of 1798, included the Aliens Act, the Alien Enemies Act, and the Sedition Act (which need not concern us here, as it applied to everyone). The Aliens Act gave the president unlimited power to order out of the country any alien "whom he shall judge dangerous to the peace and safety of the United States." The Alien Enemies Act, which would have gone into effect only if the United States had declared war, empowered the apprehension and confinement of male enemy aliens fourteen years of age and older.[15]

In the event, the Aliens Act was never enforced against anyone, although some who felt threatened left the country and others went into hiding. When the turn of the political tide put the Jeffersonians in power in 1801, they rolled back most, but not all, of the Federalist antialien legislation. The Naturalization Act of 1801 restored the residence requirements of 1795 (but not of 1790), and five years has remained the period of residence until this day. The registration requirements of that law were repealed and aliens did not have to register again until 1940, when the United States again felt threatened from abroad. The Aliens Act simply expired, as it had a two-year limit and was never renewed. Not until the passage of the Internal Security Act of 1950 was a president given power, in peacetime, to incarcerate persons, and that statute applied to citizens as well as aliens. Nor were the Jeffersonians immune from a xenophobia of their own. During the War of 1812, Congress amended the Nationality Act to provide that no British alien who had not declared the intention to become a citizen before the war began could be naturalized—a provision that was repealed after the war was over.

American Nationalism

One can make too much of the xenophobia and prejudice of the revolutionary era and the early republic. The victory in the Revolution produced in the new nation an overwhelming feeling of confidence that all battles could be won, all obstacles overcome. And most Americans were concerned, as Jefferson put it, with "the population of these states" and, to that end, welcomed immigration. When immigration was opposed, it was largely on ideological rather than ethnic or religious grounds. Federalists opposed radical immigrants from England, France, or Ireland; Jeffersonians in Congress, unduly concerned about the migration and settlement here of exiled nobility from France, got a provision put into the 1795 Naturalization Act requiring an applicant for citizenship to foreswear any hereditary titles of nobility.

The very circumstances of the Revolution tended to defuse religious prejudice, at least for a time. Not only were the Americans allied with Catholic France, but they wished to gain the support of French Catholic Quebec. Canada was given pride of place in the Articles of Confederation: It could join the new nation at any time. Any other new adherent was dependent on the agreement of nine states. When the Continental Congress sent a delegation to Canada in a vain attempt to get it to join in the Revolution, its leaders had the wit to persuade the only Catholic signer, Charles Carroll of Carrollton, to join it, and he persuaded his cousin, Father John Carroll, later to become the first American Roman Catholic bishop, to join them. The priest was reluctant, noting in a memorandum that:[16]

> I have observed that when the ministers of religion, leave the duties of their profession to take a busy part in political matters, they generally fall into contempt, and sometimes even bring discredit to the cause in whose service they are engaged.

Later in the Revolution, when the French fleet used Philadelphia as a home port, priests became a common sight on its wharves and streets. On two occasions in Philadelphia, the French minister had Te Deums said in the small Catholic church to celebrate the third anniversary of the Declaration of Independence and the victory at Yorktown. Many members of Congress, even some from pope-hating New England, attended. If they were not more tolerant, at least their public demeanor was.

In addition, the natural-rights philosophy that inspired so many of the leaders of the revolution tended them toward tolerance and nonsectarianism. In the Declaration, Jefferson spoke of "Nature's God"; the more politically minded draftsmen of the Articles of Confederation spoke of "the Great Governor of the World"; and Washington in his first inaugural spoke, appropriately enough for the Father of his Country, of "the benign Parent of the Human Race." These statements by leaders were paralleled by legislative actions. The Virginia and Massachusetts Bills of Rights (1776 and 1780), the Virginia Statute of Religious Liberties (1786) and, above all the "no establishment" clause of the First Amendment ("Congress shall make no law respecting an establishment of religion, or prohibiting the free exercise thereof"), all set a laudable and unprecedented standard of religious liberty—though it was one many or most Americans did not really believe in.

In the revolutionary period and the years immediately afterward, the tolerance and nonsectarianism engendered during and after the struggle for independence were not seriously challenged or put to the test. Relatively few immigrants came—perhaps an average of ten thousand a year from the time of Yorktown to the end of the War of 1812, and the vast majority of these were Protestant Britons. The rapid population growth of those years—it grew more than two and a half times between 1790 and 1820 (3.9 to 9.6 million)—was almost entirely due to natural increase. And, because the nation annexed so much territory during the period, the number of persons per square mile barely increased, going from 4.5 to 5.6.

In these circumstances early American nationalism could be, and was, almost entirely positive. It did not attack—as later aspects of it would—ethnic, national or racial groups or single out certain ideological strands as "un-American." It was, as Philip Gleason has cogently argued, characterized by three elements: its ideological quality, its newness, and its future orientation. Because it was ideological, it did not stress national or ethnic origin or any denomination, although there was, both in rhetoric and in the laws of many states, a distinct bias toward Christianity. The obvious newness of American nationality—the closeness of most Americans to the great events of their past—lent it an essentially positive aspect, while its future orientation—the conviction of most Americans that the future would not resemble but greatly outstrip the past—only served to strengthen its positiveness.[17]

Much of this would change in the decades after 1820. Immigration would increase to reach its highest incidence in the mid-1850s. The bulk of that immigration would not be English but Irish and German, and much of it would be Catholic. In addition, the nation would face, in the 1860s, the greatest crisis of its existence. Under these pressures, one strain of American nationalism would turn ugly and produce, not the sporadic nativism and xenophobia of the colonial period, but a full-blown movement that exalted bigotry to a matter of principle.

Shortly after the end of the War of 1812, John Quincy Adams (1767–1848) caught the spirit of early American nationalism nicely. Writing to a European, Adams, who had as much intimate knowledge of Europe as any American of his generation, made this point:[18]

> That feeling of superiority over other nations which you have noticed, and which has been so offensive to other strangers, who have visited these shores, arises from a consciousness of every individual that, as a member of society, no man in the country is above him; and, exulting in this sentiment, he looks down upon those nations where the mass of the people feel themselves the inferiors of the privileged classes, and where men are high and low, according to the accidents of their birth.

And we should not forget that such sentiments as Adams describes were not restricted to American natives. Many immigrants quickly took on the ideological accents of their adopted land. As my mentor, Theodore Saloutos, one of the pioneers of American immigration history, used to say, when an immigrant becomes an American patriot, he is often not satisfied to become a 100 percent American but tries to be a 200-percenter. Adams, in the letter quoted above, described how he thought immigrants should behave, echoing, in some ways, the remarks of Crèvecoeur at the end of the Revolution.

> [Immigrants] must cast off the European skin, never to resume it. They must look forward to their posterity rather than backward to their ancestors; they must be sure that whatever their own feelings may be, those of their children will cling to the prejudices of this country.

This hyperassimilationist ideal, as we have seen, was never realistic; immigrants generally died in the skins they were born with. And, while most immigrants were, almost by definition, future oriented, their visions of both the present and the future would be modified significantly by the cultural environment of their past.

PART II

THE CENTURY OF IMMIGRATION (1820–1924)

6

Pioneers of the Century of Immigration: Irish, Germans, and Scandinavians

In the next five chapters we will examine the "nineteenth century"—really the period from the end of the War of 1812 to the passage of the National Origins Act of 1924. In doing so one of the great shibboleths of American immigration history—the distinction between "old" and "new" immigrants—will be ignored. The classic description of the differences between the two groups runs something like this. The old immigrants, persons from the British Isles and northwestern Europe who came before the 1880s, were very much like the settlers of the colonies and were relatively easy to assimilate. The new immigrants, persons from southern and eastern Europe who came after the 1880s, were of very different ethnicity (many late-nineteenth and early-twentieth-century writers used the term *races*) who spoke strange languages and worshiped strange gods—that is, they were not Protestants. While some aspects of the description are congruent with historical reality (the sources of American immigration did change drastically in the course of this period, as table 6.1 shows), the notion of polar differences suggested by the words *old* and *new* is not accurate. In the course of these chapters we will see that there are similarities and differences over time both between groups and within groups. The change in migration from Catholic Ireland from largely male to largely family to largely female, for example, is as striking as the differences between most immigrant groups. And it could be argued that perhaps the most fundamental difference of all—whether immigrants were rural, as most colonial and early national period immigrants had been, or urban, as so many of the immigrants since the 1820s have been—is one that cuts across the artificial "old" and "new" lines. Scandinavians and Japanese were mainly rural, as were many Germans; Irish, Chinese, and Italians were mainly urban. And

Table 6.1

Immigrants by Region of Last Residence, 1820–1920 (percentages)

	1820–60	1861–1900	1900–20
Northwestern Europe	95	68	41
Southeastern Europe	—	22	44
North America	3	7	6
Asia	—	2	4
Latin America	—	—	4
Other	2	1	1

if we look at such things as family structure, occupation, and education, similar discrepancies in the supposed old-new dichotomy will appear.

It also must be emphasized that while the reasons for the changes in sources of immigration are to be found chiefly in changing circumstances in Europe and Asia—and later in Latin America and Africa—changes in the settlement and occupational patterns are due chiefly to changes in the American economy and society. Swedish immigration, for example, was at first highly rural, with settlement concentrated in the wheat belt of the west-north-central states—Minnesota, the Dakotas, and Nebraska. By the 1880s and 1890s, most Swedes were settling in urban areas such as Chicago and Jamestown, New York. If it was primarily old-stock Americans who settled the traditional western frontier (immigrants were there, too, however), it was immigrants who were the most numerous on the urban frontier. The great changes in this long century are industrialization and urbanization. The United States, historian Richard Hofstadter liked to say, was born in the country but moved to the city. At the first census in 1790 only 5 percent of Americans were in urban areas, some of which were as small as 2,500 persons. The 1920 census would show that a bare majority of Americans lived in urban places, many of them in very large cities. Much earlier, however, a majority of foreign-born persons lived in cities, most of them in very large cities.

Not only did the sources of immigration and the settlement patterns of the immigrants change, but the volume of immigration in absolute and relative numbers soared, as tables 6.2, 6.3, and 6.4 indicate. Illegal immigration, except by Chinese after 1882, was not statistically significant during this period, but it is clear that many immigrants were not counted for a variety of reasons. In the earlier part of our century the counting was inefficient, particularly for those who came from or via

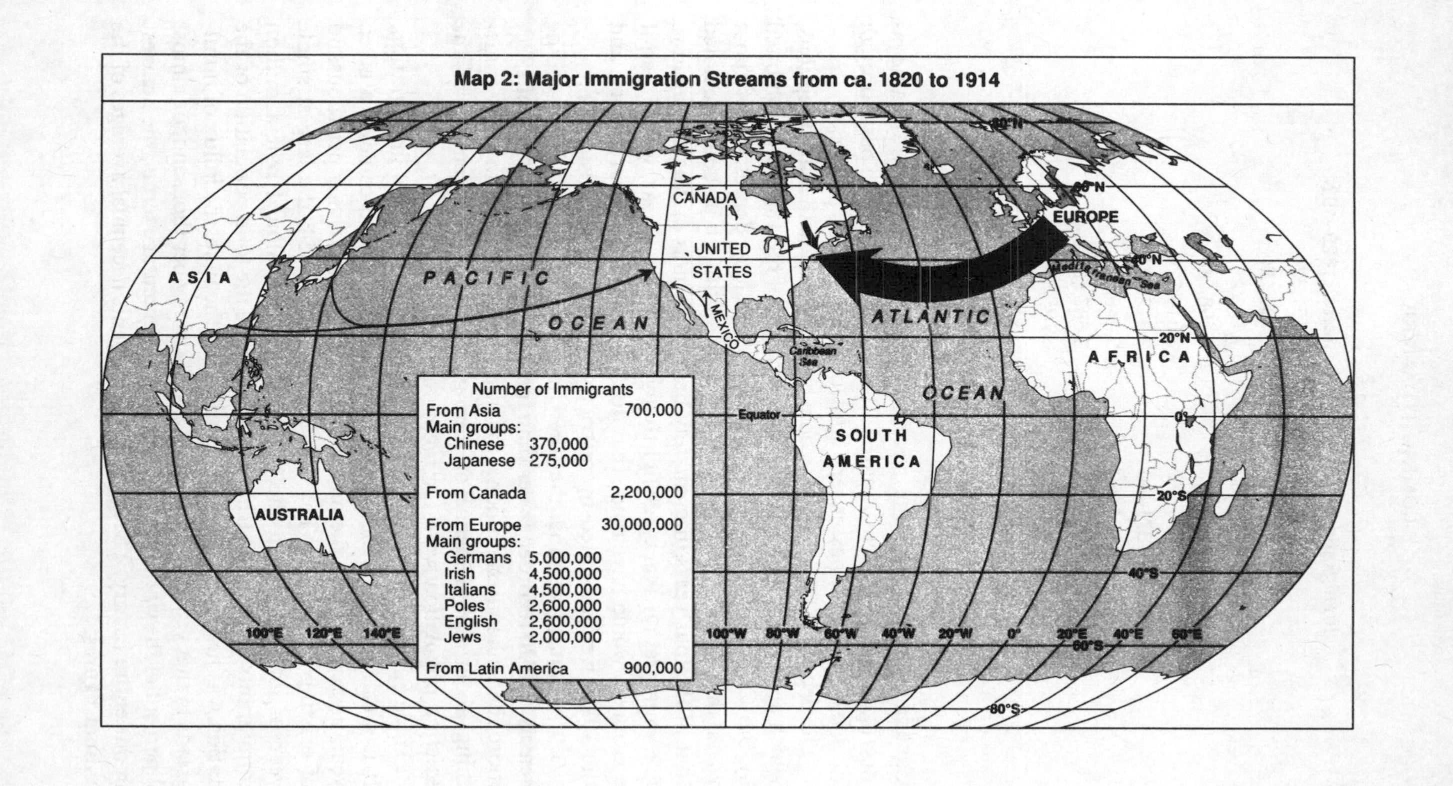
Map 2: Major Immigration Streams from ca. 1820 to 1914
ASIA
PACIFIC
OCEAN
CANADA
UNITED
STATES
MEXICO
EUROPE
Mediterranean Sea
ATLANTIC
OCEAN
Caribbean Sea
AFRICA
SOUTH
AMERICA
AUSTRALIA
Equator
60°N
40°N
20°N
0°
20°S
40°S
60°S
80°S
100°E
120°E
140°E
100°W
80°W
60°W
40°W
20°W
0°
20°E
40°E
60°E
Number of Immigrants
From Asia 700,000
Main groups:
Chinese 370,000
Japanese 275,000
From Canada 2,200,000
From Europe 30,000,000
Main groups:
Germans 5,000,000
Irish 4,500,000
Italians 4,500,000
Poles 2,600,000
English 2,600,000
Jews 2,000,000
From Latin America 900,000

Table 6.2

Immigration to the United States, 1820–1924

1820–30	151,824
1831–40	599,125
1841–50	1,713,251
1851–60	2,598,214
1861–70	2,314,824
1871–80	2,812,191
1881–90	5,246,613
1891–1900	3,687,564
1901–10	8,795,386
1911–20	5,735,811
1921–24	2,344,599
Total	35,999,402

Canada. Later, during the steamship era, only those who came as steerage passengers were counted: It was they who had to pass through Ellis Island. Cabin passengers landed in New York proper.

The decennial figures show a constant increase, with four exceptions, which are easily explained. The immigration of the 1860s was inhibited by the Civil War; that of the 1890s, by the long depression of those years; that of the 1910s, by World War I—although it should be noted that in the four years immediately before the guns of August began to fire, immigration averaged slightly over a million a year, which, had it continued, would have made that decade the richest in our history; and the 1920s were inhibited by restrictive immigration legislation.

But we must remember that the United States was growing, too: The census of 1820 recorded fewer than ten million Americans; that of 1920, more than a hundred million. It is important, therefore, to know something about the incidence of immigration. What percentage of the existing population were the incoming immigrants?

Table 6.3 attempts to indicate the numerical impact of immigration. It is a calculation of the number of immigrants who came in a given year, as expressed by the number of immigrants arriving per thousand of existing population. Thus, for example, in 1854, the largest single year of immigration in the antebellum years, almost half a million immigrants entered the United States. Since the total population of the nation was just over twenty-five million at the time, the influx of immigrants in that year was at a rate of about twenty per thousand, or almost 2 percent of the whole population, and in percentage terms, the heaviest in our entire history. The table below presents decennial averages of the annual figures.

Table 6.3

Rate of Immigration per Thousand, 1820–1924

1820–30	1.2
1831–40	3.9
1841–50	8.4
1851–60	9.3
1861–70	6.4
1871–80	6.2
1881–90	9.2
1891–1900	5.3
1901–10	10.4
1911–20	5.7
1921–24	5.3

But immigration is cumulative, and one more table is necessary to give a well-rounded view of the demographic impact of immigration on the American people. (I hate to burden the reader with all these figures, but in demographic matters, if you don't get the numbers right, you just don't know what you are talking about.) The following table shows the foreign born as a percentage of the total population at each census between 1850—the first year that the census recorded and tabulated that information—and 1920.

One notices the remarkable steadiness of those figures between 1860 and 1920: a period in which the United States grew from 30 million to 105 million, changed from a predominantly agricultural nation to a basically industrial one, from a rural to an urban nation, from a relatively isolated nation concerned chiefly with its own affairs to a world power deeply if reluctantly engaged in Europe and Asia. In an era in which almost *everything* changed, the incidence of foreign born, citizen and alien, in our population remained a constant one in seven. This will

Table 6.4

Foreign Born as a Percentage of Population, 1850–1920

1850	9.7
1860	13.2
1870	14.0
1880	13.3
1890	14.7
1900	13.6
1910	14.7
1920	13.2

be an important figure to keep in mind, especially when we deal with anti-immigrant movements, both in this period and in the more modern era. As we have noted, immigrants were *not* distributed evenly but went, in the final analysis, where the jobs or other economic opportunities were and where they perceived economic growth to be the strongest, although few of them would have couched their thoughts in those precise terms.

We should also note that this period, or most of it, fell in the century of relative international peace between the end of the Napoleonic Wars and the outbreak of World War I. This was the last period of unchallenged European global hegemony, and one in which European overseas migration to places other than the New World—chiefly the Antipodes and South Africa—first becomes statistically important, and when working-class immigration from Europe first reaches South America. Improvements in transportation not only quickened the pace of immigration but also, with the invention of the prepaid ticket and the development of cheap and reliable facilities for transferring funds abroad, it became easier for immigrant families to stagger, and pay for, their own migration. And, finally, the period also witnessed the beginnings of large-scale migrations out of Asia, as first Chinese and Indians, and then other Asian ethnic groups first become involved in the world labor market, initially as part of a new system of indentured labor and then as free immigrants.

The Irish

If you ask an Irish American who is descended from nineteenth-century immigrants why his or her ancestors emigrated, more often than not the answer will contain some reference to the potato famine or merely the famine. While, as we shall see, the famine—or what Cecil Woodham-Smith called the "great hunger"—was a major factor in the late 1840s and early 1850s, we must look much deeper than that into the Irish past to understand the segments of Catholic Irish migration that have come out of Ireland in great numbers since the 1820s. Such emigration was important before the famine began and continued long after its effects—but not its memory—had disappeared from the scene.[1]

Until the very end of our period (1921) Ireland was an unfree nation and one largely unchanged by the Industrial Revolution. And, alone of the nations of Europe in the nineteenth century, Ireland suffered a massive loss of population: The census of 1841 found about 8.2 million Irish; that of 1851, about 6.6 million. That loss of 1.6 million persons in a decade—more than 17 percent of the population—can be attributed, in large part, to the famine. But the losses did not end there: In 1891 the population stood at 4.7 million, some 57 percent of what it had been half a century earlier. Continued immigration, late marriages, and an unusually high number of persons who never marry have kept population growth very low. In the nine decades since 1891 a total of only three hundred thousand persons have been added to Ireland's population—a little over three thousand a year. In the 1890s alone almost four hundred thousand Irish emigrated to the United States, and, it must be noted, when Irish migrated, they stayed. Although about one nineteenth-century immigrant to the United States in three did not remain, only about one Irish immigrant in twelve returned home. (Only Jews had a lower return rate.)

Altogether, about 4.5 million Irish immigrated to the United States between 1820 and 1930 according to American statistics. During the same period perhaps half a million went to Canada, another 350,000 to Australia and New Zealand, and another 60,000 to other overseas destinations for a rough total of more than 5.4 million overseas emigrants. In addition, over the years uncounted numbers of Irish went to Britain; and some of those listed as British in the American statistics were second or third-generation Irish. Similarly, a large percentage of the Irish who went to Canada eventually came to the United States; some did so almost immediately. Much smaller but still significant numbers of Irish Americans were drawn to Canada as part of the process that scholars describe as the mingling of the Canadian and American peoples. It is clear, then, that the depopulation of Ireland was *not* just a case of "America fever" and that any precise enumeration of either Irish emigration or immigration is impossible. It is also clear that the American immigration data somewhat understate Irish immigration.

But, even if we take the American data at face value, the impact of the Irish on American immigration is enormous. Between 1820 and 1860, Irish were never as few as a third of all immigrants, and in the 1840s they were nearly half (45.6 percent). In the twenty years *after* the Civil War, not usually thought of as a heavy period of Irish immigration, they still made up more than 15 percent of the recorded entrants. Perhaps the best single indicator of Irish incidence is the census of foreign born in the United States. In 1860 there were just over four million foreign born: nearly two-fifths—38.9 percent—were Irish. In 1900 there were more than ten million foreign born: Nearly a sixth of them—15.7 percent—were Irish.

From the Irish point of view the emigration may be divided into three periods: the years before the famine, that is to 1844; the famine years and their aftermath, roughly, 1845–55; and the postfamine migration, which goes on until the onset of the Great Depression, Ireland not being adversely affected by the Immigration Act of 1924.

PREFAMINE MIGRATION

The migration of the prefamine years laid the basis for an Irish Catholic American ethnic group. Relative prosperity in Ireland during the Napoleonic Wars had led to a rapid and dangerous population growth—from 4.8 million in 1791 to 8.2 million just fifty years later—which made Ireland the most densely populated "country" in Europe. This rapid growth caused agricultural holdings to shrink steadily, as small tenants rented to subtenants, and more and more of the population became impoverished. With no industrial sector of any size, emigration was the sole source of relief. The emigrants of those years, increasingly drawn from the Catholic population, were not the poorest of the poor. The migration at first was heavily male and, to a very large degree, was associated with the existing lumber trade between the British Isles and eastern Canada.

The British Passenger Acts attempted to deflect immigration from the British Isles to Canada rather than the United States by making it much more expensive to travel to the latter. Instead of the four or five pounds a fare to New York would cost in those years, the rate to the Canadian Maritime Provinces was sometimes as low as fifteen shillings (there were twenty shillings to the pound). In addition, Canada-bound ships left from every seaport in Ireland and were both much more convenient for Irish immigrants and much cheaper than making the twelve- to fourteen-hour crossing of the Irish Sea to Liverpool, the chief port of the immigrant trade proper. But there were few economic opportunities in Canada, and the curious combination of patterns of trade and anti-American British legislation produced what Marcus Lee Hansen called "the second colonization of New England," a colonization that was largely Irish. Immigrants quickly discovered that they could get cheap transportation south from Canadian ports or, if they lacked money, as was often the case, they could walk. This became well known to both captains and emigrants. When the master of the ship *Ocean,* sailing from Galway to New Brunswick in 1835, advertised for immigrant passengers, he pointed this out, adding (with a bit of blarney) that "those living on that line of road being very kind to Strangers as they pass." Although the road led to Boston, many Irish found work and settled along the way, and Hansen pointed out that one can trace the Irish migration route by the pioneer Catholic churches established in Maine in those years.[2]

Hansen also argues that, because of the section's need for labor, New England would have attracted immigrants in any event, but since there was little direct trade between Boston and Britain, those immigrants would have been funneled through New York and have been varied. It was, according to Hansen, the prosaic timber trade that made New England so heavily Irish, and—we might add—that eventually produced the line of Irish American politicians whose most illustrious scions have been the Kennedy brothers. The facts of growing Irish predominance in New England's population are undisputable. New England, which had been the most homogenous of American regions, had become by the 1840s heavily foreign born. It contained more than a fifth of all the Irish born in the country in 1850—the first year for which we have such data—but only 2 percent of the German born, the only other foreign group of comparable size. Table 6.5 shows the number of Irish immigrants from 1820 through 1924.

Table 6.5

Immigrants from Ireland, 1820–1924

DECADE	NUMBER	PERCENTAGE OF TOTAL IMMIGRATION
1820–30	54,338	35.8
1831–40	207,381	34.6
1841–50	780,719	45.6
1851–60	914,119	35.2
1861–70	435,778	18.8
1871–80	436,871	15.5
1881–90	655,482	12.5
1891–1900	388,416	10.5
1901–10	339,065	3.9
1911–20	146,181	2.5
1921–24	71,865	3.1
Total	4,578,941	12.7

Writing of New England's metropolis, Boston, a community with "no room for strangers," Oscar Handlin puts it this way:

> The great waves of European migration [in the years before 1865], with one exception, caused scarcely a ripple in the placid stream of the city's life. Only one country directed a dislodged population

> to a city where no promise dwelled; elsewhere events promoted the departure of those only who could choose their destination more prudently.[3]

Boston's population figures bear out the dominance of the Irish immigrants. In 1850 Boston had a population of 136,881: 35 percent of that population, 47,933 persons, were foreign born. Nearly three-quarters of the foreign born, 35,287, or 74 percent, were natives of Ireland, and surely some of Irish ethnicity had been born elsewhere (Britain, Canada), not to speak of the growing second generation. By 1880 the numerical ascendancy of the Irish born, although declining, was still impressive. Ireland still accounted for 56 percent of Boston's 114,796 foreign born—64,793 persons.

Most of the half million Irish who migrated to the United States between the end of the War of 1812 and the famine did not come in family groups. Before 1820 at least three thousand Irishmen were employed in digging in the Erie Canal in New York State, and from then through the post–Civil War decades Irish muscle created much of the American infrastructure: other canals, then railways and urban amenities of all kinds: streets, sewers, water works, and so on. Large numbers of these men sent remittances home even out of the meager wages they drew. The development of the prepaid ticket made it easier for the Irish to bring relatives and even friends over after them, and although the terms did not then exist, they were probably the first group to practice *chain* or *serial migration* on a large scale. And while single men predominated, and, most women who came in that period either joined male relatives or came in family groups, single women did come, and they too sent remittances. And many who came wondered why anyone stayed at home. Margaret McCarthy wrote home to after arriving in New York in 1850:

> My dear Father. . . . Any man or woman are fools that would not venture and come to this plentyful Country where no man or woman ever hungered or ever will and where you will not be seen naked.[4]

Although written just after the Civil War, the following excerpt from an account by an Irish priest who visited America—while idealized, sentimental, and conforming to the stereotyped role that the church prescribed for women—nevertheless describes with some accuracy an important ignored segment of the Irish American community, the working women.

> To better the circumstances of her family, the young Irish girl leaves her home for America. There she goes into service, or engages in some kind of feminine employment. The object she has in view . . . protects her from all danger, especially to her character: that object . . . is the welfare of her family, whom she is determined, if possible, to again have with her as of old. To keep her place or retain her employment, what will she not endure?—sneers at her nationality, mockery of her peculiarities, even ridicule of her faith, though the hot blood flushes her cheek with fierce indignation. At every hazard the place must be kept, the money earned, the deposit in savings-bank increased; and though many a night is passed in tears and prayers, her face is calm, her eyes bright, and her voice cheerful. One by one, the brave girl brings the members of her family about her.[5]

The anti-Irish feelings Father Maguire refers to—which will be discussed at length in chapter 10—surfaced in want ads as early as 1830. The *New York Evening Post* of September 4, 1830, for example, contained:

> Wanted. A Cook or Chambermaid . . . must be American, Scotch, Swiss or African—no Irish.

Much of such discrimination was sheer bigotry, but there was also a widespread and accurate belief that standards of cleanliness and hygiene in Ireland were not of the highest and that young Irish women were not likely to know much about the techniques necessary to run a middle-class American household. But they learned and, often reluctantly, were accepted, and by midcentury "Bridget" had taken an essential role in thousands of northeastern homes. In Boston, for example, Handlin notes that at least 2,227 Irish women worked as servants in 1850. The great reluctance of native-born Americans—and this, of course, included second-generation children of immigrants—to work as servants ("I'd starve first") created an endemic servant problem in the United States from the era of Andrew Jackson to the Great Depression. It was a problem whose solution was almost always immigrant "help," who—when not Irish—were usually German or Swedish immigrant women and, in the Far West, Chinese and Japanese men. The exception was Afro-Americans: they were the one native-born group that could not afford to turn down domestic service.

These and other employments of most of the Irish were urban in a society that was still highly agrarian. In 1830 the United States was only 9 percent urban; by 1850 this had increased to only 15 percent. The Irish were concentrated in the Northeast, which was the most heavily urbanized region, but even there the urban percentages were quite small: 14 percent and 27 percent for 1830 and 1850. Many writers, from contemporary Irish American intellectuals such as Thomas D'Arcy McGee (1825–68) to historians such as Carl Wittke, have bewailed the fact that so few Catholic Irish became farmers. These complaints are unrealistic. In the first place, most of the Irish did not have the capital necessary to buy a farm or prepare a frontier holding for agriculture. In the second place, their experience in Ireland ill prepared them for American agriculture. When, in the second half of the nineteenth century, Catholic philanthropists tried to establish Irish immigrant settlers on donated railroad land in Minnesota, the result was unsuccessful for just those reasons.[6] As a wharfman wrote to Boston's leading Catholic newspaper

> Some critic in next week's *Pilot* may tell me, why don' I "go West." They say a man requires money to take him there and then requires something to start with.[7]

The fact of the matter was that few Catholic Irish were well prepared for either urban or rural success in America: almost none had trades to ply and few had much more than the rudiments of an education. Their transition to growing American cities was clearly difficult as were those of most immigrants: Even if one rejects Oscar Handlin's value-laden "uprooted" as a synonym for most nineteenth-century immigrants, the traumas that accompanied so much of the immigrant adjustment cannot be ignored, either for the Irish or any other group.

By 1845 Irish immigration was growing and would undoubtedly have continued to grow at a quickening pace in normal circumstances. But the great famine which began in that year and its aftermath influenced not only Irish immigration, but also the whole Irish American community, for decades to come.

THE FAMINE YEARS

In Ireland the burgeoning population had the inevitable effect of reducing in size the already small holdings of Irish farmers. Thus economic deterioration was steady, although minor improvements were made in the political situation of Irish Catholics. Those willing to swear fealty to England could buy, bequeath, and inherit property, those with a certain income could vote, and, thanks to the agitation led by Daniel O'Connell, the Emancipation Act of 1829 allowed Irish Catholics to be members of Parliament and to hold any civil office short of becoming lord lieutenant (governor) of Ireland or lord chancellor (chief justice) of England.

As their plots grew smaller and smaller and rents grew higher and higher, more and more Irish farmers sold their grain and came to subsist largely on the potato. The potato, itself a immigrant *from* the New World, part of what Alfred Crosby has called the "Columbian Exchange," was easily cultivated and took little labor and practically no equipment. An acre and a half of potatoes could feed a family of six. It has been estimated that a third of the Irish poor—much of the population—ate almost nothing else, and it formed the bulk of the diet of many more. The greatest drawback of the potato was its susceptibility to disease, particularly the fungus *Phytophthora infestans,* commonly known as the potato blight. Blight was no stranger in Ireland: It had struck at least twenty times in the century and a quarter before 1845. So, when another outbreak began in October 1845, no great hue and cry was raised. It would mean, of course, more hunger and misery, but these were not strangers in Ireland either. Thus began the last great peacetime famine in western European history.

The next year, 1846, saw the worst outbreak of the blight; nearly the entire crop was destroyed, leaving a stinking, rotting mass. The twentieth-century Irish writer Sean O'Faoláin said, "The land blackened as if the frown of God had moved across it." Most Irish cotters (small farmers) were reduced to eating the potatoes they normally would have kept for seed, so that in 1847, a year in which the blight abated, only about a sixth of a normal potato crop was even planted. Concurrently disease, the inevitable partner of famine, began to wreak havoc on the weakened population. Epidemic diseases, chiefly those borne by ticks and lice, such as typhus and relapsing fever, raged.

To say that the reaction of the British government was inadequate is to understate the situation greatly. Although close students of the matter can differentiate between the policies of the two prime ministers, the Tory Robert Peel and the Liberal John Russell, each was utterly incapable of dealing with the situation. Like Herbert Hoover nearly a century later, their ideological predilections kept them from taking the one step that would have relieved, at least in the short run, most of the suffering: feeding the hungry. The potato blight was unavoidable; but the Great Famine, in the words of Professor Kerby Miller, was "largely the result of Ireland's colonial status and grossly inequitable social system." Both British prime ministers honestly believed that the Irish needed to be more enterprising—a classic case of blaming the victim—and that they should go to work to earn money to buy imported food. Some public works jobs were provided, but these were too little and too late. Even worse were the views of Charles Trevelyan, the treasury official in charge of Irish relief. He was largely concerned that too much relief would damage the character of the Irish people, demoralize them, and make them dependent, and he was sure that what he called Ireland's "great veil" was not famine but deficient moral fiber: "the selfish, perverse and turbulent character of the people."[8]

The census of 1851 showed clearly the results of natural disaster and human mismanagement; there were about two and a half million fewer people in Ireland than there would have been under normal conditions. About half, 1,000,000 to 1,500,000 human beings—perhaps a sixth of the population—died from a combination of hunger and disease. An equal number emigrated, most to America. Our concern will be the latter.[9]

Although emigration was viewed as an escape, for many it was no escape at all. Even before the famine years, the number of Irish lost at sea had been quite high. The small timber ships on which so many traveled were not designed for the transportation of people, and many were old and unfit for the heavy seas of the North Atlantic. In the single year 1834—a particularly stormy time—at least 17 immigrant ships sank with the loss of 731 passengers. By midcentury and after, the ships were larger and safer and subject to some regulation by both the British and American governments, but when there was a major accident, as is the case with today's jumbo jets, the carnage was fearful. In 1848 a fire aboard the British sailing ship *Ocean Monarch* took 176 lives; the worst disaster of the immigrant trade was in the age of steam in 1858, when a fire—ironically, caused by a health measure, as hot tar being used to fumigate the steerage—ignited a blaze on the Hamburg-Amerika Line's iron steamship *Austria* that took the lives of 500 emigrant passengers. And, in a final example, four years earlier the iron-hulled steamer *City of Glasgow* sailed from Liverpool with 480 emigrants and was lost without trace. But the great killer of immigrants was disease, and no emigrants were more susceptible than the weakened Irish poor of the famine years.

Typhus, cholera, dysentery, and what was called "ship fever"—in the mistaken belief that shipboard conditions caused the epidemics—were the great killers. We now know—and medical authorities at midcentury were beginning to realize—that these diseases did not originate at sea but were brought aboard by either passengers or crew. Once aboard the conditions on the crowded and unsanitary ships were ideal for the propagation of disease. In the famine year of 1847—the worst year in terms of mortality—perhaps 100,000 men, women, and children embarked for Canada from British ports. Some 17,000 died at sea and another 20,000 died of disease after landing, mostly along the shores of the St. Lawrence. At just one place, the quarantine station at Grosse Isle off Quebec City, between mid-May and early November 1847, 8,691 persons were admitted to a hospital whose normal capacity was 200; 3,228 died. Conditions at Grosse Isle defy description; during the latter half of 1847 "only" 850 of the 7,000 admitted to New York's new quarantine hospital died. Nor did the horrors end in 1847. Kerby Miller estimates that in the cholera year of 1853, 10 percent of the 180,000 Irish emigrants died at sea.

All told, in the famine years something more than two million Irish went overseas. Most of them, nearly a million and a half, came to the United States; a third of a million went to Canada, and many of those came sooner or later to the United States; perhaps a quarter of a million settled in Britain, and thousands of others went to Australia and elsewhere. The total emigration was about a quarter of the prefamine population. More people left Ireland in the eleven years 1845–55 than in its previous recorded history.

Almost all historians writing about the American Irish agree that the famine years left enduring scars on the Irish and on Irish American psyches, exacerbating, in many ways, attitudes that were already there. While not all scholars agree with the conclusions of Kerby Miller on this matter, his notions should be taken into account. He argues that:

> First, both collectively and individually the Irish—particularly Irish Catholics—often regarded emigration as involuntary exile, although they expressed that attitude with varying degrees of consistency, intensity, and sincerity. Second, this outlook reflected a distinctive Irish worldview—the impact of a series of interactions among culture, class, and historical circumstance on Irish charac-

> ter. Finally, both the exile motif and its underlying causes led Irish emigrants to interpret experience and adapt to American life in ways which were often alienating and sometimes dysfunctional, albeit traditional, expedient, and conducive to the survival of Irish identity and the success of Irish-American nationalism.

Views of national and ethnic character, even in the hands of a careful scholar like Miller, always run the risk of blending into stereotype. Certainly the cultural baggage that immigrants brought with them can never be ignored and there are observable differences in the collective behavior of American immigrant and ethnic groups. But it seems to many that views like Miller's go too far. Were he correct, it seems to me, the experiences of the Canadian Irish and the Australian Irish would be more similar to those of the American Irish than they are. The view presented here will always stress—perhaps too much—the role of the American environment, the impact of the material and social conditions of American life on the experience of the immigrant and in formation and development of American ethnic groups.

Those circumstances, as indicated, made the Irish highly urban, and their experiences prefigure, in many ways, those of very different immigrant groups who began to come in the late nineteenth century. The Irish were concentrated in cities, but not all cities. In 1870, for example, Irish-born persons constituted almost 15 percent of the population of the fifty largest American cities. In nine of those cities Irish born were more than a fifth of the population. All of them were in New England, New York, and New Jersey. (In rank order they ranged from Lawrence, Mass.—some 26 percent—through Troy, N.Y.; Boston and Lowell, Mass.; Jersey City; New York City; Fall River or Worcester, Mass.—to Hartford, Conn., which was just above 20 percent.) As we shall see, a similar listing for the other very large group of urban pre–Civil War immigrants, the Germans, will produce a list of mostly midwestern cities.

Wherever they lived in urban America in the middle decades of the nineteenth century, large numbers of Irish were at the very bottom of the economic structure, overrepresented as common laborers and domestic servants and as residents in various municipal institutions—poor houses, jails, and charity hospitals. Whenever Irish and blacks were present in significant numbers, significant competition between them developed, sometimes murderously, as in the draft riots in New York City in July 1863. In the antebellum South it was widely believed that Irish should be employed in dangerous, high mortality jobs rather than risking the loss of valuable Negro slaves. New Orleans's New Canal, for example, one of the great Southern public works in the 1830s, was largely dug by immigrant Irish laborers, who died in great numbers during the four years of construction. Whether the traditional figure of 20,000, as expressed in song, is accurate, no one can say:

Ten thousand Micks, they swung their picks
To dig the New Canal.
But the choleray was stronger 'n they,
An' twice it killed them all.[10]

But the concentration of Irish workers at the bottom of society, though pronounced, can be overemphasized. The most widely read book on urban Irish immigrants, Oscar Handlin's study *Boston's Immigrants,* which set the pattern for urban ethnic group biography, may well have been an examination of the worst case. Boston, as Handlin makes clear, was not a magnet for immigrants: Irish were, in a way, trapped there in large numbers. In 1850 Irish born made up a third of the working population of Boston (they actually outnumbered natives of Massachusetts there, 14,595 to 13,533). In 1880 they represented less than a quarter, but of course, by that time, many of the native born who made up almost three-fifths of the city's work force were Irish Americans, but they cannot be easily identified in the census. Very high percentages of Irish were in just two occupational categories: laborers and servants. These categories employed 63.7 percent of Irish immigrants in 1850 and 49.6 percent in 1880. For native born the figures are 7.1 percent and 11.4 percent, while for the next largest immigrant group of the antebellum years, the Germans, the data show 14.8 percent and 11.9 percent. Yet, even in Boston, at the height of the famine immigration, more than a third of Irish-born workers were not in the lowest occupational ranks and by 1880 a slight majority were not. In 1850 Irish born were represented in every one of the sixty-two occupational categories Handlin uses, although, to be sure, very thinly in some of them. Only one of Boston's thirty-two undertakers was Irish born, but nearly a tenth of its physicians were and, even at that early date, just over a tenth of its police.

In other cities the Irish fared somewhat better, or perhaps *less badly* is the proper term to use. In Philadelphia, for example, as Dennis Clark has written, "there was simply a better chance for Irishmen to compete," largely because of more rapid growth in the City of Brotherly Love. But even in Philadelphia the overrepresentation of Irish born among laborers was quite high, although lower than in Boston. In Clark's census samples of one Philadelphia neighborhood, he calculates the percentage of male Irish-born laborers in three successive censuses, 1850, 1860, and 1870, as 34.4 percent, 22 percent, and 25 percent. And, because he is using a sample of actual census returns, he can create figures for native-born Irish Americans, 12 percent, 12 percent, and 8 percent respectively. (Clark does not discuss either servants or working women.)[11]

In the Far West—which of course, was out of reach for famine immigrants and other poor Europeans—the Irish experience was entirely different, as the English scholar R. A. Birchall has shown.[12] In San Francisco—which in 1870 was almost half foreign born and had the highest percentage of immigrants of any large American city—Irish were more than 15 percent of the population, and they were largely members of what modern sociologists would call the lower middle class. Of course, the presence of a large Chinese population tended to "promote" the status of all whites. Many who came were not workmen at all, but entrepreneurs. James Phelan (1818?–1900) is the kind of Irish immigrant not usually discussed. He emigrated to New York and moved on to Cincinnati, where by 1848 he had a wholesale grocery business worth perhaps $40,000. Drawn to San Francisco by the gold rush, he made his first fortune selling provisions to miners and then became a real estate magnate and one of the city's first millionaires. His son, James D. Phelan, became a United States senator and a leader of the anti-Asian movement. For a while father and son had offices in the modern "Phelan block" that the father built, where, as the story goes, the elder Phelan was once asked by the building's cigar counter operator why he smoked five-cent cigars although his son smoked twenty-five-cent ones. "Because," Phelan snapped back, showing a mind-set that only Protestants are supposed to have employed, "he has a millionaire for a father and I don't."[13]

The massive Irish immigration of the 1840s, not surprisingly, utterly transformed the Roman Catholic Church in America. The twenty-five thousand Catholics of 1790 had probably grown to about 100,000 by the end of the War of 1812. By 1860, there were three and a half million, and Catholics were the largest single denomination in the United States, although still a small minority of the population. As late as 1830 the leadership of the American church was largely French or French trained: Of ten bishops in the nation then, six fitted that description. French and French-trained priests, many of them exiles from France, were a bulwark of the church in the early national period. Anti-Irish sentiment was not restricted to natives and Protestants. English-born and French-trained James Whitfield, fourth archbishop of Baltimore, wrote a fellow cleric in 1832 about a vacancy in an American see:

> If possibl[e] . . . let an American born be recommended and (between us in strict confidence) I do really think we should guard against having more Irish bishops. . . . This you know is a dangerous secret, but I trust it to one in whom I have full confidence.[14]

In the next twenty years, however, what can be called the Hibernization of the American Roman Catholic Church was well begun: The major struggle was not between Irish (and later Irish Americans) and French, but between them and Germans (and later German Americans).

What was crucial, in the final analysis, was not the appointment of Irish or Irish American bishops, although that did happen, with Irish-born John Hughes, bishop from 1838 to 1850 and from 1850 to 1864 archbishop of New York, playing perhaps the key role. What was crucial was more than a million Catholic immigrants, most of them Irish but with a substantial minority of Germans, who came between the end of the War of 1812 and 1860. They and their children made up the vast majority of the church membership, and the Roman Catholic church had become an immigrant church.

Although it is widely believed that Irish Catholic immigrants were, almost ipso facto, devout Catholics, Jay Dolan has demonstrated that this was not the case. He quotes with approval a midcentury New York priest who wrote in his diary that "half our Irish population here is Catholic merely because Catholicity was the religion of the land of their birth." Many had never received communion and were ignorant of the basic tenets of their religion. "It is clear," Dolan concludes, "that all Catholics did not come to the United States in sound spiritual condition," and this assertion could be made, in varying degree about most immigrants from Catholic countries. While the theme of "preserving the religion of the immigrants" has been a major motif of American Catholic history, to a very great degree the Church also had to be concerned with instilling the fundamentals of religion in populations that were largely ignorant of them. This is not to say that Irish were indifferent to the church. Their ignorance was often coupled with a fierce loyalty. They identified with the church and, as we shall see in our discussion of nativism in chapter 10, an attack on it was seen as an attack on them, and vice versa. The church, after all, had been their only bulwark against the hated English, albeit a not very effective one, and loyalty, in the Irish culture, was placed quite high among the virtues.

The impact of immigration on Catholicism can be seen by merely enumerating the Catholic churches in New York City. The first church, Saint Peter's on Barclay Street, was founded in 1785. Only one more was established before 1826. Six were set up in the period 1826–36; twelve in the 1840s; ten in the 1850s; and two more in 1860–63. Of the thirty-two parishes, seven were German, one was French, and the rest were essentially Irish. The struggle between Irish and German Catholics for control of the church began in New York as early as 1808 and highlighted what would be an ongoing problem for the American church: Immigrants, and sometimes their children, wanted to worship in their own language. Even when the church authorities were willing to meet this demand, they were not always able to find priests who commanded immigrant languages. In fact, the church was chronically short of priests and other religious, and importation was the only possible solution as American seminaries—the first of which was Saint Mary's in Baltimore established by French Sulpicians in 1791—could never fully satisfy the demand.[15]

IRISH IMMIGRATION, 1860–1930

Nothing so deflates the notion that most American Irish are descended somehow from famine immigrants than a look at postfamine migration figures. As Table 6.5 shows, more than 2.6 million Irish came in the decades after 1860, an absolute majority of all Irish immigrants. The drain from Ireland was fairly constant: About 13.9 percent of Ireland's population emigrated to the United States in the 1850s, and about 12.6 percent did so in the 1880s. But their incidence in the United States changed greatly. The Irish immigrants of the 1850s were more than a third of all arrivals, while those of the 1880s were just an eighth. Similarly, the nearly one million Irish born in the 1850 census were almost 43 percent of all foreign-born, while almost twice that number in 1890 constituted only 20 percent. In other words, the arrival of even larger numbers of immigrants from elsewhere helped to mask the coming of large numbers of Irish. The new Irish tended to settle where Irish pioneers had established sizable urban enclaves, which contributed to their relative "invisibility."

In addition, when in 1880 the Census Bureau first began to enumerate the second generation—under the awkward if precise rubric of "native-born of foreign or mixed-parentage," "mixed" meaning one foreign-born and one native-born parent, it found more than 3.2 million second-generation Irish, making the Irish American community more than 5 million persons. By that time there were also significant numbers of third- and subsequent-generation Irish Americans, but the published census only shows them as "native born of native parentage" so we have no gross data.

But if we look behind the numbers, other significant differences appear in the new pattern of Irish immigration. Much of the very early migration had been heavily male; in the famine decades, migration was largely a family affair, although to be sure, many families arrived serially in chain migration and others were mutilated by the high mortality of those years. Males still predominated, but not by a very large margin. In the post-famine years, and particularly after 1880, more females came from Ireland than males. The best data on Irish emigration compiled by Kerby Miller, show this very clearly. From 1851 to 1880, males outnumbered females among Irish emigrants to all destinations by a ratio of 1.14 to 1. From 1880 to 1910 the ratio is 0.98 to 1, showing a slight female preponderance. Similarly, Archdeacon's chart of ethnic "maleness" shows Irish-born Americans as the only group with a female majority, 53.6 percent.

The reasons for this are to be found in Ireland, not America. While the position of women in Irish society was always quite disadvantaged, their place in postfamine Irish society clearly worsened. The decline of cottage industry lessened their economic contribution and independence while the development of arranged marriages and an increasingly important dowry system made marriage more and more difficult. Ireland had the highest age of marriage in Europe, and about a quarter of Irishwomen who stayed home never married. Before the famine about 20 percent of all Irish wives were at least ten years younger than their husbands; by the early years of this century about half were. And in Ireland—in contrast to Irishwomen who migrated to England or America—women did not live longer than men. All of this made the emigration of women—particularly young, single women—a salient characteristic of Irish postfamine emigration, and one that grew more pronounced with the passage of time. One early immigrant, Mary Brown from County Wexford, writing from New York just before the Civil War, advised a friend to come over to a land "where thers love and liberty." To be sure, the life of immigrant women was filled with hard work at menial occupations, but compared to conditions in Ireland the fact that "good girls to do house work in respectable families can readily get from one and a half to two dollars per week and good board and food" made such work seem quite attractive, especially at first.

In addition, underlining the nonfamily character of Irish immigrations an increasing proportion of Irish immigrants were young adults, female and male. In the 1850s through the 1870s, persons from fifteen to twenty-four were about 45 percent of all Irish emigrants, and those were fairly evenly divided by gender. From the 1880s into the 1920s, members of that age group were nearly 60 percent or more, with young women taking a larger and larger share; by 1921 young women outnumbered outgoing young males by two to one. Children fourteen and under, who had comprised more than a fifth of all emigrants at the end of the famine period, were, after 1890, less than 10 percent of those departing. Old people simply did not come: Persons over fifty-five represented a fairly constant 1 percent of both genders.

Irish and Irish American residential patterns continued to be urban and were still clustered in the northeast. Of the eleven large cities with the highest incidence of Irish born in 1910, ten were in that region, seven of them in New England. San Francisco, again, was the exception. But looking at the data in this way tends to disguise the fact that large numbers of Irish and Irish Americans were settling out of the region of initial concentration. In 1910 the Irish immigrants who made up a mere 3 percent of the population of Chicago, for example, actually outnumbered their fellow immigrants in the Irish American metropolis, Boston, by about two to one, even though Boston's Irish immigrants amounted to a tenth of its population.

The occupational structure of the Irish American population was also changing slowly but subtly. In the middle decades of the nineteenth century, as American cities were undergoing rapid growth and beginning to develop an infrastructure and creating the governmental machinery and personnel necessary to run it, the Irish and their children got in on the ground floor. The Irish policemen and firemen are not just stereotypes: Irish all but monopolized those jobs when they were being created in the post–Civil War years, and even today Irish names are clearly overrepresented in those occupations. Irish workmen not only laid the horsecar and streetcar tracks, they were also the first drivers and conductors. If the first or immigrant generation worked largely at unskilled and semiskilled occupations, their children increasingly worked at skilied trades. By 1900, when Irish American men made up about a thirteenth of the male labor force, they were almost a third of the plumbers, steamfitters, and boilermakers. Those who worked in industry found themselves lifted up into boss and straw-boss positions as common laborers and were, more and more, immigrants from southern and eastern Europe—Italians, Slavs, and Hungarians. Yet, at the turn of the century, very large numbers of Irish American men found themselves in unskilled or semiskilled jobs. In 1900 about a fifth of all male workers of whatever generation—25 percent of the Irish born, 17 percent of the second generation—had such jobs.

Not surprisingly, occupational mobility varied from place to place, as several studies of different American cities have shown. In 1890 Boston was, for most Irish born, still a place where little "promise dwelled": almost two-thirds of them were in unskilled jobs, while in Chicago the same could be said about only one third. A decade earlier the same could be said for nearly half of Philadelphia's Irish, including nearly a third of the native born, while in 1900 it was true of more than half of Detroit's Irish born and more than a quarter of its second-generation Irish. For white males born in America the national rate in unskilled jobs in 1900 was about 15 percent.

Although working Irish American women—like all American women workers—enjoyed much less upward mobility than did their brothers, husbands, and sons, some did improve their status. Irish American women workers were overwhelmingly concentrated in domestic service, laundry work, and in the least-well-paying jobs in the New England textile industry. At the turn of the century more than half of all working Irish immigrant women were servants, as were nearly a fifth of second-generation workers. Englishwomen of both generations then comprised an absolute majority of New England's servants and two-fifths or more of those in most eastern cities. Like most American female workers at that time, Irish American married women "with husband present," to use the Census Bureau's category, usually left the labor force at marriage or after becoming pregnant.

Irish Americans soon developed networks of kith and kin that helped to smooth the way for later immigrants. "Should your Brother Paddy Come to America," wrote an Irish American workingman to a relative in Ireland, "he can rely on his Cousins to promote his interests in Procuring work." The networks were not just familial. Irish trade unionists and leaders and Irish politicians illustrate nicely the intimate connection that can exist between economics and politics. Even today, one of the crucial points of friction between ethnic American skilled workers and affirmative action programs is the strong feeling that a son or a brother's son should be able to inherit the father's or uncle's union card—and the job that goes with it.

The contributions of Irish Americans—along with those of other ethnics from the British Isles—to the American labor movement are hard to overestimate. From an early second-generation Irish American like Terrence V. Powderly (1849–1924), longtime grand master workman of the Knights of Labor and later Commissioner General of Immigration, to the fiery immigrant Michael J. Quill (1905–66), leader of New York City's transport workers, Irish men have been significantly overrepresented among American labor leaders, particularly in the American Federation of Labor which organized almost exclusively among skilled workers. And although Irish Americans were concentrated in the skilled trades in the East, pockets of Irish American trade unionism existed in the Far West in places as disparate as Butte, Montana, and San Francisco.[16] Conversely, Irish American women were not particularly prominent among female trade unionists, most of whom came from the needle trades where Irish were not well represented, but there were some prominent radicals, the most noted of whom was the American communist leader, Elizabeth Gurley Flynn (1890–1964).

Even more well known and notorious were the ways in which Irish Americans adapted to and changed American politics, especially urban ethnic politics, which they practically invented. The Irish American identification with the Democratic party, which has a putative history going back to Irish support for Thomas Jefferson, was really a product of the political coming of age of Irish American communities in the mid- and late-nineteenth century. Although it was not universal—in Philadelphia, for example, Irish politicians and voters were an integral part of the corrupt Republican machine that dominated local politics—it was nearly so. And most other Catholic ethnics followed the Irish into the Democratic party. Thus, in the 1920s, North Carolinian Josephus Daniels (1862–1948), one of FDR's political mentors, could explain that the Democratic party was made largely of Southerners, Catholics, and Jews, "none of whom can be elected President."

Irish American urban politics is almost universally associated with "bad" political machines, the most venal of which, and the only one that readers of history textbooks ever hear about, was New York's William Marcy Tweed (1823–78). In writing about Tweed, textbook writers usually mention his greatest journalistic antagonist, the cartoonist Thomas Nast (1840–1902). They sometimes point out that Nast created two of the most potent symbols of American politics, the Tammany tiger and the Republican elephant, and popularized the third, the Democratic donkey. They do not note that Nast also popularized one of the great ethnic slurs of American journalism, the depiction of the Irishman as a stupid brute with simian characteristics.

More typical than Tweed was George Washington Plunkitt (1842–1924), whose as-told-to memoir spoke frankly of "honest graft," graft based essentially on inside information—precisely where a street car line was going to go, for example—rather than on outright theft. Plunkitt also makes it clear that the politicians performed real services for their constituents: handing out jobs, giving free turkeys at Thanksgiving or Christmas, or interceding with the authorities when a constituent ran afoul of the law.[17] In an age when respectable politicians, the so-called good government types—what regulars called "goo-goos"—were opposed to spending public money on anything even slightly resembling what we would today call social service or welfare, the political boss was as close as nineteenth-century America came to the "welfare state." The critics of the Irish machine politicians also usually ignore the fact that the machines sometimes produced real reformers, of whom New York's Alfred E. Smith (1873–1944) and German American Robert F. Wagner (1877–1953) are the best known. In addition, Irish Americans were willing to support liberal national Democratic candidates, such as Woodrow Wilson and Franklin Roosevelt. The Irish Americans and other ethnic voters who gave their votes to the bosses and their machines knew what they were doing. They were voting their interests, supporting those they perceived to be their friends and opposing those they knew were their enemies. That the Irish played ethnic politics so successfully and for so long, often dominating local politics long after the numerical preponderance on which that dominance was originally based had disappeared, has been attributed to the Irish genius for politics. No one who has even glanced at the history of politics in Ireland can believe that the Irish are, somehow, inherently good at it. What the successful association between Irish and politics in the United States should be attributed to, I believe, is the achievement of getting in on the ground floor, of numbers, of ethnic concentration, and of organization. What the Irish gained in politics, it can be argued, may have cost them dearly in other realms, such as social mobility and advancement in the arts and professions. Perhaps. But it is difficult to envisage a scenario in which Irish immigrants and their children eschewed politics and still came out as well as they did.

German Americans

Superficially, the immigration of Germans during the peak immigration years seems similar to that of the Irish. Between the 1830s and the 1880s Germans were never less than a quarter of all immigrants. They were nearly a third of all foreign born in 1860, 30.9 percent, and a quarter in 1900, 25.7 percent. Between them, Germans and Irish were almost seven out of ten foreign born in the former year, 69.8 percent, and more than four out of ten in the latter, 41.4 percent. Never again would two ethnic groups so dominate immigration. But that, and the fact that both groups were European, were about the only similarities. Germans spoke a foreign language, represented three broad confessional groups—Protestants, Catholics, and Jews—and had a varied pattern of distribution in the United States, participating significantly in both urban pursuits and agriculture. Germans more consistently migrated in family groups and had a return migration rate more than half again as high as that of the Irish, 13.7 percent as against 8.9 percent. Germans were drawn to the United States largely by economic reasons, but they were not fleeing from a national disaster or from a stagnant economy; they were, rather, often persons dislocated or threatened by a vigorous, if uneven, economic growth. Germans and Irish both had ethnic pride, but while Irish Americans could have little identification with the government of Ireland, that "most distressful nation," most German Americans took particular pride in Germany's achievements, especially after the creation of the German Empire, the Second Reich, in 1871. Mack Walker's observation about nineteenth-century German immigrants, that they came not to establish something new but to reestablish something old, could never be made about the Irish.[18] The cultural apparatus created in the United States by Germans

Table 6.6

Immigrants from Germany, 1820–1924

DECADE	NUMBER	PERCENTAGE OF TOTAL IMMIGRATION
1820–30	7,729	5.1
1831–40	152,454	25.4
1841–50	434,626	25.3
1851–60	951,667	36.6
1861–70	787,468	34.0
1871–80	718,182	25.5
1881–90	1,452,970	27.7
1891–1900	505,152	13.7
1901–10	341,498	3.9
1911–20	143,945	2.5
1921–24	148,102	6.3
Total	5,643,893	16.4

in the nineteenth and early twentieth centuries dwarfs that of any other ethnic group.

As noted earlier, there are special problems in trying to decide who is German. Prior to 1871, there was no German nation, but only a collection of separate German states, which in 1871 became unified under Prussian leadership. The chart above follows the American data which generally treat as German all persons immigrating from the post-1871 boundaries of Germany. This means that German speakers from Switzerland, Austria, and the Austro-Hungarian Empire, so on, are not included. My maternal grandmother, for example, a German-speaking Jew from Hungary who arrived in 1900, was almost certainly listed as a Hungarian although she read a German-language newspaper for the rest of her life. In that sense the data understate Germans and Germanness. On the other hand, some German states had ethnic minorities—Poles in Prussia, Danes in Schleswig-Holstein for example—who, if they came to the United States, were almost certain to be counted as Germans. This tends to overstate German immigration. Which factor is more significant no one can say. This counting by nationality rather than ethnicity continues to create problems in American immigration data. Many recent refugees and immigrants from Vietnam and other parts of Southeast Asia are ethnic Chinese who so identify themselves to the census taker, but who are recorded in the immigration data as citizens of the nation of their birth.

Although some German immigrants headed to Russia and other less-developed parts of central and eastern Europe (large numbers of their descendants would eventually come to the United States) and others migrated to such destinations as Argentina, Brazil, Canada, and South Africa, once the postcolonial migration of Germans to America had got under way in the mid-1830s, about 90 percent of all German immigrants came to the United States. The rhythms of their movements were largely affected by the fluctuations of the American business cycle, although, to be sure, economic and political events within Germany also had their effect.

Religious motivations, so important in establishing the colonial American migrations, were much less significant in the nineteenth century. Some Old Lutherans from Prussia were impelled to come by the unification of their church with the Reformed Church in the 1830s, and small handfuls of religious radicals, mostly pietists, founded new societies in Pennsylvania and, as the century progressed, in Ohio, Iowa, South Dakota, and the prairie provinces of Canada, chiefly Alberta. Among Catholics, Bismarck's *Kulturkampf* (cultural struggle) against them shortly after unification was a push factor.

Politics was even less significant as a pushing force. The Metternichian reaction after the defeat of Napoleon and the sometimes violent repression of liberalism in the 1830s did send some activists abroad. Even the great upheavals following the failed revolutions of 1848 made a very small numerical contribution to total emigration, and the legend of the forty-eighters in the German American community has been greatly inflated. It is true that some exiles did flee to the United States after 1848 (many also went to London, including the most famous exile of them all, Karl Marx, but at most there were a few thousand of them). One or two—most notably Carl Schurz (1829–1906), who became a general in the Union Army, represented Missouri in the Senate, and served as Rutherford B. Hayes's secretary of the interior—did have significant American political careers. But as opposed to the Irish, not many Germans were attracted by American politics. Despite all the ink that has been spilled about the failure of an Irish Catholic to be elected president until 1960, it is almost never remarked that no member or descendant of the larger German American community was elected until 1952, and Dwight Eisenhower was elected largely because he had led the armies that defeated Germany. The Kaiser's ambassador to the United States before and during World War I, Count Johann Heinrich von Bernstorff (1862–1939) tells us in his memoirs that his illusions about the potential political power of the German American community were shattered when he visited Milwaukee, the large city with the highest proportion of Germans, and found that it had an *irisch Bürgermeister* ("Irish mayor"). In addition it should be noted that a few German Marxists came to the United States, as did, even more notoriously, some German anarchists. After several of their number were hanged—on no real evidence—for the Haymarket bombings in Chicago (1886), the German radical became, for a time, a stock figure in American social novels, most notably William Dean Howells's *A Hazard of New Fortunes* (1890). The surviving anarchists were eventually pardoned by another German immigrant prominent in American politics, Illinois governor John Peter Altgeld (1847–1902), who was also a very successful businessman.

GERMAN AMERICAN OCCUPATIONS

It was in the economic sphere that the motivations for most German immigrants were to be found. It was the economics of prosperity, not of poverty, that impelled most of them. What students of socioeconomic change now call "modernization"—industrialization, urbanization, and the whole complex of political and social changes usually attendant with them—was nowhere on the Continent more apparent than in Germany from the mid-nineteenth century onward. These changes in the traditional structure of society made it increasingly difficult to maintain former ways of life. When population growth had been slow or nonexistent it was often possible for sons to succeed fathers, whether in ownership of land or in skilled trade. Increasingly, this became difficult under the new conditions, and, faced with the necessity of moving to a new job, many Germans found emigration a rational alternative. Others, having initially made an internal move—say from a German farm or village to a German city—would then make a further move overseas. Modernization not only increased the pressures for leaving, it also created speedier and more convenient ways to emigrate. Steamboats and railways made it easier to move within Germany, as did the gradual removal of internal barriers to the flow of people and goods. The development of German ports and a German merchant marine was also significant. As noted, in the Colonial Era much German emigration was through Holland; in the early nineteenth century Le Havre, the major continental port for the cotton trade, was a key exit point, especially for emigrants from southwestern Germany. At midcentury, many Germans came to America by taking a complicated but competitively priced route from Holland to Hull, crossing England by rail, and embarking from Liverpool. In the later nineteenth century, north German ports predominated. First Bremen—which originally took immigrants as return cargo in the tobacco trade, most often landing them at Baltimore—and then Hamburg developed as major ports in the immigrant trade. They served immigrants not only from Germany but also from central and eastern Europe and took them chiefly to New York.

Within the United States the pattern of German settlement was in stark contrast to that of the Irish. Whereas the Irish settled almost exclusively in large cities, only a minority of Germans did. Between 1860 and 1890 about two-fifths of the German-born lived in cities of twenty-five thousand or more, a figure considerably higher than that of native-born Americans. And, of course, second-generation German Americans had a propensity to move to the city, like Theodore Dreiser and his heroine Carrie Meeber. A list of the nine large cities in 1870 in which Germans had the highest incidence has only one city—New York—in common with the similar list given for the Irish earlier. The other heavily German cities ranged from Milwaukee—where German born were about a third of the population—through Cincinnati, Buffalo, Saint Louis, Chicago, Cleveland, Toledo, Dayton, and Detroit. Within those cities Germans tended to cluster in ethnic enclaves—Kleindeutschland (little Germany) in New York and Over the Rhine in Cincinnati—which replicated the culture of the homeland, or tried to. Historical geographers sometimes speak of the German triangle, whose three points were Saint Louis, Cincinnati, and Milwaukee, and within which in the late nineteenth century was contained an absolute majority of the German born in the United States. Some, mostly those who came through Le Havre, landed in Gulf ports and, largely eschewing the slave South, went up the Mississippi or by rail to somewhere in the triangle. Others came overland from East Coast ports along the route of the Baltimore and Ohio, the Pennsylvania, the Erie, and the New York Central railroads.

Much larger numbers of Germans than of Irish worked at skilled trades: In 1870 more than a third of German-born workers (37 percent) had such jobs. The lager beer industry, for example, employed German capital and German skilled labor and catered to German customers in Cincinnati, Milwaukee, and Saint Louis. Germans were also prominent as bakers, butchers, cabinetmakers, cigar makers, distillers, machinists, and tailors. German-born women, on the other hand, were less likely to enter the labor force, as defined by the Census Bureau, than were other American women, immigrant or native born. They were underrepresented in factory and clerical jobs and concentrated largely in the service sector as bakers, domestic workers, hotel keepers, janitors, laundry workers, nurses, peddlers, saloon keepers, and tailors. Many women worked either in family-owned businesses or businesses in the German American community.

The traditional argument has been that because so many immigrant women were housewives or otherwise employed at home, they were less likely to learn English and become acculturated as quickly as immigrant men. Kathleen Conzen, writing about German immigrants in Milwaukee in the decades before the Civil War, has suggested that sometimes the opposite was the case. Noting that it was quite common for immigrant girls as young as eleven or twelve to be hired out to Yankee families as day helps, she suggests that "domestic service not only augmented family incomes and provided nest eggs, but it brought an important segment of the immigrant population—the future mothers of the second and third generations—into intimate contact with middle-class American home life." They quickly learned English, adopted American dress, and with it, undoubtedly, American attitudes.

Many young men, in a heavily German city like Milwaukee, by contrast, lived and worked solely among German speakers.[19]

About the urban Germans of the nineteenth-century the stereotype exists of the beer-swilling, fat, and contented exponent of gemütlichkeit. One academic opponent of the German American way of life, quoted by Carl Wittke, complained, from far-off Göttingen, that "wherever three Germans congregated in the United States, one opened a saloon so that the other two might have a place to argue."[20] It is true that saloons and beer gardens, where families went to drink beer, socialize, and listen to music, proliferated in German American neighborhoods and cities: There were two thousand places where drinks were sold in Cincinnati in 1860, more than one for every hundred residents. The German attitude toward drink was one of the crucial points of friction between Germans and some of their neighbors, between continental and Puritan ways of spending a Sunday, between a relaxed attitude toward alcohol and a crusade against it. As we shall see in chapter 10, it was not only an issue in the long struggle over Prohibition but also, during World War I, a way of discrediting everything German.

About one German-born person in four was engaged in agriculture in 1870, and at that date Germans were more than a third of all foreign-born farmers. This does not count most Pennsylvania Germans, who were by this time almost all native born. German farmers settled in large numbers in the German triangle and elsewhere in the Midwest and in Texas, while there were scattered German agricultural settlements in the Far West, such as the one in Anaheim, California. The stereotype of the industrious, tight-fisted, and successful German farmer, reluctant to give up Old World techniques and customs and more persistent in his settlement than most of his neighbors, was carried over from the image of the Pennsylvania Germans. What validity it has relates to those German farmers who settled in places with a high density of other Germans. In such circumstances what an early student of immigration called "old world traits transplanted" were most likely to persist. However, the myth, also carried over from colonial times, that German farmers somehow knew how to choose good land has been exploded. Kathleen Conzen, a close student of rural Germans, has written that:

> Germans were not especially gifted in the choice of land. They settled on what was available, valuing access to market and to nearby German urban settlements. . . . Even in clustered settlements, Germans planted the locally prevailing crops, and almost

everywhere abandoned Old World village settlement patterns for the dispersed farmsteads of America.[21]

Germans did tend to persist more than their mobile neighbors and tended to invest more heavily in buildings and other improvements. This has also produced, in some rural-smalltown areas, the persistence of the Germans language into the fourth and fifth generations. Sometimes, in such areas, German Americans were able to obtain political control and tried to impose their culture on others. In an extreme example of this, as late as 1888 English-speaking persons in a few Missouri school districts had to struggle to stop the local boards from using German as the basic language of instruction in the public schools. And when I was in college in Texas in the 1950s, I had fellow students from the area around New Braunfels who spoke of not learning to speak English until they went to public school. This kind of language persistence, it must be emphasized, was the exception rather than the rule. But the continued presence of German Americans in agriculture, combined with their heavy urban concentrations, is one of the factors that sets off the Germans not only from the Irish but from most other immigrant groups.

DIVISIONS AMONG GERMAN AMERICANS

During much of the nineteenth century, divisions among Germans seemed more significant than those between German Americans and other groups. These divisions were based on geography, on ideology, and on religion. The first two were most apparent before 1871, when the push for German unification tended to unite most but certainly not all German Americans in feelings of pride in their fatherland and its achievements. Initially, German immigrants tended to identify themselves as Bavarians, Württembergers, Saxons, and so on, although intellectuals and those who were politicized yearned for some kind of German unification. Most of these were liberals of one kind or another, who dreamed of a more-or-less democratic Germany. Even so, when unification did come on Bismarckian, autocratic terms after the wars of unification, all but the most ideologically committed German Americans rejoiced: Liberals and conservatives, as well as the more numerically important apolitical, were united in a feeling of pride.

Religious differences were more enduring. Most German immigrants were Protestants, with Lutheranism by far the most numerous denomination; perhaps a third of German immigrants were Catholics, and perhaps 250,000 were Jews. Within the Lutheran community in the United States there was considerable friction. Nineteenth-century German Lutheran immigrants found that the existing German Lutheran churches in the United States had developed what, to them, were unwelcome tendencies. Most had been Americanized enough so that English was used for all or part of the services. Even worse, doctrine had been liberalized. The older churches and their offshoots, established by immigrants who had come before the Revolution, had come closer to Reformed and even Anglican churches and in many instances had adopted preaching styles similar to that of the Methodists. These trends were, not surprisingly, more pronounced in the cities than in the country. In New York and Philadelphia, for example, Lutheran bodies had adopted new constitutions in which all reference to the Augsburg Confession had disappeared. The result was, eventually, schism. By 1847, under the leadership of a recent immigrant pastor, C. F. W. Walther, whose enemies called him "the Lutheran pope of the West," the newer Lutheran arrivals who wished to maintain the old-style doctrine had organized the Missouri Synod. Over the years it has remained the bulwark of the more conservative American Lutherans, regardless of where they live.

Catholic Germans were allowed national parishes (something that would be given only reluctantly to later-nineteenth-century immigrant groups, such as the French Canadians, and denied to many Mexican immigrants in the twentieth century) but were swamped by the numbers of Irish immigrants in eastern dioceses, such as New York, where the clerical leadership was almost invariably Irish or Irish American. The United States has always been a net importer of clergy and religious, with the seminary at Maynooth, Ireland, being one of the major sources. In much of the Midwest, German and later German American bishops and eventually archbishops came to prevail, but in their struggles with the Irish and Irish American bishops for control of the American church, the Germans lost. This conflict can be overstated and, it must be emphasized, German-Irish conflict never produced a significant schism in the church, as did a similar conflict initiated by Poles and Polish Americans. Nevertheless it did persist and in some ways persists to this day. The differences have come to be expressed primarily in social and political rather than doctrinal terms. As a rule German and German American prelates have been and are more liberal on social and political questions. It is no accident that the member of the hierarchy with whom Franklin D. Roosevelt felt most comfortable was the German American archbishop of Chicago, George William Cardinal Mundelein (1872–1939). However, it must be noted that not all German Catholics were social liberals, not all Irish Catholics social conservatives. The leading social activist among the church leadership in the first half of the twentieth century, for example, and the moving spirit of the National Catholic Welfare Council's Social Department, was Monsignor John A. Ryan (1869–1945). Ryan's liberalism was so patent that, when he clashed with the Irish-Canadian-American priest, Charles E. Coughlin (1891–1979), the latter sneered at the "Right Reverend New Dealer," a nickname a subsequent friendly biographer of Ryan used to title his book.

The struggle for control within the American church came to a head toward the end of the nineteenth century. Despite a growing number of German-speaking clergy—one count records 2,882 by 1892—and large numbers of German national parishes—705 in 1869 and 1,890 in 1916, the latter figure representing about 10 percent of all American parishes—German and German American cultural nationalists continued to strive for greater and more independent roles for Germans within the church and more German and German American bishops. In 1886 the German-born vicar-general of the diocese of Milwaukee, acting on behalf of a number of German-speaking priests in his own diocese and in Cincinnati and Saint Louis, memorialized the Vatican about the lack of sympathy toward the German language and customs evinced by Irish-born and Irish American bishops. Four years later, the Saint Raphael Society, a German lay society founded for the protection of German Catholic emigrants and the preservation of their religion, forwarded a memorandum to the Vatican advocating that Germans and other Catholic ethnic groups be represented in the hierarchy and that each, whenever possible, should be ministered to in its own language by priests of its own nationality or ethnicity. It did *not* advocate—as is sometimes alleged—that the American church be organized ethnically rather than geographically, although there were those who did advocate such a step. This movement was often called Cahenslyism for Peter Paul Cahensly (1838–1923), a German merchant and philanthropist who sponsored the Saint Raphael Society—and who never even visited the United States. The Vatican temporized, but in a way that did not challenge Irish and Irish American dominance in the church. It recognized, in practice, the national parishes that had long existed but refused to consider any change in the method of appointing American bishops, which would, of course, have infringed somewhat on the prerogatives of the pope, forcing him into a sort of ethnic affirmative action.

But the German-Irish divisions were not polar, and on some issues some bishops and priests split along other than ethnic lines. In the related but separate controversy over the so-called Americanist heresy, for example, the German clergy of Milwaukee and the Irish clergy of New York were on the same side, but that involved doctrinal rather than ethnic matters. In any event, as time went on, the dilution of urban German American enclaves and the gradual abandonment of the use of the German language in sermons meant, effectively, the dissolution of most German parishes except in the most heavily German cities. And although much of this abandonment took place in the hysterical atmosphere of World War I, it was an inexorable process already under way by then.

The immigration of German Jews, beginning in the 1820s, had a transforming impact on the American Jewish community similar to that of Irish and German Catholics on the American Catholic community. The initial Sephardic synagogue in New York, Shearith Israel, was the only one there until 1825, when German-speaking Ashkenazic Jews founded B'nai Jeshurun. By 1835 there were ten synagogues in New York. At the outbreak of the Civil War, Jewish communities once confined to the coastal littoral had spread across the continent in more than 150 places; however, the East Coast continued to attract most Jewish immigrants and their descendants. In New York, where the Jewish population was perhaps 500 in 1825, there were an estimated 40,000 by 1860. Many families that originally settled elsewhere were eventually attracted to Gotham. The Lehmans, for example, who founded the Wall Street firm of Lehman Brothers came to New York from Alabama, having established themselves as merchants there. Of the Midwestern centers of German American Jewry, none was more important than Cincinnati, where there were perhaps 3,300 Jews in 1850 and 10,000 a decade later. In the earlier year the nation's first Jewish hospital was established there. In the Far West, the Gold Rush drew many Jews to California in general and to San Francisco in particular. Among them was Levi Strauss (1829–1902) who emigrated from Bavaria to New York in 1848 but by 1850 had set up a dry goods business in Sacramento and later developed for the miners who were his customers the Levi's that have become the international uniform of young people.

Initially, most German Jews came from southwestern Germany, although soon large numbers also came from Posen (now Poznan), in Prussian-occupied Poland. Within Germany most Jews had lived in small cities and market towns and were artisans and petty traders and cattle dealers, although a few could properly be called merchants, and a very few were bankers. The same kinds of economic disruptions that impelled German migration generally, also spurred the German Jews who had the additional impetus of wanting to avoid special taxes and discriminatory legislation which impinged on their most basic civil rights, such things as the right to marry and reside in certain places.

While much German Jewish migration was of families, sometimes very large families, such as the first American Guggenheims who came with an even dozen children, perhaps more often there was chain migration. Joseph Seligman (1819–180), the founder of the New York banking house which bore his name, migrated from Baiersdorf, Bavaria, as a teenager in 1837 and sent for his two eldest brothers two years later. By 1843 eight more brothers and sisters and their widowed father had been brought over. While these two families were more numerous and became more prominent than most, their modes of immigration paralleled that of the majority.

German Jewish migration was part and parcel of the German migration of the period. It is not always easy or even possible to differentiate between them and other Germans. The federal immigration and census records merely give place of birth; they do not list religion. We are able to say what we do about German Jews largely because of the existence of synagogue records. The American Jewish community has also been particularly assiduous in collecting documents and writing family, institutional, and communal accounts that range from genealogies and mythmaking filiopietism to the scientific history still practiced by Jacob Rader Marcus (born 1896) at Cincinnati's American Jewish Archives and to the sophisticated "number crunching" of statistically minded contemporary social scientists. But while a historian of Germans can construct a table from census data showing that in 1860, in Milwaukee, 38 percent of German-born household heads were skilled workers as opposed to 19 percent of similar Irish-born persons, careful historians writing about German Jews are forced to resort to such statements as "a New York merchant's reference guide for 1859 listed 141 wholesale firms with Jewish names."

Apart from imprecision, there are other obvious pitfalls in using names. While the chances of someone named Aaron Rabinowitz being anything other than Jewish approach zero, many German surnames were and are shared by Jews and Gentiles. My grandmother, for example, was named Lustig: There are a dozen Lustigs in the Hamburg telephone book for 1989, apparently none of them Jewish. In addition, many individuals changed their names or had them changed by officials in the process of emigrating and settling. When such an individual became well known, as did August Belmont (1816–90), who changed his name from Schönberg when he left Frankfurt in 1837 and became an Episcopalian and an important banker and politician, historians have become aware of the change. But there must have been countless cases of "leakage" of which we are not aware, some occasioned not by an intent to obscure origins, as in Belmont's case, but by settlement in a place without any established Jewish institutions. Such problems of ethnic identification and misidentification are endemic to American social history. We have already seen how Irish Catholic indentured servants could "disappear" in the colonial South.

But there are characteristics other than religion that set the German Jewish immigration somewhat apart from that of other Germans. One distinction is in the rate of return migration. While one in seven or eight of all German emigrants returned to Europe—perhaps as many as seven hundred thousand persons—very few German Jews did. (The traditional figure for all Jewish return migration is less than 5 percent.)

Other differences involve patterns of settlement and occupation. Very few German Jews were engaged in agriculture, which also meant that, unlike other Germans, hardly any lived in rural areas. A relatively high percentage settled in New York City—about a fourth of the total—with a consequent smaller concentration in the Middle West. In addition, a larger percentage of German Jews settled outside the major centers of settlement than did Germans generally.

Relatively few German Jews were artisans, and most of those who worked with their hands were tailors. But it was in the retail trades that a disproportionate number of German Jews found their economic niche in America. Many began as itinerant peddlers who walked the dusty roads of rural and small town America with packs of notions and other small goods on their backs. It was a trade which many had plied in Europe and with which Americans were familiar, as the Yankee peddler had long been a fixture of American rural life. In a number of Jewish families, the first son established in America would found or take over a retail or wholesale establishment, and send later arriving sons, or other relatives, or *landsleit* (fellow townspeople from Europe) on the road. In a number of instances these early traveling salesmen established branch stores. This was especially true in underdeveloped American regions, such as the South, Southwest, and Far West. Although the merger mania of the 1970s and 1980s has now erased most of the old names, most of the major stores in these regions were founded by German Jews: Rich's in Atlanta, Sakowitz's in Houston, Goldwater's in Phoenix, and Meier and Frank in Portland, Oregon, to name just four. Sometimes the great success came in the second generation: Julius Rosenwald (1862–1932), the guiding spirit of the great mail-order house Sears, Roebuck, was the son of a German Jewish peddler who settled in Springfield, Illinois.

At the very top of the German Jewish occupational ladder stood the few handfuls of bankers associated with German Jewish banking houses. (Almost no Jews were then hired by any other American banks.) Some, like the Schiffs and the Warburgs, came from families that had already been established bankers in Frankfurt and Hamburg respectively; others, like the previously mentioned Joseph Seligman, came here relatively poor. Most German Jewish businessmen, of course, did not make spectacular successes, but more of the first generation entered into business or other middle-class occupations than any other nineteenth-century immigrant group.

It is difficult today, in the aftermath of the Holocaust, to realize just how German most German American Jews were. The oft-quoted credo of the immigrant Bernhard Felsenthal (1822–1908), a Ph.D. from the University of Munich and rabbi of Chicago's Zion Synagogue, eloquently expresses attitudes shared by many, particularly among the elite.

> I am a Jew, for I have been born among the Jewish nation. Politically I am an American as patriotic, as enthusiastic, as devoted an American citizen as it is possible to be. But spiritually I am a German, for my inner life has been profoundly influenced by Schiller, Goethe, Kant, and other intellectual giants of Germany.[22]

The final major distinction between German Jews and other German immigrants was that, as Jews, they faced a double prejudice, since, in addition to being foreigners, they were not Christians. For reasons of contrast, and because certain things are best understood in a comparative setting, I am deferring a discussion of both anti-Semitism and the developments within American Judaism until chapters 8 and 10, when the migration of Jews from Eastern Europe and nativism will be discussed.

Not all major distinctions among German Americans were confessional. To many in the nineteenth century the greatest division seemed to be between the church Germans—this term seems to have been used for Christians—and club Germans. The one group, mainly of peasant origin and largely but not exclusively settled in rural America, made the church the hub of their noneconomic life. The other group, predominantly urban in America, was involved in a whole gamut of secular activities, from musical societies to shooting groups. Although many of the second group were freethinkers or agnostics, most of the membership espoused a religion, but not as a primary associational activity. (Jews were sometimes members of these clubs; they also had clubs of their own.)

GERMAN AMERICAN CULTURE

Indispensable for most cultural institutions that were intended to endure beyond the immigrant generation was some way of ensuring that the second and subsequent generations learn and use the ancestral language—what scholars now call language maintenance. As has been noted, the basic requirements for such maintenance in the United States are numbers, desire, and a high degree of segregation verging on isolation. In some few rural areas these all occurred, but in no city of any size did Germans constitute even a bare majority. Beginning with parochial schools, largely but not exclusively Lutheran and Catholic, Germans eventually turned to the public schools and political action in an attempt to make German instruction in all subjects available when enough parents wanted it. In such public schools, English might be taught as a special subject as if it were a foreign language, which, of course, it was and is to many young children of immigrants raised in essentially monolingual homes. And in many parochial schools, English was not taught at all.

Beginning in 1839 a number of states, starting with Pennsylvania and Ohio, passed laws enabling (or in some cases requiring) instruction in German in the public schools when a number of parents—often but not always 50 percent—requested it, and these laws were copied, with inevitable variations, in most states with large blocs of German settlers. The Ohio law authorized the setting up of exclusively German-language schools. In Cincinnati this option was exercised so fully that there were, in effect, two systems, one English, one German, and, in the 1850s, the school board recognized the right of pupils to receive instruction in either German or English. In Saint Louis, on the other hand, the use of bilingualism was a device to attract German American children to the public schools. In 1860 it is estimated that four of five German American children there went to nonpublic schools; two decades later the proportions had been reversed. In Saint Louis all advanced subjects were taught in English. So successful was the integration that even before the anti-German hysteria of World War I, German instruction, as opposed to instruction in the German language, was discontinued.

The backlash against German instruction began at the end of the 1880s, when a number of states enacted statutes that, in effect, regulated the use of German. The best known of these was Wisconsin's Bennett Law of 1890, modeled on an Illinois enactment of the previous year. The former simply provided that English be taught in all schools in the state for at least sixteen weeks a year and that schools not teaching reading, writing, arithmetic, and United States history in the English language were not to be regarded as schools under Wisconsin law—which meant that attendance in them would not satisfy the state's compulsory education law. Supposedly a newly elected Republican governor called for the law after discovering that in 129 Lutheran parochial schools in the state there was no teaching in English at all, and the same held true in many Catholic parochial schools. The Bennett Law caused great controversy in Wisconsin, leading to the governor's defeat at the next election and the law's repeal. Most laws in other states, such as Massachusetts and Texas, simply mandated the use of English in public schools. In American high schools in the early twentieth century, German was the most-taught modern foreign language, studied by perhaps a fourth of all high school students.

The anti-German hysteria of World War I, however, put an end not only to German instruction everywhere, but to most teaching of German as a foreign language, not only in high schools, but in some colleges and universities as well. What happened in Nebraska illustrates the tendency well. There, in 1913, a state law, backed by a number of German American groups, required that every public high school and urban public schools at any level provide instruction in any modern European language requested by the parents of at least fifty pupils. During the law's five-year life apparently German was the only language requested by the requisite number of parents. In 1918 the law was repealed by unanimous vote after the governor denounced it as pro-German and "un-American." The next year the war was over but the hysteria was not. Every tenth bill dropped into the legislative hopper in Lincoln in 1919 concerned the language issue. One legislator put it nicely: "If these people are Americans, let them speak our language." What finally emerged was a statute forbidding any teacher in any kind of school—public or private—from teaching any subject in any language but English. It also forbade the study of any foreign language—

including Latin!—before grade eight. An appeal to the Nebraska Supreme Court failed, but the opinion suggested that language instruction outside regular school hours might be possible. The 1921 legislature countered by making instruction in a foreign language illegal at any time.

A teacher in a Protestant parochial school, Robert T. Meyer, who was fined twenty-five dollars for violating the law, appealed. He lost in the state courts, but in *Meyer* v. *Nebraska* (262 U.S. 390) the Supreme Court of the United States overturned the law, holding that the law unreasonably interfered with the "liberty" of parents and guardians to direct the economic upbringing and education of their children and thus was in violation of the Fourteenth Amendment. Although the *Meyer* case was a milestone in civil liberty, and a precedent for the even more important case of *Pierce* v. *Society of the Sisters* (268 U.S. 510), which three years later overturned an Oregon law that attempted to abolish all private schools, the damage done to the German-language schools by then (1925) was irreparable. Where there had been hundreds of thousands of students in German-language parochial schools before the war, by 1927 there were about thirty-five thousand in the schools run by the Missouri Synod—which was probably a majority of all such students—and by 1936, only about half that. In the public schools, the destruction was total: No German-language instruction program survived, and, in addition, the teaching of foreign languages in American high schools had suffered a setback from which it has yet to recover. And, of course, the demise of instruction in German hastened the death of most of the other cultural institutions of German America.

A mere listing of the major cultural enterprises of German America between 1850 and the outbreak of World War I gives some notion of their scope which was and remains unrivaled by any other American ethnic group. The German American press, which traces its origins to ephemeral sheets established by Benjamin Franklin (1732) and Christopher Sauer (1739), was clearly the largest and most influential segment of the American foreign-language press and may have been the best edited as well. It tended, in the nineteenth century, to have a liberal cast and at one time perhaps a majority of all German papers were edited by forty-eighters, one reason that it became traditional to overstress their incidence in the emigration. There was, however, a wide spectrum of opinion expressed in the papers, which, in good nineteenth-century fashion for all American newspapers, were very much the products of their individual editors. It has been said that the only issue in the whole century on which the press was unanimous was the Franco-Prussian

War, when even forty-eighters cheered a victory that produced, in the final analysis, the antithesis of their former dreams for a democratic Germany. There were about eight hundred German newspapers in the 1880s, at which time about four out of five foreign-language papers in the United States were German.

It must be remembered that, however great the volume of the German American press, most German Americans did *not* read German newspapers regularly, even assuming multiple readership for each copy. At their peak the seventy-four daily newspapers, the first of which was Cincinnati's *Volksblatt* (1836), had a circulation of perhaps three hundred thousand at a time when there were nearly two million German-born persons in the United States and as many or more of the second generation. The nearly four hundred weeklies had a combined circulation of more than a million, but since many were church and club publications and since a large number of readers probably subscribed to more than one, their numbers should not be cumulated with the daily circulation.

Other aspects of high German culture also flourished in America to a degree that outstripped those of any other ethnic group. Autobiographies, histories, novels, poems, short stories, and even an eleven-volume *Deutsch-Amerikanisches* encyclopedia, all flowed from the pens of German American authors in enormous quantity, but none is regarded by present-day literary critics as having lasting esthetic value. Only during the Nazi period, when a number of already-established German writers fled to America— Nobel laureate Thomas Mann was the most notable—were significant contributions to German literature made by writers in America.[23]

The contributions of German Americans to American musical culture have been of permanent value. In Milwaukee, which liked to think of itself as the German Athens in America, the local Musical Society was able to stage a production of Albert Lörtzing's opera *Zar und Zimmermann* before an audience of eight hundred in 1853 when the city's population was not yet 30,000. Even more important were the *Sängvereine* (singing societies), which blossomed wherever there was a German American community of any size. *Sängerfeste* (singing contests) evolved into national competitions: In 1900 in Brooklyn, for example, 174 associations with six thousand singers competed. This tradition, although largely a thing of the past, survives in some localities as an annual concert or concert series, such as Cincinnati's May Festival, with professional soloists and orchestra and a local nonethnic chorus. Germans are also intimately connected with the development of American symphony orchestras, both in cities like Philadelphia and Chicago, which had large German American populations and a place like Boston where a German immigrant, Carl Zerrahn (1826–1909) served as director of its Handel and Haydn Society for four decades. Later such outstanding German-born conductors as Theodore Thomas (1835–1905) and the Damrosches, Leopold (1832–85) and his son Walter (1862–1950), enriched the musical life of New York and the nation by taking orchestras from the metropolis on tour. Other German Americans were important patrons of music, including, perhaps most notably, Otto Kahn (1867–1934) who underwrote the deficit of the Metropolitan Opera for years.

The German American theater, while leaving no enduring institutions, once flourished almost everywhere there were German settlements. Not only in eastern centers but as far west as St. Paul there were professional German theaters. For Minnesota alone, one scholar has calculated that there were over 1,600 performances of German-language plays between 1857 and 1890, an average of almost one a week in a state not long removed from its frontier period. In Cincinnati, and many other German American centers, regular German theater did not survive World War I. Today German-language theater exists only on college campuses and in sporadic tourist performances such as the annual open-air staging of Schiller's *William Tell* in New Glarus, Wisconsin, originally a Swiss American settlement.

The reasons for this cultural flowering in America are not to be found merely in the numbers of German Americans, but in the cultural and intellectual attainments of so many of the immigrants who came. These middle-class largely urban types, many of them products of German gymnasiums and universities, provided the leadership and the stimulus. At the same time, there developed a great deal of German American cultural arrogance about the superiority of their culture no so much vis-à-vis other immigrant groups but over and against general American or "Yankee" culture. The forty-eighter quoted by Carl Wittke, who refused to throw away "the intellectual achievements of a thousand years . . . for the culture of the primeval forest," was more eloquent than most, but not atypical of German American cultural leaders' views. It was, admittedly, an arrogance that had some basis in fact. Other cultural frictions between German Americans and "Yankees" included such matters as Prohibition and the German or continental Sunday as opposed to the Puritan Sunday. These all helped produce considerable tensions between German Americans and some of their neighbors even before World War I raised the curtain on the last act of what John

Hawgood called *The Tragedy of German America.*

There were times and places when one and another exponent of German American *Deutschtum* fantasized, Walter Mitty-like, about a German colony or at least a German state in America. Realists in the German American community, such as Carl Schurz or Gustave Koerner (1809–1896), understood that this could never be. German America contained within itself the seeds of its own destruction, and even without the cultural terrorism of the era of World War I its decline was inevitable simply because the adaptation of most Germans and their descendants was so successful. The British churchman, Dean Robert W. Inge, once quipped, "In religion, nothing fails like success": The same can be said about American ethnic communities. If the cultural disappearance of German Americans has been often overstated—some institutions do survive in places like Milwaukee and Cincinnati—German Americans, as an effective ethnic group in contemporary America, just do not exist.

Scandinavians

Compared with the masses from Ireland and Germany, the immigration from Scandinavia (Norway, Sweden, and Denmark) is significantly smaller. Predominantly rural, focused on the farther Middle West and the Great Plains, and almost totally Protestant, it provides interesting contrasts with the others. At the same time, there are decided and instructive differences between the flows from each Scandinavian country in terms of size, incidence, internal origin, and motivation.

As table 6.7 shows, a little more than half of the 2.15 million Scandinavians were Swedes, almost a third Norwegians and a seventh Danes. But these are small countries, and in some cases the numbers above are quite high in incidence of total population. The 176,000 Norwegians who emigrated in the 1880s for example, represented more than nine percent of that country's population, and in no decade between 1861 and 1910 did the loss fall below 4.5 percent. In Denmark, by contrast, the decennial loss in that period never reached 4 percent. Sweden's loss was intermediate, reaching a peak in the 1880s of seven percent. The millions from Germany from 1820–1924 were never as much as 3 percent of the total population in one decade, while the heaviest losses, between 10 percent and 15 percent consistently, are for Ireland. The only other major sending nation in that period to exceed the Norwegian percentage in the 1880s was Italy after 1900 whose losses were in the 10 percent range.

Table 6.7

Immigrants from Scandinavia, 1820–1920

DECADE	SWEDEN	NORWAY	DENMARK	SCANDINAVIA
1820–30	94[a]		189	283
1831–40	1,201[a]		1,063	2,264
1841–50	13,903[a]		539	14,442
1851–60	20,931[a]		3,749	24,680
1861–70	37,667	71,631	17,094	126,392
1871–80	115,922	95,323	31,771	243,016
1881–90	391,776	176,586	88,132	656,494
1891–1900	226,266	95,015	50,231	371,512
1901–10	249,534	190,505	65,285	505,324
1911–20	95,074	66,395	41,983	203,452
Total	1,116,239	695,455	300,036	2,147,859

NOTE: If one regards Scandinavian migration as a function of emigration from Europe and measures it in five-year periods beginning in 1851, in only one period, 1866–70, was it as much as 10 percent of the total. It was 11.4 percent. For the period 1851–1915 it was 5.1 percent.

[a]Until 1861, immigrants from Norway and Sweden were not counted separately. These 36,129 immigrants are not included in the total for either country but are included in the total for Scandinavia.

On the other hand, the rhythms of emigration from all three countries are quite similar. Perhaps 125,000 Scandinavians migrated to America before the Civil War. When immigration got properly started after 1865 some two million came between then and the outbreak of World War I. The volume of emigration from all three countries peaked in the 1880s, declined in the 1890s, and rose somewhat in the next decade (quite sharply in the case of Norway), and suffered a decided decline in the 1910s. And, for a final comparison, in all three countries the overwhelming majority went to the United States: In the period 1871–1925, 97.7 percent of all Swedes, 95.6 percent of all Norwegians, but "only" 87.9 percent of all Danes who emigrated did so. And about half of those who did not go to the United States went to Canada. Another similarity is that, unlike the Irish and the Germans, there are few problems in identifying Scandinavian immigrants. It is true that some Swedes emigrated through Copenhagen and that after 1864, when North Schleswig was lost to Germany, some Danes emigrated through Hamburg, and, finally, that the American statistics before 1930 include Icelanders with Danes. But none of these aberrations distorts the statistics very much. There were, for example, probably no more than five thousand Icelanders in the whole period. Other comparisons and contrasts will appear in our discussion of the individual national movements.

Because the Scandinavian countries had become quite bureaucratized even in the nineteenth century, we know relatively more, statistically, about their emigrants than we do about those from elsewhere. Perhaps the most remarkable "statistics" were those relating to "intelligence" created by Swedish pastors. Since the time of Ravenstein, as we have seen, scholars have assumed that immigrants generally had higher levels of intelligence than the populations from which they came. Nativists, on the other hand, have generally assumed that immigrants, or many of them, were of substandard intelligence. Among the duties that the state church of Sweden assigned to its pastors was the conducting of an annual intelligence evaluation of their parishioners with regard to literacy and general knowledge. The pastor did not administer examinations; intelligence testing had not yet been invented. He simply made his evaluation of every name on his parish register based on his knowledge of that person. Thus we have a uniform if not "scientific" evaluation of the intellectual accomplishments of the Swedish population dating back in some cases to the eighteenth century. The Danish scholar, Kristian Hvidt, describes the results this way:

> A comparison of those Swedes, rural as well as urban people, who emigrated later on, with those who remained in Sweden, reveals that everywhere, irrespective of the corner of the country chosen, emigrants generally made higher marks than the rest of the population. They were brighter in school, had a wider picture of the world, and were the kind of persons to whom it would occur to leave their habitual surroundings. It is equally clear that the persons who contributed to migration within Sweden, from village to town, were also placed intellectually above those who never left the place where they were born. But the actual emigrants abroad seem to have had the highest intellectual level; a fact that says something about the loss to their native country that large-scale immigration caused, and which also contributes to an understanding of the immense economic expansion that occurred in the major countries of immigration, especially in the United States.[24]

SWEDES

The Swedes were not only the largest contingent, they were also, not surprisingly, the most varied in both origin and settlement.[25] Those who are familiar with some or all of Vilhelm Moberg's trilogy *The Emigrants* (1949–59), about lower-middle-class farmers, or the outstanding film of that name based on its first segment (arguably the finest social history in a commercial movie), must remember that these deal only with the beginnings of large-scale Swedish emigration and settlement. There is no traceable connection between the Swedish migration in the Colonial Era and that of the nineteenth century. The earliest immigrants from Sweden in the nineteenth century were not farmers at all, but middle-class individuals and adventurers—not a few had been soldiers—and technicians who came as passengers on ships that brought Swedish iron ore to the United States. Prior to 1846 there were perhaps a thousand such migrants. Then, in the later 1840s and early 1850s immigrants of the kind Moberg wrote about came, small farmers hungry for more and better land along with some seeking broader religious freedom than the austere Church of Sweden (Lutheran) was willing to allow. These often traveled in parties. This kind of emigration continued, with minor variations, into the 1870s. The majority of Swedes who came came in family groups. But in the 1880s the sources of emigration within Sweden began to change, so that rural laborers who had never been landowners began to predominate. By 1900 emigrants came more and more from towns and cities rather than from the countryside, and at the same time single young men and women rather than family groups began to dominate. And, with a growing Swedish population base in America—one hundred ninety thousand by 1880—there was increasing chain migration. One evidence for this is the fact that in some years of the eighties, slightly more than half of Swedish immigrants traveled on prepaid tickets purchased in the United States. In Denmark, this figure never reached a third, and more often was a quarter or less. For the overwhelming number of Swedish immigrants economic motives prevailed.

Although the desire for religious freedom was a minor element in Swedish emigration, it is worth discussing. In the late 1840s and 1850s perhaps fifteen hundred Swedish "Janssonists" came to the United States and established a utopian colony at Bishop's Hill, in Henry County, Illinois. The name comes from their prophet, Eric Jansson, a farmer's son who defied the law by holding services apart from the state church, attacked the clergy on both religious and populist grounds, publicly burned the writings of Martin Luther, and otherwise scandalized the establishment. The Bishop's Hill settlement was, if anything, more authoritarian than the system its members had fled. Jansson ordained shared property, work, and meals and even dictated family planning. He laid out the town according to a plan of his own, which included a grid system of streets with curiously large buildings. Like most such colonies, Bishop's Hill, named for the founder's birthplace, was short lived. There was a devastating cholera epidemic in 1849, and the next year Jansson was murdered by a colonist. It endured as a cooperative colony until 1861, when it went bankrupt. Some of the colonists remained in the area as individual farmers and many prospered. The colony's importance extends beyond the numbers involved. It received enormous amounts of attention in Sweden and helped to popularize the notion of migration to America. Later in the century some eight thousand Mormon converts left Sweden, but these will be discussed in the section on the Danes who were the key group of Scandinavians who came, in William Mulder's phrase, "homeward to Zion."

Much of the economic pressure for emigration arose from a rapidly expanding population coupled with a finite amount of land suitable for agriculture. In the century before 1850, Sweden's population doubled, and it increased another 50 percent by 1900, reaching 5.1 million. The factors causing the increase include vaccination for smallpox and, as in much of the rest of Northern Europe, the introduction and cultivation of the potato. A system of partible inheritance left more and more farm families on smaller and smaller plots and on less and less arable land. By 1870, 48 percent of the farm population was totally landless, and the number was growing. Added to that, the late 1860s saw severe famine in Sweden—the last that country has experienced—which, in one year, caused the death of twenty-two Swedes in every one thousand. The result was a surge in Swedish immigration to the United States, a surge that is partially masked by the decennial data in table 6.7. About 150,000 Swedes came in the twenty years after 1861; two-thirds of them came between 1868 and 1873.

These Swedes settled largely in the wheat belt of the Midwest. The U.S. Census for 1910 shows nearly a fifth of all Swedish immigrants in Minnesota, just over a sixth in Illinois, and about a fourteenth in New York, with 43.4 percent in just those three states. Later many Swedes and second-generation Swedish Americans would migrate, along the lines of the Northern Pacific and Great Northern—and later along highways—to the Pacific Northwest. (By 1930 the state of Washington was second only to Minnesota in its number of Swedish ethnics, a distinction it still holds.) But the percentage concentration of Swedes in Minnesota can be deceiving: The 1930 census, for example, showed that there were more first- and second-generation German Americans in Minnesota than there were first- and second-generation Swedes, some three hundred thirty thousand to two hundred eighty thousand. (Norwegians were third with about two hundred fifty thousand.)

Yet the traditional picture of the Swede as a farmer in the Upper Midwest, although accurate as far as it goes, must be balanced by the recognition that a significant minority were urbanites right from the start, and that one American city, Chicago, became the second largest Swedish city in the world. In 1900 about 150,000 Swedish Americans were nearly 9 percent of the Windy City's population. Ulf Beijbom, the historian of Swedish Chicago, has traced the evolution and migration of the ethnic enclave, Swede Town, from the near North Side to the South Side and the West Side and indicated its eventual dispersal into suburbia. But the picture he draws of the mid- and late-nineteenth-century urban poor is a far cry from the favorable stereotype of neat, prosperous, and independent Swedish American farmers. One perhaps overly censorious Swedish community leader in the 1850s wrote of:

> the dirty beds, the unappetizing food, served in stoneware dishes on the trunk lids, saw both adults and children in coarse, dirty clothes, which not even the lowest laboring people in America would want to use.[26]

Cholera epidemics, like the one that devastated Bishop's Hill, raged almost every summer, affecting poor immigrants the worst, killing some even before they arrived in Chicago. When the doors of one boxcar packed with immigrants were unlocked, four dead bodies were found. Almost 3 percent of all Chicagoans died of cholera in the summer of 1851. In one three-block area on the North Side every individual died, 332 persons, most of them immigrants. During the time such conditions existed, the Swedes of Chicago were largely wage workers. Not only Swedish men but Swedish women played prominent roles in the labor force, the former largely as laborers, skilled and unskilled, the latter as domestic servants or textile workers. As Stina Wiback wrote from Chicago in a New Year's letter in 1871:

> A Woman here can support herself and her husband rather well on her earning, without his earning anything, for that is what I have been doing ever since I came here.

Despite hardships and straitened circumstances typical of working people in urban America, the confidence of persons like Stina Wiback that they could better themselves was not misplaced. Most of Chicago's Swedes—or most of those who survived—could, after a decade or so, move to a somewhat better neighborhood, setting off an escalatorlike movement, typical of urban areas, in which as one group moved out of a neighborhood and presumably up, another immigrant group—often but not always a more recent one—moved in behind it. In Chicago, Swedes generally seemed to rank behind the Germans, in occupation, income, residence, and so on. In turn, the Germans were behind the "Americans," and generally ahead of the Irish. While some of the well-known antipathy between Swedes and Irish can be laid to sectarian differences between Lutherans and Catholics, others certainly stem, in Chicago anyway, from their proximity on the ethnic escalator. Yet the hostility was not total. At the turn of the century, when incoming Italians began to press both Swedes and Irish, the traditional foes joined forces in an unsuccessful attempt to stem the newcomers' advance. What had been called Swede Town became Little Sicily. Often a symbolic event in the struggle for a neighborhood, the equivalent of a military salient, came when the new group managed to build or secure a church of its own in formerly enemy territory. In this instance it occurred with the erection of Saint Philip's Roman Catholic Church at Oak and Cambridge Streets in 1904.

The ethnic enclave, a place where the language and the customs of the old country were transplanted, however inexpertly, was a typical development of most American ethnic groups wherever their numbers reached a certain critical mass. Though today, ahistorically, they are called ghettos and are often viewed as a bad thing, in the past these enclaves provided an important transitory phase for millions of urban immigrants. If, in some instances, these enclaves survived long enough to serve as a brake on the pace of acculturation, they nevertheless provided an important way station for immigrants on the road to fuller integration into the larger streams of American life. Swede Town was no exception. Writing in 1890, the radical Swedish American journalist Isador Kjellberg remembered nostalgically:

> The liveliest section of the busy Chicago Avenue shows, its entire length, a large mass of exclusively Swedish signs, that Anderson, Petterson, and Lundstrom were here conducting a Swedish general store, a Swedish bookshop, a Swedish beer saloon . . . and so on. And wherever one goes one hears Swedish sounds generally, and if one's thoughts are somewhat occupied, one can believe one has been quickly transported back to Sweden.

The near universality of such a reaction can be seen by comparing the remarks of a Chinese describing Chicago's Chinatown nearly forty years later:

> It is only in Chinatown that a Chinese immigrant has society, friends and relatives who share his dreams and hopes, his hardships, and adventures. Here he can tell a joke and make everybody laugh with him; here he may hear folktales told which create the illusion that Chinatown is really China.[27]

Other Swedish urban concentrations—smaller in total size but larger in incidence than Chicago's, occurred in Worcester, Massachusetts, and Jamestown, New York.

Whether rural or urban, Swedes established large numbers of institutions, including churches, schools and newspapers, and magazines. The Swedish churches, all Protestant, were dominated by Lutheranism. In 1860 Swedish American pastors formed the Augustana Synod (from the 1530 Augsburg Confession) jointly with Norwegians. The latter soon left, and by the early twentieth century perhaps a fifth of Swedish Americans were affiliated with Augustana Synod churches, another tenth with other ethnic churches, and the rest were in non-Swedish churches or were unaffiliated.

Augustana College in Sioux Falls, South Dakota, and Gustavus Adolphus College in Saint Peter, Minnesota, were the first two Swedish denominational colleges, created in 1860 and 1862. The success of Swedes and certain other European ethnic groups in founding enduring institutions of higher education is remarkable, particularly in the face of the failure of the larger and, in other ways, better-organized Germans to do so. Perhaps the Germans were too conscious of the German university and realized that they could not possibly duplicate such an institution in America. But the founding of small, denominational immigrant colleges like Augustana was an act of courage and perhaps arrogance, similar to the one that had launched Harvard more than two centuries earlier.

Attitudes about other sides of life in America often reflected those of the church, which, as in Sweden, were puritanical and advocated an austere way of life. Lutheran support of temperance and later Prohibition helped associate Swedes with the Republican party. And there were inevitable conflicts between the conservative church leaders and liberals and radicals, both inside and outside the church. These conflicts were often expressed in the burgeoning Swedish American press, much of which was under religious control. Altogether more than eleven hundred Swedish-language newspapers and magazines were published in America, none of which ever claimed a circulation of as much as a hundred thousand, and most of which had tiny circulations and short lives. Perhaps ten thousand books were published in the Swedish language, many of them by the synod's Augustana Book Concern, established in 1883. By 1920 the firm's English-language titles were three times as numerous as those in Swedish and after 1937 all were in English. This outpouring has produced little that contemporary critics value: The great Swedish immigrant novel—Moberg's trilogy—was written by a Swede who visited the United States only *after* he published it. Those Swedish Americans who produced notable prose, for example, Carl Sandburg or Nelson Algren, did so as American rather than Swedish American writers.

Swedes have been relatively successful at politics, helped not only by their geographic concentration but also by the fact that, unlike the Germans, they were almost all in one political party, the Republican. (In the late nineteenth and early twentieth century, there may have been more Swedish socialists than Democrats.) This began to change after World War I, when an unpopular war was followed by severe economic problems in the wheat belt, so that many Swedes became radicalized and supported the Non-Partisan League and the presidential candidacy of Robert M. La Follette. The Great Depression created a strongly Scandinavian Farmer-Labor Party in Minnesota, which eventually merged into the Democratic Party. Swedish and other Scandinavian names are still prevalent, and in some places in the wheat belt still dominate state and local ballots.

NORWEGIANS

In 1979, 4.1 million Americans reported themselves to be of at least partial Norwegian ancestry, a figure roughly equal to the entire present-day population of Norway. Nowhere else on the European continent was the pressure of population on arable land as strong as in Norway. Norway was under Danish rule from the late fourteenth century until 1814, when Denmark, part of Napoleon's continental system, was forced to cede it to Sweden. Norway, which had been undergoing a nationalist revival, used the opportunity to declare itself independent on May 17, 1814 (a date still celebrated by Norwegian Americans). After brief hostilities with Sweden, it was agreed that the king of Sweden should also be the king of Norway. This tenuous union endured until 1905, when the Norwegians got a king of their own. Although Scandinavians can generally understand one another, Norwegian and Swedish are separate languages, while Danish and Norwegian are so similar that the term *Dano-Norwegian* is often used.[28]

Although Norway is not one of Europe's smaller nations—its 125,000 square miles make it slightly larger than New Mexico—only about 3 or 4 percent of its land was tillable. The population grew 50 percent between 1801 and 1845, when it reached 1.3 million. This created impossible conditions in Norway's rural areas, where two-thirds of the population still lived, neither industrialization nor urbanization having made much progress there, at least by general European standards. Soon an absolute majority of Norway's rural population was landless. Thus the outpouring of Norwegians, which was largely a migration from the Norwegian countryside to the American countryside, where many of the landless got land and the landowners got larger and more fruitful farms in the United States.

But, in addition to land hunger and population pressures, religious intolerance played a part in the origins of nineteenth-century Norwegian migrations. In 1825 a much-heralded group of fifty-two religious dissenters in the small sloop *Restauration,* sometimes called the Norwegian *Mayflower,* sailed from Stavanger to New York, having been persuaded to come by Cleng Peerson (1783–1865) a dissenter and would-be colony promoter. The group settled near Geneva, in western New York state, but the community soon dissolved. Peerson, who had been a member of the largely Swedish community at Bishop's Hill for a while, continued to promote communitarian settlements for the rest of his long life. He eventually established a small Norwegian settlement in Bosque County, Texas, where he died. A number of other essentially communitarian settlements by Norwegians, the most famous of which was subsidized by the concert violinist, Ole Bull (1810–80) in the early 1850s in Potter County, Pennsylvania, on one hundred forty thousand acres of heavily wooded land. All failed in short order.[29]

The bulk of Norwegian immigration came after the Civil War and comprised almost entirely families and individuals rather than groups. Yet, to a very high degree, Norwegians settled in rural American regions that became heavily Norwegian American, a "nation within a nation," as Peter Munch calls them. Even more highly concentrated than the Swedes, 57.3 percent of the Norwegian-born persons in the United States resided in three states: a sixth in Wisconsin, a quarter in Minnesota, and an eighth in North Dakota, according to the 1910 census. There and elsewhere, Norwegians had a tendency to settle in compact enclaves, with persons from the same areas in Norway often settling together. For about a quarter century after the Civil War, Norwegian settlement was almost all rural; after that, as was the case with contemporary German and Swedish immigration, more of the immigration was of individuals rather than families and, while still predominantly rural in both origin and destination, more and more migrants stayed in cities. The most distinctive concentration of Norwegian-born persons in any city during the peak immigration years occurred in Brooklyn, where groups of professionals and artisans connected with the maritime trades—brokers, ship chandlers, and so on—formed a distinct ethnic enclave that catered initially to Norway's large merchant marine and later to the ships of other nations. In the later 1870s an estimated sixty Norwegian-flag vessels arrived in the port of New York every week. In addition an estimated fifty thousand Norwegian seamen immigrated to the United States "unofficially" by simply walking off their ships in New York and other ports between 1870 and 1914. Perhaps 40 or 50 percent of these eventually returned to Norway, but Norwegian seamen played an important role in the American merchant marine on both coasts and on the Great Lakes as well. It is not surprising that the most prominent Norwegian-born leader of the American labor movement was the West Coast seamen's leader, Andrew Furuseth (1854–1938).

Like the Swedes, the overwhelming majority of Norwegians were Lutherans, but for nationalistic, religious, and ideological reasons they organized themselves into a variety of separate synods, some of which reflected divisions brought over from Norway, such as high church/low church, and others that represented splits created by American conditions, such as the debate over slavery. An affiliation of the Norwegian Synod with the German Missouri Synod produced a variety of schisms, some of which concerned slavery, widely condemned by most Norwegian Americans. When the pastors of the Synod were challenged, they insisted that while they did not condone the abuses of slavery they could not find an explicit condemnation of the institution in the Bible. One of the chief reasons for the affiliation with the Missouri Synod was to enable Norwegian Americans to get pastoral training at Concordia, its Saint Louis seminary. (The proliferation of "Concordias" among Lutherans is really evidence not of agreement, or concord, among them, but of disagreement. Jacob Rader Marcus has noted similarly that the use of the word *shalom*—"peace, or concord"—in the names of so many Jewish American synagogues reflects pious hopes for unity rather than unity itself.) One of the results of schism was the founding of

Luther College (1861) in Decorah, Iowa. By 1917 the three main Norwegian Lutheran synods had merged to form the Norwegian Lutheran Church of America, which claimed nearly half a million members in the early 1920s.

Also like the Swedes, the Norwegians founded a number of church-affiliated institutions of higher learning. In addition to Luther College, denominational colleges included Augsburg (1869) in Minneapolis; Augustana (1889) in Sioux Falls, South Dakota; Concordia (1891) in Moorland, Minnesota; and Pacific Lutheran (1894) in Parkland, Washington. The largest and most prestigious Norwegian-founded institution, Saint Olaf's (1874) in Northfield, Minnesota, began without church affiliation but gained one before the end of the century. It is now one of the higher-ranked liberal arts colleges in the United States.

Norwegian American politics, as indicated above, became strongly Republican as early as 1860. But, both because of their location in the more marginal areas of the wheat belt and because of strongly egalitarian and populistic tendencies they brought from Norway, more Norwegians became agrarian radicals sooner than most other Midwestern Scandinavians. Large numbers of Midwestern Norwegians joined the Populist Party and its forerunners in the 1880s and 1890s, and Norwegian Americans were a major element supporting A. C. Townley's socialistic Non-Partisan League, which began in heavily Norwegian North Dakota.

Along with other ethnic groups, Norwegian Americans established an ethnic press totaling some eight hundred different publications; perhaps the largest was the Iowa weekly *Decorah-Posten* which had more than forty thousand subscribers all over the U.S. Norwegian American writers, the most noted of whom is Ole E. Rølvaag, (1837–1931) perhaps the most influential immigrant to create literature in America in a non-English tongue. Rølvaag would have put it another way:[30]

> We can call these works . . . emigrant literature, but then we give the child a wrong name. For they are not that: they are American literature in the Norwegian language.

At the other end of the ideological spectrum stood Hjalmar Hjorth Boyesen (1848–1895), who said that once he had set foot in America he spoke Norwegian only when absolutely necessary (since he was a professor at Cornell and Columbia universities for most of his life that would not have been very often) and in his novels urged that his fellow immigrants do the same and leave Norway and things Norwegian behind them. This dichotomy illustrates nicely what the leading scholar of Scandinavian-American literature, Dorothy Burton Skårdal, has called "the divided heart." Rølvaag, the Norwegian fisherman's son who got most of his education in America and had hoped that the Norwegian language could be passed on to the second and third generations, understood before he died that this was not to be. Characteristically, the author of *Giants in the Earth, Peder Victorious,* and *Their Father's God,* which were first published in Norway and then translated for publication in the United States, didn't blame the younger generation but the American environment. He put it this way in 1922:

> Again and again [second-generation Norwegians] have had impressed on them: all that has grown on American earth is good, but all that can be called *foreign* is at best suspect. Many of our own people have jogged in the tracks of the jingoists. "Norwegian church service? Why should there be Norwegian church service in America? No, talk English. . . . No full blooded American can be expected to want to belong to a Norwegian church!" . . . The young are extremely sensitive in matters of honor, and much more so in their patriotic honor! It has been—and to some extent still is—a point of honor to be able to prove that nothing *foreign* hangs about one's person. Under such conditions how could anyone expect that young people should show only enthusiasm for their forefathers' tongue—that would be to expect the impossible.

DANES

Although the Danes are, by far, the smallest of the three Scandinavian groups, we know a great deal more about their social background than we do for any other national emigrant group. This is because the Danish police recorded information about Danish emigrants between 1869 and 1914 in "58 thick volumes [of] handwritten registers," which Kristian Hvidt, now head librarian of the Danish parliamentary library, discovered and utilized in his *Flight to America* (1975).[31] The registers contain seven kinds of information about emigrants:

1. Year and month of departure
2. Sex
3. Traveling alone or in a group
4. Occupation

5. Age
6. Place of last residence
7. Destination

The registers contain some 300,000 names. Hvidt has concentrated on the period 1868–1900 and computerized the information about 172,022 persons recorded as leaving Denmark, which Hvidt estimates as 90 percent of all Danish emigration in those years. These registers represent a unique source in the history of European emigration. Some notion of the singularity of Hvidt's findings can be gathered by glancing at Bernard Bailyn's description of the second largest collection of names, 9,868 departures for the Western Hemisphere from Britain between December 1773 and March 1776:

> This register of emigrants was so complete, the range of ancillary sources that could be associated with it so vast, and the geographical scope involved so huge that a kind of cross-sectional analysis of the peopling process, or a significant part of it at least, seemed to be emerging.[32]

Without in any way denigrating the importance of the British emigrant registers, or the prizewinning work that Bailyn wrote using them, they contain only about 3 percent of the number of names in the Danish registers and cover a period of time only 5 percent as long.

A very few Danes emigrated to America before the onset of mass immigration. Some came to Dutch New Netherland, the most prominent of whom, Jonas Bronck (died 1643), bought the large area north and east of Manhattan now known as the Bronx. Some Danish Moravians settled among their German coreligionists in Pennsylvania and were largely absorbed by them. Scattered Danes appear in the revolutionary and early national periods—most notably the blacksmith from Copenhagen, Peter Lassen (1800–59), who blazed one of the early trails to California, which, along with a magnificent California mountain, is named for him.

Among the first large groups of Danes in nineteenth-century America were converts to Mormonism, who came, in William Mulder's phrase, "homeward to Zion," to the valley of the Great Salt Lake.[33] Mormonism was nowhere in Europe more successful than in Denmark: Nearly twenty thousand Danish Mormons emigrated, most of them in the second half of the nineteenth century. The reasons for this special relationship are to be sought both in Utah and in Denmark. When, in

October 1849, the church leaders in Salt Lake City decided to proselytize in continental Europe, they sent one missionary to work in Germany and Austria, one to France and Italy, and three to Scandinavia—two to Denmark and one to Sweden. The man in Sweden was expelled after a few days and joined his colleagues in Denmark. The Danish authorities did not interfere with the Mormons' recruiting efforts, although physical attacks on Mormon meetings and other persecutions all too familiar to the missionaries were prevalent. Another advantage the Danish missionaries had was that the Book of Mormon had been translated into Danish before the first missionaries arrived in Denmark. Of the twenty-five thousand Mormon converts in Scandinavia more than half were Danes, and more than half of Danish converts emigrated, emulating the words of the Mormon hymn:

Oh Babylon, Oh Babylon, we bid thee farewell
We're going to the mountains of Ephraim to dwell.

Most of the converts were of relatively modest socioeconomic circumstances. They were able to come to America because of a unique Mormon institution, the Perpetual Emigration Fund. Established in 1850, it operated until 1887, when, as part of an antipolygamy crusade, the U.S. government annulled its charter and seized its assets. The fund made it possible for some fifty thousand poor European converts to emigrate by financing their passage. Careful records were kept, and the advances were supposed to be repaid later, but many were not. This has been called the first "travel now, pay later" scheme, but the whole system of indentured servant-redemptioner migration accomplished the same thing from quite different motives. There was never enough money in the fund to accommodate all of those who wanted to come. In Denmark poor families who wished to emigrate would assemble around the missionary once a year to learn how many of them could leave in the year to come. (The Mormons, who were superbly organized, ran the most efficient "travel bureau" for immigrants that ever existed.) Not surprisingly, most Mormon migration was family oriented. Hvidt's figures for the period 1872–1914 show that among Danish emigrant Mormons, women slightly outnumbered men—32.2 to 29.3 percent—while 38.5 percent were children. Among other Danish emigrants of that period, men outnumbered women by 52.0 to 30 percent, with children comprising only 18 percent.

If Mormon group emigration was well organized and generally tidy, the same cannot be said for other Danish group migration of the era.

There were several socialist colonies, none of which endured. Perhaps representative was the one founded in April 1877 in Ellis County, Kansas, near Hayes City. Its chauvinistic historian has written:

> There were 18 colonists, some married, some unmarried. They at once set to work to build a log cabin with separate apartments for the married and the unmarried. Tools and stocks were purchased. The men worked "like Hell." The women quarreled. And the naked prairie—save for the abundance of buffalo bones, rattlesnakes, prairie dogs, owls and an occasional soldier—seemed so unresponsive to the demands for a better society that the colonists could stand it for no longer than six weeks.

But the overwhelming majority of the three hundred thousand Danish immigrants came for economic reasons. For the period before 1900 about 56 percent were from rural areas; after that, following the pattern we have seen elsewhere on the Continent, some 52 percent came from cities and towns. In terms of age, the Danish data verify what students of immigration have always assumed: that it was primarily a phenomenon of the young. Between 1868 and 1900 an absolute majority, 55.5 percent, were between fifteen and thirty years of age. Another 20 percent were children, leaving less than a quarter—22 percent—thirty years old and over. (The ages of the remaining 2.5 percent are unknown.)

As we have seen, the preponderance of adult immigration—except for an occasional group like the Mormons—was male. This created problems of sex distribution in both the sending and receiving countries. The British Women's Emigration Association was formed to encourage female emigration to lessen the "excess" number of women in England. In Dorpat (today Tartu, Estonia), a German professor named Rauber argued that the ancient legend of the Amazons was really Greek government propaganda to encourage surplus women to emigrate. Rauber's theory probably owes as much to the brothers Grimm as to scholarship, but does illustrate nicely late-nineteenth-century concerns. In Denmark the emigration of ninety-six thousand adult men and of sixty thousand women caused the already predominantly female sex ratio to increase. In 1840, 494 persons in every thousand were male; by 1911, the figure was 485. The difference may seem small and insignificant, but it meant a significant increase, in every part of Denmark, of women who would never marry and never have children.

In receiving countries, the opposite was true. In the United States, the proportion of women among the foreign born fell steadily between 1860–90, from 46.7 percent to 45.7 percent, and its effects were felt particularly in foreign-born communities. An itinerant Danish clergyman in the United States reported that he often heard complaints that Danish girls who had been brought over to work for Danish families seldom stayed long, as they married Danish young men. He reported that one employer had even stipulated that she wanted a servant both "old and ugly," but even that servant got engaged in half a year. The shortage, of course, somewhat improved the labor market for young women. A Danish politician who visited the United States in 1887 wrote in a Copenhagen paper:

> The demand for Scandinavian girls, particularly Danish ones, is enormous. [A New York employment broker] told me that I was welcome to send 3000 girls any day I pleased. Inside a few days he would promise to get them all work at a beginner's wage of 7 to 12 dollars. Girls who are skillful at housework and dairywork might obtain 25 dollars. Working in the fields is unknown for women in the States.

Again, as we have seen elsewhere, the tendency as the nineteenth century wore on was for more and more individuals and fewer families to emigrate, even if we include serial migration for the latter. Over the whole period, 1868–1900, only four Danish immigrants out of ten were in family groups; the other six were unmarried men and women at roughly a two-to-one male ratio. But if we break these figures down by decades we find that in the 1870s 43 percent of Danish emigrants were in families and in the 1890s only 29 percent.

But statistics tell us nothing of the often-agonizing decisions that families had to make about emigrating. Hvidt gives the following case histories which indicate the range of problems faced. Neils Jensen Neilsen got married shortly after being demobilized following the war of 1864. He, his wife, and their growing family failed to make a go of a small urban enterprise and went to live on the twenty-five acre Neilsen family farm in return for keeping Neilsen's father as a pensioner. As a son later told the story:

> It came to quarrels between my parents . . . my father was fed up with life as a small farmer and wanted to emigrate. But my mother, who knew how fond the Neilsen family were of the bottle, feared

> moving to a strange country with the risk of being left alone to fend for her children. Finally when my grandfather died, she let herself be persuaded into leaving. The little house and most of their belongings were sold, tickets were bought, and reluctantly she set out. The first stop was Vejle, and that was as far as they got. It so happened that they were met by the agent who had sold them their tickets. He explained that the berths he had booked for them had also been sold in some other town at the same time. Consequently my father and his family had to wait for the next ship to sail. Of course, my father was furious, there he was with his whole family and with only the barest necessities in a small town where he had to find a place to stay for several weeks. At this point my mother intervened and seized the opportunity to change my father's mind, with the result that he told the agent to go to hell. The ticket money was refunded and we all settled in Vejle. . . . My father, who gradually realized that all he could expect was the miserable existence of a worker, felt he had been cheated all his life. He had had his chance to see what America offered, and it had been wasted. Occasionally he felt bitter, and now and then friction would break out between my parents.

The other case involved an actual emigration.

> In the summer of 1871 the family received a visit from two persons recently back from the United States. The visitors gave an ecstatic description of the wonders of America. The father . . . had for years been thinking of emigrating while the mother had been stubbornly against the idea. The morning after this visit the eldest [of five sons] aged 14 told his parents firmly, "As soon as I have been confirmed I shall go to America." This made a strong impression on the mother, who then became convinced of the need of the whole family to leave together. If she refused to go she would have to face the risk of being left alone in Denmark with a dissatisfied husband and all her children in America.

But whether they went in families or as individuals, the overwhelming majority of Danish emigrants, as we would expect, came from the lower ranks of society. The following table, from Hvidt, ignores dependents, listing only those with occupations.

Whatever their status, Danes were less concentrated than other Scandinavians or, indeed, most other immigrant groups. In the 1910

Table 6.8

Danish Emigrants with Occupation, 1868–1900 (percentages)

Occupation	Percentage
Independent farmers	3.4
Rural laborers	43.2
Shipping and fishing	1.5
Commerce and professions	7.8
Craftsmen including apprentices	18.5
Domestic and industrial workers (urban)	25.6
	100.0

census, the state with the largest number of Danes, Iowa, had only just over 10 percent, and the addition of the next two, Wisconsin and Minnesota, produced just under 30 percent. Partially because of this lack of clustering, the intermarriage rates for Danes are higher than for other Scandinavians. The same census showed that while 72 percent of the American-born children of Swedish immigrants had both parents born in Sweden, the figure for Danes was only 57 percent. In addition, the Danes who married out had a greater tendency to marry non-Scandinavians than did either Norwegians or Swedes.

Danes did not form the great number of ethnic religious institutions as did the other Scandinavians, and very large numbers of Danes seem to have been unchurched. One estimate is that only about a third of the Danes in America were members of any church. Not surprisingly, the great schism among Danish Lutherans in the United States was, in part, over attitudes toward assimilation and the ethnic heritage. The conservative wing, commonly known as the Grundtvig faction, felt that Danes should settle together, establish Danish schools, perpetuate the language. The rival, or Inner Mission, wing was opposed to attempts to foster Danish culture in America. As one of its leaders put it in 1888:

> We should serve ourselves and our children poorly by doing all in our power to prevent them from becoming Americanized. . . . Even if the Danish language is lost to our posterity, they might retain all that is good and true in the Danish character.

Each wing set up its own theological seminaries, which have evolved into Dana College at Blair, Nebraska, and Grand View College in Des Moines.

One of the other remarkable things that the Danish data show is just how many of the emigrants came with specific destinations in mind. About a third of the emigrants left Denmark with tickets that took them only to New York, a place where relatively few Danes stayed, perhaps 3 or 4 percent. Another 10 percent had tickets to Chicago, another destination from which many branched out. The rest—56 percent, or some 88,000 persons—had tickets to some specific American destination. These are clearly immigrants who had connections, were part of an immigrant stream, who had been invited to America by someone they knew or knew of. Some 20 percent of all Danish emigrants, perhaps 35,000 persons, traveled on prepaid tickets between 1868 and 1900. (For Norway the figure may have been as high as 40 percent, for Sweden 50 percent.) The majority purchased their own tickets. In addition we know that relatively large sums of money were transferred from the United States to Denmark via postal money orders, which, since they could not exceed $50, were virtually useless for business purposes. In the mid-1890s these remittances hit a peak of $800,000, just before the panic of 1893, rising again to that level in 1900, to $1.6 million in 1906, and hitting $2.4 million before the outbreak of World War I. How much was to help finance emigration, how much to assist relatives left behind, no one knows. At the same time, it should be noted that there was a steady stream of remittances the other way. An early peak of $400,000 just *after* that panic suggests that the stay-at-homes increased their help to many immigrants, but the balance of payments from the United States was always higher.

Was There an Old Immigration?

The groups that we have talked about in this chapter, along with large numbers of Britons, constitute what historians of immigration have long called the old immigration. It seems to me that, for a whole variety of reasons, it is time to retire that hoary concept. In the first place, if the old immigration from northwestern Europe was succeeded by the new immigration from southern and eastern Europe, which was ended by the Immigration Act of 1924, what are we to call the millions of immigrants who have come in since World War II? Much of the immigrant canon was established during the years between the wars. It was only in 1927 that Marcus Lee Hansen published his pathbreaking article, "Immigration as a Field for Historical Research," in the *American Historical Review.* In its very first paragraph he talked about the process of settlement "from its beginnings in 1607 to its virtual close in 1914." Similarly, writing twenty-eight years later, John Higham in his classic *Strangers in the Land* wrote about immigration as something that had ceased.[34] The 3.3 million legal immigrants of the 1960s, the 4.5 million of the 1970s and the more than 6 million of the 1980s make the old nomenclature and the assumptions on which it rested more and more inappropriate.

In addition, I would argue, it never was an appropriate way to structure rational discussion of the immigrant process. Immigrants came, as we have seen, largely for economic gain. Some, like the Irish, were largely pushed at one time, pulled at another. Most immigration flows adapted to changes in both the home country and in America. What changed was not just the sources from which immigrants came, but the nature of opportunity in America. The relative filling up of the American West, the end of the frontier announced by the census, meant that the attraction of the United States for the land-hungry was lessened and, as we have seen, the nature of immigration from Northwestern Europe changed. The conditions of the poorest immigrants, the persons farthest down, has always been deplorable whether one is talking about hapless indentured servants in the colonial South, Swedish poor in mid-nineteenth-century Chicago, Italians and Poles in industrial America, or Puerto Ricans and Cambodians in today's inner cities. Differences within groups have at times been as important as differences between groups.

And, finally, the old typology between old and new simply assumed that immigrants were Europeans. Chinese, who came in the era of the old, and Japanese, who came in the period of the new, were treated as exotic exceptions when they were not written off as sojourners and thus not immigrants at all. By the 1960s, however, Asians and Latin Americans constituted just over half—52 percent—of legal immigration, and in the first half of the 1980s that figure zoomed to more than four-fifths—83 percent.

Thus, in discussing Southern Europeans, Eastern Europeans, and East Asians in the next three chapters I am not conscious of crossing any particular historical watershed. The demarcation points of this section—the ending of the Napoleonic Wars and the Immigration Act of 1924—were each, in their way, external to the immigration process, although each, of course, had broad influence upon it.

7

From the Mediterranean: Italians, Greeks, Arabs, and Armenians

As we have seen from our examination of northwestern European groups, the nature of immigration from those countries began to change toward the end of the nineteenth century from largely family groups to single individuals and from immigrants heading for rural economic opportunity to those heading toward urban economic opportunity. Conditions on both sides of the Atlantic caused these changes. In Europe greater and more rapid economic growth—stronger in the West than in the East—attracted migrants from within and emigrants from without to industrializing cities such as Berlin, Stockholm, Paris, and London. In the United States similarly, the cities rather than the prairies drew larger and larger proportions of new Americans. In 1920, when a bare majority of Americans lived in urban areas, three-quarters of the foreign born did. Immigrants from Russia, Ireland, Italy, and Poland were the most urbanized of the larger groups, with rates of urbanization of from 88.6 to 84.4 percent. And the disparity between the urbanization of immigrants and natives is greater than the bare data indicate. The Census Bureau counted places of twenty-five hundred and above as "urban territory." Thirty percent of the urban population—more than sixteen million persons—lived in places with populations between twenty-five hundred and twenty-five thousand. These were overwhelmingly native born, while immigrants tended to live in large cities rather than in the small towns that made up so much of America.

As the steam-powered transportation revolution reached into all but the most remote corners of Europe it provided millions of peasants the practical means of leaving their native villages. The creation of steamships, first of iron, then of steel, intended for the immigrant trade not only reduced the time required for an Atlantic crossing to a matter of days rather than weeks but also made transportation of immigrants into a modern big business. The major European passenger companies—no American line had a significant share—set up vast networks of ticket agencies. The Hamburg-Amerika line, for example, had more than three thousand of them in the United States in the 1890s, most of them sidelines for ethnic entrepreneurs in the process of "making it." Thus the use of prepaid tickets, purchased in America for the use of the immigrant, increased sharply. In the early 1890s perhaps one immigrant ticket in three was prepaid; just after the turn of the century, two in three were.

Emigrant ships still left from dozens of European ports, but by 1907 just four ports—Naples, Bremen, Liverpool, and Hamburg—accounted for more than three European immigrants out of five. Immigration peaked in that year (July 1, 1906–June 30, 1907) as more than a million and a quarter immigrants were recorded entering the United States. Table 7.1 shows the numbers leaving from each of those ports and indicates the main nationalities involved, as these ports, except for Naples, were funnels for much of Europe and exported more foreigners than natives.

The great shipping lines, including North-German Lloyd, Cunard, and Hamburg-Amerika, under the pressure of competition and of increasing protective legislation, created modern assembly lines to funnel immigrants onto their ships. Cunard Line representatives met trains in Liverpool and escorted immigrant passengers to its dormitory enclave,

Table 7.1
Emigrant Ports, 1907

PORT	NUMBER OF EMIGRANTS	MAJOR ETHNIC GROUPS
Naples	240,000	Largely Italians; also Greeks and "Syrians."
Bremen	203,000	Poles, Czechs, Croats, Slovaks, and other Slavs.
Liverpool	177,000	More than half Irish and other British; many Swedes, Norwegians, and Eastern European Jews.
Hamburg	142,000	Eastern European Jews and Scandinavians predominate, but some from every part of Central and Eastern Europe.

NOTE: Total immigration to the United States: 1,285,349; Europe—1,199,566; Americas—41,762; Asia—40,524; Australasia—1,989; Africa—1,486; other—22.

where they were housed, fed, and medically examined while waiting for the next ship. Those non-Germans who departed from German ports had a double examination. After the cholera epidemics of 1892—more than nine thousand died in Hamburg alone—the German government subjected migrants from the East to medical examinations, which included baths and fumigation. Those going on to Hamburg, where the shipping magnate Albert Ballin had established a huge immigrant depot, got a second set of baths and fumigations: Men and boys had their heads close cropped and received a chemical shampoo, while women and girls had their hair combed with fine-toothed metal combs. The Amerika village, which could accommodate four thousand persons at a time, was isolated behind brick walls, and emigrants in transit were forbidden to venture beyond them. Inside everything was hygienic. There were even kosher kitchens for observant Jews. Amenities such as daily band concerts and a library were also provided. As regimented as all this sounds—and was—it was a vast improvement on the midcentury conditions, which left immigrants to fend for themselves against those who sought to prey on their inexperience. The ships, too, were cleaner and better run, especially the larger liners, which had begun to substitute "third class" for steerage and actually provided some privacy for immigrant passengers. But even in the twentieth century, not all ships were modern; ships operating out of the Mediterranean—French, Italian, and Greek—probably had the worst conditions. One journalist who traveled as an immigrant from Naples in 1906 reported:

> How can a steerage passenger remember that he is a human being when he must first pick the worms from his food . . . and eat in his stuffy, stinking bunk, or in the hot and fetid atmosphere of a compartment where 150 men sleep, or in juxtaposition to a seasick man?

To similar complaints defenders of the status quo, such as Joseph Chamberlain, the British cabinet officer responsible for overseeing conditions on passenger vessels, had responded a quarter century earlier, that improving bad conditions would cause transatlantic fares to rise beyond the ability of immigrants to pay, and, in addition, argued that immigrants weren't used to any better conditions and that steerage berths were as roomy as peasant cottages or cramped tenements. But, as we have seen, not only did the conditions improve but the price of tickets actually went down, at times of rate wars, to two pounds (ten dollars) from Liverpool, ten days' pay for a common laborer in America.[1]

The immigrants who came to America after 1880 were increasingly from southern and eastern Europe, although, as table 7.2 demonstrates, British, Irish, German, and Scandinavian immigrants continued to come. In the heaviest decade of American immigration, 1901–10, these countries sent more than 1.5 million immigrants; but those sizable numbers were dwarfed by the 6.5 million who came from other parts of Europe: 2 million Italians and very large numbers of Poles and Eastern European Jews, neither of whom can be enumerated with any precision. The obvious lacunae in this table, taken from U.S. immigration statistics, are the absence of Poles and Jews, as well as peoples who came in smaller but still substantial numbers. The Austro-Hungarian and Russian Empires, and, to a lesser degree, Germany, contained subject peoples who did not get recorded in the American data, which listed only national origin, not ethnicity.

Italians

Between 1880 and 1920 more than 4.1 million Italians were recorded as entering the United States. No other ethnic group in American history sent so many immigrants in such a short time. Prior to the 1870s only scattered thousands of Italians had come and, unlike most other emigrant groups, large numbers of Italians had ventured outside of Europe—to South America and to North Africa—before and while large numbers were coming to the United States. Up to 1900, perhaps two out of three Italians who crossed the Atlantic did not come to the

Table 7.2

European Immigration to the United States, by Country, 1901–1910

COUNTRY	NUMBER	PERCENTAGE
Austria-Hungary	2,145,266	26.6
Italy	2,045,877	25.4
Russia	1,597,306	19.8
Britain	525,590	6.5
Germany	341,498	4.2
Ireland	339,065	4.2
Sweden	249,534	3.1
Norway	190,605	2.4
Greece	167,519	2.1
Other European	453,780	5.6
Total European	8,056,040	

United States: Most of them went to Argentina and Brazil. Thus the Italians, like the Chinese, had a strong migratory tradition before the major movement, called by one scholar "the Italian emigration of our times," set in.[2] The figures in table 7.3 show only one aspect of the Italian transatlantic migration, that which brought them to the United States. It does *not* show the return migration, which was quite high among Italians, having been estimated at anywhere between 30 percent and nearly half. Thomas Archdeacon has hypothesized that while Italian immigration between 1899 and 1924 was more than twice as heavy as Jewish immigration—3.8 million as opposed to 1.8 million—the number of permanent immigrants from each group was more nearly equal—2.1 million as opposed to 1.7 million. This was because perhaps fewer than five percent of Jewish immigrants returned to Europe. In the U.S. Census of 1890, for example, 182,000 persons born in Italy are recorded, even though some 307,000 had immigrated in the previous decade alone. And, of course, many of the returning emigrants—*ritornati* as the Italians say—reemigrated to the United States. We also know that large but unknowable numbers of Italians participated in temporary migration—within Italy, within Europe, or overseas to South America or North Africa—before migrating to the United States.

It was a Genoese who "discovered" America, and many other Italian explorers participated in making it better known to Europeans—the Florentine, Amerigo Vespucci, who gave America its name; the Venetian, Giovanni Caboto, who "discovered" New England and is usually disguised as John Cabot in American history books; and another Florentine navigator, Giovanni da Verrazano, who first sailed into New

Table 7.3

Italian Immigration to the United States, 1820–1920

YEARS	NUMBER
1820–30	439
1831–40	2,253
1841–50	1,870
1851–60	9,231
1861–70	11,725
1871–80	55,759
1881–90	307,309
1891–1900	651,893
1901–10	2,045,877
1911–20	1,109,524
Total	4,195,880

York Harbor and for whom the magnificent bridge across the Narrows is named. In addition to these seamen, Italian soldiers marched across the American Southeast with De Soto and La Salle, and Italian priests, such as the Franciscan Marco da Nizza and the Jesuit Eusebio Francesco Kino, were among the most important of the mission fathers who explored and established settlements in what became the American Southwest. All told several hundred Italians participated in the exploration and conquest of North America, most of them in the service of either France or Spain. Yet no New Italy was founded to parallel a New Spain, New France, or New England. This was largely because there was no Italy in a national sense until the mid-nineteenth century.

Italy, in some of whose northern cities the civilization of the Renaissance was born in the thirteenth century, became a classic case of uneven development in modern times. The northern cities, although not politically unified, continued to develop culturally and economically. In the center, under papal denomination, a slower but significant modernization evolved. South of Rome—in the Mezzogiorno and Sicily—development was retarded and grinding rural and urban poverty was the lot of the vast majority of the inhabitants. It is no surprise that, eventually, most Italian emigrants came from the south, but initially almost all of them came from the north and center.

During the colonial and early national periods in America, most of the few Italians who came were skilled artisans, like the Venetian glassblowers who were brought to Jamestown in 1622 or relatively well-to-do individuals such as one Peter (surely Pietro) Alberti who is recorded as a landowner in New Amsterdam. As was true for so many other groups, the first large group immigration was impelled by religious motives. In 1656 a party of more than a hundred Waldensians—Italian Protestants—was brought to the colonies by the Dutch who settled them in New York and Delaware.

During the late eighteenth and early nineteenth centuries a surprising number of Italian intellectuals came to the United States. Politically the most significant figure was Philip Mazzei (1730–1816), a physician and scholar who was persuaded by Benjamin Franklin to go to Virginia and attempt to establish some typically Italian agricultural products there: silk culture, wine grapes, and olives. He settled near Thomas Jefferson and is believed to have influenced him and Thomas Paine and, after the Revolution began, served as an American agent and fundraiser in Europe. (Jefferson had a continuing relationship with Italian immigration. He wrote to Italy to obtain some Italian masons to help in building Monticello and, as president, recruited a number of Italian musicians to form the nucleus of the United States Marine Band.) There were a number of Italian soldiers and soldiers of fortune in the revolutionary armies, one of whom, Colonel Louis Nicola, proposed a throne for Washington. More significant were the activities of the Piedmontese Giuseppe Maria Francesco Vigo (1747–1836) who became one of George Rogers Clark's chief lieutenants in the conquest of the trans-Appalachian West.

Apart from Jefferson's bandsmen, many Italians played an important part in the musical life of the new nation, the most illustrious of whom is Lorenzo da Ponte (1749–1838), Mozart's librettist for *Don Giovanni, The Marriage of Figaro,* and *Così fan Tutte.* Born a Venetian Jew, da Ponte converted to Catholicism at age fourteen, became a priest and then renounced his vows, and managed to be expelled from several cities and countries for both political and moral reasons. The United States, where he came in 1805, was the end of the line. He supervised the building of the first American opera house, was a professor of Italian (unpaid) at Columbia University, and engaged in such unmusical activities as running a whiskey distillery in Sunbury, Pennsylvania. Italian performers, impresarios, and even an occasional composer—for example, Pietro Mascagni—ranged all across the nineteenth-century United States, even in such presumably unmusical places as Cheyenne, Wyoming, where an Italian traveler reported hearing an Italian opera company perform *Lucia di Lammermoor:*

> The touching melodies of Donizetti heard in a still wild region where only a few years before ferocious Indians like Sitting Bull and Crazy Horse roamed, left an indelible impression upon my mind.[3]

In the plastic arts, too, Italians made an important contribution. Both the first U.S. Capitol in Washington, which was burned by the British in 1814, and its successor were decorated by Italian painters and sculptors. The most famous and controversial of these was Constantino Brumidi, who has been called, mostly by those with undeveloped esthetic sensibilities, "the Michelangelo of the United States Capitol." Over the protests of American artists who wanted the commission, he painted the huge frescoes in the dome of the Capitol, working 180 feet above the floor when he was more than seventy years old. Other works are in the corridors and on the walls of various parts of the Capitol, notably a fresco of Washington at Yorktown in the dining room of the House of Representatives. More prosaically, the granite quarries of

Vermont were exploited with the help of Italian stonecutters who formed one of the first Italian American trade unions.

Among other prominent Italians to come before the great movements of immigrants in the later nineteenth century were priests and political exiles. Most notable among the former were Giuseppe Rosati (1749–1843), who was, for a time, bishop of both New Orleans and Saint Louis, and the Milanese Dominican, Samuele C. Mazzuchelli (1806–64), who organized the Catholic Church in Wisconsin and Iowa and designed many public buildings, including the State Capitol in Des Moines. A different kind of Italian religious leader was Sabato Morais (1823–97), the Livorno-born Sephardic rabbi and friend of Mazzini, who came to the United States in 1851 to serve Philadelphia's Mikveh Israel synagogue and became an abolitionist and a founding father of Conservative Judaism in America. Italian exiles were fairly common during the Risorgimento (1848–1870), and included, for a time, the liberator Giuseppe Garibaldi (1807–82), who supported himself in New York by manufacturing candles. On more than one occasion he used a dubious claim of American citizenship to gain or maintain his freedom. (This ploy was used by a later nationalist patriot, Sun Yat-sen, who sometimes claimed to have been born in Honolulu for the same reason.)

After 1849 there was a regular exile community in New York, large enough to start several newspapers, none of which survived for more than a few years. Nor were all of the political exiles *prominenti* or upper class. From time to time various Italian states simply deported captured or suspected revolutionaries, many of whom took passage on American ships. One such group of exiles from Bologna in 1856 found work in coal mines near Scranton, Pennsylvania. By the time of the American Civil War there were enough exiles and other Italians in New York to form a "Garibaldi Guard," a kind of International Brigade that fought in the Union Army and drew volunteers from many nationalities, although it was officered almost exclusively by Italians and Hungarians.

Most of the Italians who came here were artisans, merchants, businessmen, professional people, musicians, actors, waiters, and seamen. The 1850 census, the first to record foreign born, enumerated 3,645 persons born in Italy. They lived in every state and territory but Delaware and New Hampshire, and almost half of them resided in the South, with the largest concentration anywhere in Louisiana, in and around New Orleans. (Ironically, in 1891 New Orleans would be the scene of a lynching of Italians, but the antebellum Italians were well accepted and, when war came, marched off to join the Confederacy along with their American-born neighbors.) It is also interesting to note that, just a year after the California gold rush began, more than two hundred Italians were reported in California, most of them around San Francisco. Settled largely by Genoese, San Francisco remains an important center for Italian Americans, and, as late as 1870 it and New Orleans still rivaled New York as a mecca for Italians. In that year's census New York City (Manhattan) had 2,749 Italians; San Francisco, 1,622; and New Orleans, 1,571. In the latter two the incidence of Italians was much, much higher.

Although most Italians in the United States, then and later, were urban dwellers, there *were* rural colonies, perhaps the earliest of which was a successful enterprise begun in 1868 in Bryan, Texas, comprising chiefly Sicilians who had worked as section hands on a local railroad. By 1900 there were five hundred Italian families there. A Piacenzan, G. F. Secci de Casali, the proprietor of the first Italian American newspaper, helped organize agricultural colonies of Italians near Vineland, New Jersey, in 1874. The men worked in factories while the women and children did most of the truck farming. By the early twentieth century nearly a thousand Italian families owned land in the vicinity. (Vineland later became the site of a Jewish agricultural colony as well and, during and after World War II, provided agricultural employment for "relocated" Japanese Americans and Japanese Peruvians.)

But most of the Italian presence in agriculture was in the Far West, particularly in California. There Italian immigrants introduced new crops and techniques, and some established businesses that have survived to become some of the major "factories in the field" that dominate California agriculture. From the 1880s immigrants from Genoa, Turin, and Lombardy began to enter California's then small grape-growing, winemaking industry. An early Italian American banker, Andrea Sbarboro (1839–1923) helped establish the cooperative Italian–Swiss Colony vineyard and winery at Asti, California. The vineyard started with five thousand acres. By 1897 it produced more wine than barrels could be supplied for, so the group built a five hundred thousand-gallon cement wine tank that may have been the largest in the world. Its completion was celebrated by a dance for two hundred persons inside the tank. Farther south, in the Central Valley, the Di Giorgio brothers, Joseph (1874–1951) and Rosario founded the Di Giorgio Fruit Corporation, which became the largest shipper of fresh fruit in the state. The Di Giorgios owned more than forty thousand acres, and bought and canned the produce of others under the S & W label. The company became one of the most notorious exploiters of labor in California, as John Steinbeck and others interested in reforming Californian agricultural labor practices often pointed out.

That Italian immigrant farmers were helped by immigrant bankers like Sbarboro is not surprising; in fact, there are consistent patterns of such relationships in most American immigrant groups. Most immigrant and immigrant-oriented bankers were and remained small fry in the banking world. (The German-Jewish bankers mentioned earlier were often bankers before they came here and were not immigrant oriented.) The typical immigrant "banker" of whatever ethnicity was likely to be an entrepreneur who performed all kinds of services for his clientele and was able to do so in their common language. He sent money to the old country, served as a ticket agent, and often went bankrupt, by either mismanagement or outright fraud, as did many nonimmigrant banks in America. The largest and most traumatic such collapse was the bank supervised by Archbishop John B. Purcell (1800–83) of Cincinnati, which failed with a great loss of immigrant's savings. Immigrants, who came to America ignorant and/or distrustful of banks in general, often had their prejudices confirmed by their experience here. The great exception among "immigrant" bankers was Amadeo Pietro Giannini (1870–1949), the son of immigrants from Liguria, who founded the small Bank of Italy in San Francisco's North Beach, a largely Italian enclave, in 1904. Tradition has it that he won the confidence of many by being the first banker to resume payments after the earthquake and fire of 1906. Whatever the reasons, its growth was phenomenal and, under its later name of Bank of America—adopted in 1928 as Mussolini's unpopularity began to rub off on anything named for Italy—became one of the largest banking empires in the world.

Beginning in the 1880s, the size and character of Italian immigration changed. The three hundred thousand immigrants of the 1880s were more than three times as many as had come in all American history; the six hundred thousand of the 1890s doubled that, and the more than two million of the first decade of the new century represented a further tripling. Put another way, the forty-four thousand Italian born of 1880 represented about six of every thousand foreign born in the nation; by 1920 the more than 1.6 million Italian born represented 117 in every thousand foreign born. (Italians ranked well behind natives of Russia and Germany; the three nationalities comprised two-fifths of the nation's foreign born in 1920.) These Italian immigrants were heavily male—males outnumbered females three to one—and were quite likely to return. Perhaps a majority of all the Italians who came here never intended to stay. What is important is not that fact, but that so many *did* stay.

The settlement patterns for Italians changed with the mass immigration from the Mezzogiorno. Some 97 percent of Italians in the four and a half decades after 1880 migrated through the port of New York, and vast numbers stayed there. By 1920 there were almost four hundred thousand Italian immigrants in New York City, nearly a quarter of all the Italian foreign born. Yet, although some authors persist in writing about New York's Italians as if they all lived below Washington Square, the fact is that as early as 1920 fewer than half lived anywhere in Manhattan; more than a third lived in Brooklyn and a tenth in the Bronx, with some 7 percent in Queens and Richmond. Similarly in Chicago, the neighborhood around Hull-House, habitually referred to as Italian, was in fact only about one-third so. What gave these neighborhoods, and the enclaves of other ethnic groups, their particular flavor were the predominant ethnic businesses in each. Despite much loose talk about ethnic ghettos, only racially segregated neighborhoods, those for blacks and Chinese primarily, came anywhere close to 100 percent uniformity. Counted by region, Italians were concentrated in the Middle Atlantic and New England states; despite a large Italian population in Chicago, they were not heavily represented in the Middle West. In the Far West there was a concentration in California, where in 1920 Italians were the largest foreign-born group, comprising more than a tenth of that population. And although relatively few immigrants now ventured to the South, there was a continuing Italian presence in and around New Orleans.

Counted by occupation, Italians in the United States were found largely in manual labor. As early as the 1890s commentators were noting that the Irish no longer built the railroads and paved the streets; Italians did. This concentration continued into the twentieth century and soon extended into the building trades. Perhaps the classic Italian immigrant novel, Pietro di Donato's *Christ in Concrete* (1939), movingly describes the life of a construction worker and his family as seen through the eyes of his son Paulie, obviously di Donato himself. Other Italian occupations included vending. The pushcart became one of the stereotypes of Italian American life, as did what must have been a relatively rare occupation, that of organ grinder with monkey. So strong was this image that when one showed up in faraway Prescott, Arizona, the young Fiorello La Guardia recalled vividly his embarrassment when his schoolmates noted that both the musician and he were Italians.[4]

Many Italians, especially in the early years of mass migration, got their jobs through ethnic labor contractors, called *padroni*. Similar arrangements were made by other immigrant groups, including Chinese, Japanese, Mexicans, and, as we shall see later in this chapter, Greeks. The labor contractor was most often himself an immigrant, but one who had mastered enough English to mediate for his fellow countrymen. The *padrone,* of course, often exploited the workers shamelessly, and in some industries, such as longshoring and hotel trades, kickbacks of salaries and other practices persisted long after the need for an intermediary translator had passed. (Such practices were not exclusive to immigrant labor contractors, it must be added.) A combination of muckraking exposé, more scientific management in most industries and, perhaps most important of all, increasing chain migration in which relatives and friends already here found jobs for the new immigrant, made the system less important for Italians as the new century progressed. A special and notorious form of the *padrone* system recruited and brought to America large numbers of young boys, mainly Italians and Greeks. The Italian boys were largely street musicians and acrobats, while the Greeks were fruit and candy vendors and, above all, shoeshine boys.

Wages were indeed low, and living conditions in the crowded urban tenements were appalling, but the notion one gets that the Italians were somehow poorer and more disadvantaged than other groups probably has more to do with their number and their exposure during the first age of investigative journalism than with their special plight. The occupants of the lower steps of the American ethnic escalator have always lived in deplorable conditions. Italians tended to send their children to work early: One often-cited study shows that some 90 percent of Italian American girls over fourteen were working in New York City just after World War I, fourteen then being the school-leaving age. Not surprisingly, few Italian youth in those years attended high schools: According to Humbert Nelli fewer than 1 percent of Italian American youth were enrolled in high schools on the eve of World War I.

Unlike most other immigrant groups, Italian Americans founded relatively few communal and fraternal organizations. According to Rudolph J. Vecoli, the leading student of Italian Americans, this is due to their individualism, an individualism that expresses itself as intense attachment to family—especially in South Italian culture and its American offshoots. Along with this went the extreme parochialism that caused Italians to regard themselves as citizens of a village or a town, not a nation. This has been called *campanilismo,* the notion that the true *patria,* native land, extended only as far as the sound of the bells in the local *campanile,* or bell tower. In line with this view of the world, Italian immigrants, even in large American cities, tried to reform village and small town clustering. In an important article Vecoli identified seventeen such clusters within the Italian neighborhoods of Chicago. Nor were such migration chains limited to large cities. William Chazanof and his students at SUNY, Fredonia, have published a pamphlet tracing the migration of a large number of families from one Sicilian village, Valledolmo, to one western New York village, Fredonia.[5] And similar stories could be written about literally hundreds of pairs of such places.

Although almost all Italian immigrants were and remained Catholics, their Catholicism was of a quite different nature than that of Irish and Polish Catholics on the one hand or German Catholics on the other. Whereas the former groups saw the church and its priests as the preservers of their embattled nationalism, Italian patriots saw priests as agents of the pope and opponents of the risorgimento, and thus opposed to Italian national aspirations. For Italians of the Mezzogiorno, the priest was often the agent of the bishop, who could be the landlord. A strong anticlerical tradition was part of the intellectual baggage that many Italians brought with them. In any event, Southern Italian Catholicism had its own kind of *campanilismo:* the special *festa,* feast day, of a particular saint or an aspect of the Madonna. (Robert Orsi has described the transportation of one such *festa* to New York in his *Madonna of 115th Street.)*

Such believers often found little comfort in the Irish-dominated Catholic church in America. An outsider, who didn't understand the religious aspects of the *festa* commented, hyperbolically, that Italians went to church only three times—to be hatched, matched, and dispatched—but Italians clearly did not attend church with anything like the regularity of most other Catholic ethnics. Nor did very many, proportionally, find religious vocations. Considering the size of the Italian American community, its representation in the American hierarchy has been amazingly low: Only with the emergence of Joseph Bernardin, the son of Italian immigrants, as cardinal archbishop of Chicago, has an Italian American played a major role in the power structure of the American church. Nor did Italian Americans provide much support for parochial schools, although it is not clear whether this illustrates attitudes toward religion or education. In any event, few

Italian immigrants, who slaved at low wages to save twenty-five or fifty cents a day, were going to pay the church for what the state provided free.

The most controversial aspect of the Italian American experience involves crime. Since most incoming immigrants enter American society at its bottom layers and live in decaying, crime-ridden neighborhoods, almost all have intimate contacts with crime and criminals early in their lives in the United States—most often as victims. For a minority of young immigrants, as the sociologist Daniel Bell has written, crime has served as a means of upward social mobility.[6] These immigrants have been of every imaginable ethnic group. Complaints linking ethnicity and crime began in the Colonial Era with talk about the crimes of Irish and German indentured servants and have continued, as witness contemporary stereotyping about Colombian drug dealers, the Chinese Triad Society, Vietnamese street gangs, or Soviet army deserters. It is instructive of the whole process that one of the first major talking pictures on the gangster theme, *Scarface* (1932), based vaguely on the life of Al Capone (1899–1947)—born in Brooklyn of Neapolitan parents—was remade in the 1980s with a Cuban rather than an Italian character as protagonist. (Ironically, a Jewish American actor, Paul Muni, played the Italian gangster, and an Italian American actor, Al Pacino, played the Cuban.)

Yet the persistence of the image of the Italian as criminal, from the Black Hand assassin with a stiletto at the turn of the century, to the Prohibition Era gangsters with machine guns, to the so-called Mafia and Costra Nostra of today, makes it one of the more enduring ethnic stereotypes. The notion of an international Mafia conspiracy centered in Palermo and dominating all Italian American crime is a foolish one. As students of the very real Italian Mafia point out, its power does not even extend into southern Italy, let alone Brooklyn, Chicago, Las Vegas, or Los Angeles. That there are criminal organizations in the United States with largely Italian American membership and that these and other criminal groups engaged in the drug trade have foreign connections are undeniable statements. But the attempts of some politicians, like Estes Kefauver, John McClellan, and Robert Kennedy, and some district and U.S. attorneys to turn the whole business of crime into an ethnic "conspiracy so vast" have been reprehensible. They are consonant with an American cultural tradition of scapegoating and imagining conspiracies that goes back to the Salem witchcraft trials and has, at one time or another, accused Negro slaves, Catholics, Freemasons, Mormons, and Jews of conspiring at the downfall of the republic.

The changing responses of Italian Americans have been instructive. Some of the earliest were defensive. As early as 1907 some in Chicago organized a White Hand Society devoted to law and order; others have simply denied that there were any such things as Italian American criminal organizations. Recently, however, more ambivalent attitudes have surfaced—attitudes that may be seen most conveniently in reactions to the film *The Godfather.* In Kansas City, for instance, local Italian American groups bought up every seat for the film's premiere but made sure that the house remained empty. On the other hand, some serious scholars have argued that Mario Puzo's book, on which the film is based, is a great immigrant novel about family solidarity. Actually, the Corleone family of the novel resembles real gangsters about as much as Paul Bunyan does real lumberjacks, but the overreaction is understandable. The whole Mafia syndrome is part and parcel of the particularly virulent prejudice that has been heaped on Italian Americans since the late nineteenth century and that will be discussed in context in chapter 10.

As Italians were relatively late to arrive and did not have a great propensity to become naturalized, their political participation was low and few Italian Americans won or were appointed to high political office. Some writers speak of Anthony Caminetti (1854–1923)—a second-generation Italian American who served in the California legislature, was elected to Congress, and then became Woodrow Wilson's commissioner of immigration—as an early ethnic politician, but he was not. Elected from tiny Amador County, which had no sizable Italian population, Caminetti claimed to be the first native son to be elected to Congress from California—thus nicely ignoring a Mexican American congressman, Romualdo Pacheco (1831–99), elected in the 1870s. Caminetti was not only a leading figure in the state's anti-Japanese movement but was an anti-immigrant commissioner of immigration during the nativist episodes of the World War I era. Fiorello H. La Guardia (1882–1947), initially elected to Congress in 1916 from a partly Italian district on New York's Upper East Side, was the first successful Italian American ethnic politician to occupy a significant position on the national political stage. In the 1930s, as mayor of New York, he became one of the nation's best-known leaders. Less prominent but more typical of Italian American ethnic politics were two New York ward bosses, James March and Paul Kelly. The first, born Antonio Maggio in 1860, came to the United States at age thirteen and was successively a labor agent for the Erie Railroad, a prosperous entrepreneur, and had an interest in an immigrant bank. From 1894 to 1910 he was the Republican leader on the Lower East Side. The second, born Paolo Antonio Vaccarelli in Naples, took his Irish-sounding name when he became a boxer. (In this era and later many boxers of various ethnic cities, including some Jews and blacks, fought under Irish names.) He was the Democratic counterpart to March and later became a vice president of the corrupt International Longshoremen's Association.[7]

Since most Italian Americans in the labor force were in unskilled and semiskilled jobs, relatively few Italian Americans were in the unionized sectors of American industry and even fewer were found in the leadership of the American Federation of Labor in the first quarter of the twentieth century. An exception must be noted for the garment workers' unions. Although the top leadership was largely Jewish, there were Italian vice presidents such as Luigi Antonini (1883–1968) and Salvatore Ninfo (1883–1960) of the International Ladies Garment Workers' Union. Ninfo, who had led a strike of Italian American subway diggers when he was just seventeen, became a vice president of the ILGWU in 1916. In the men's clothing union, the Amalgamated Clothing Workers, Anzuino D. Marimpetri was a longtime vice president. The one AFL union to have Italian presidents in this era was, appropriately, one for the semiskilled—the Hod Carriers and Building Laborers—which had at least two Italian-born presidents, including James Moreschi (1884–1970) of Chicago, who held the office from 1926 to 1948.

In the American radical movement, especially in its anarchist/syndicalist wings, Italians were of increasing importance. Gilded Age American anarchism had been largely German, as exemplified by the anarchists prosecuted and executed after the 1886 Haymarket Riot in Chicago who were mainly German immigrants, despite the *bürgerliche* (middle-class) image that German America presents. But early-twentieth-century anarchism in America spoke, and spoke eloquently, with a distinct Italian accent. Brooklyn-born Joseph J. Ettor (1885–1948) and Italian-born Arturo Giovannitti (1884–1959) were key organizers and agitators for the Industrial Workers of the World (IWW) in epic strikes of immigrant workers in Lawrence, Massachusetts (1912), and Paterson, New Jersey (1913). Giovannitti, in addition, is the only American labor leader I can think of who was a serious published poet, as witness his *Arrows in the Gale* (1914), mostly written while he was serving a prison sentence for his radical activities during the Lawrence strike.

It was two other Italian-born immigrants who came to symbolize

American anarchism for the world: Nicola Sacco (1891–1927) and Bartolomeo Vanzetti (1888–1927). An obscure fish peddler and an unknown shoemaker, they were arrested, tried, and convicted for a robbery and murder in 1920 and, after a protest campaign that swept the world, were executed by the Commonwealth of Massachusetts in 1927. While a spate of books by historians and pseudohistorians has tried and retried Sacco and Vanzetti with differing verdicts, no one can any longer even pretend that they received a fair trial. These men symbolized, both for patricians like Judge Webster Thayer, who sentenced the men he allegedly referred to as "anarchist bastards," and for millions of ordinary Americans, the dangerous ideas and deeds that foreigners could bring in. Their trial, which took place just before passage of the Emergency Quota Act of 1921, and their execution, which took place three years after Congress all but closed the golden door of immigration, punctuate and symbolize the end of an era—what I have called the century of immigration—for America.

No other Mediterranean groups came in the kinds of numbers that Italians did, but two of them, the Greeks and the largely Christian Arabs of the Levant, will be discussed at some length. Some six hundred thousand Greeks came before 1924, and perhaps one hundred thousand Arabs. The former are discussed in every history of immigration worthy of the name, the latter, until recently, have been largely ignored.

Greeks

Unlike the Italians, Greeks who immigrated to the United States generally had a fierce sense of their own Greekness, of the glories of the Greek heritage, even though—or more possibly because—that heritage had so long languished under the alien rule of another people of a different faith. This sense of distinctiveness was aided by the fact that they did not have to find a place within a church already well established in America: The Orthodox congregations they founded were ethnically Greek from the very beginning. While it is true that provincial and regional differences and animosities existed—as between Arcadians and Spartans, for example—the general Greek consciousness tended to override them. This is not to suggest that Greek Americans were unified, except in regard to their Turkish oppressors and Balkan enemies. Greek American politics and religion were highly partisan and conflict ridden.[8]

Although Greeks came early to the New World, they did so as

individuals and not as communities. Numerically significant Greek immigration did not begin until the 1890s. During the era of Greek immigration, more Greeks lived outside Greece—in the Balkans, in Turkey, in Egypt, and all along the Mediterranean littoral—than lived in it. The leading scholar of Greek Americans, Theodore Saloutos, thought it a good rule of thumb to assume that two-thirds of the immigrants from Turkey during this period were ethnic Greeks. The following table, which lists all immigrants from Greece and Turkey recorded by the INS, may be used accordingly. Thus, using the Saloutos formula that two-thirds of Turks are ethnic Greeks, we get a figure of ethnic Greek immigration from Greece and Turkey of 587,970 (370,405 + ⅔ of 326,347, or 217,565). Other parts of the Greek diaspora also sent smaller but still sizable numbers to the United States. In 1976, the INS estimated that 640,000 ethnic Greeks had come to the United States between 1820 and 1975. Many experts hold that figure to be too small. By 1975 the Greek American population was at least twice that number.

Although Greece, or some of it, achieved independence from Turkey in 1821, Greece has not been able to sustain its still-burgeoning population, and emigration has been a factor in its demographic history since the late nineteenth century. The Greeks who came to the United States in the era of immigration were very heavily male and were the only fairly large European group of which more than half returned: Archdeacon gives figures of 87.8 percent and 53.7 percent. This was not too different from the Italian American experience—the figures there were 74.5 percent male and 45.6 percent remigration—and, in fact, Greek

Table 7.4

Immigrants from Greece and Turkey, 1820–1920

YEARS	GREECE	TURKEY
1820–30	20	21
1831–40	49	7
1841–50	16	59
1851–60	31	83
1861–70	72	131
1871–80	210	404
1881–90	2,308	3,782
1891–1900	15,979	30,425
1901–10	167,519	157,369
1911–20	184,201	134,066
Total	370,405	326,347

and Italian immigration are similar in many other ways. But there are important differences.

The settlement patterns of Greeks were quite dispersed, but almost all of them pursued urban occupations. The soil had little attraction for Greek emigrants who associated it with misery and hardship in their native land. The majority during the two heavy decades of immigration of this century came from small villages in the Peloponnesus and elsewhere in Greece proper, but sizable minorities came from the Greek islands and European and Asiatic Turkey. The latter groups were almost all urban, somewhat better educated, and often created small businesses in the United States. Greeks settled primarily in the great immigrant belt of the northeastern and northcentral states, although, as was true for Italians, a large number went to California. A smaller number, almost all of them from the Greek islands, settled on the west coast of Florida, centered around Tarpon Springs, where they pursued, and still pursue, sponge fishing. (One of the American television networks uses a heavily Greek American precinct in Tarpon Springs to inform its viewers how Greek Americans voted, despite the obviously special nature of that area.)

The earliest Greek immigrants took what jobs they could find in both heavy and light industry and were never as concentrated as were Italians and Jews in the clothing industry. Railroad construction gangs, textile mills, meat-packing, and mining occupied significant numbers of immigrants. Others, as we have seen, were employed by *padrone*-organized shoeshine stands, some of which evolved into shoeshine parlors. Relatively large numbers of Greeks became small businessmen and, for reasons that are not at all clear, large numbers of these opened restaurants. These were not restaurants that featured Greek cuisine but were generally modest places that featured inexpensive general food, the kind of place often referred to as a "greasy spoon" restaurant. (Wherever there was a sizable Greek American community one or more coffeehouses where Greek men could meet and talk would be established, but that is a different matter.) The economic niches filled by immigrants in small businesses are not predictable based on what immigrants did in the old country. Few if any emigrating Greeks ran restaurants in Greece, just a few Italians ran barbershops in Italy, and no Chinese ran laundries in China. Yet, early Greek, Italian, and Chinese immigrants established such businesses and hired fellow immigrants who learned the ropes and later went into business for themselves. All three of these dissimilar businesses had common characteristics. None required large amounts of capital, each was labor intensive, and each provided services to the general low-income public. In some instances these businesses could have been interchangeable—Greeks could have specialized in barbershops and Italians in small restaurants—and it was largely historical happenstance that each European group found the niche that it did. (On the West Coast, at about the same time, Japanese immigrants with a little capital often set up small workingmen's restaurants, and, in Seattle, a number of barbershops. And in Lima, Peru, but nowhere else as far as I can discover, Japanese immigrants came to dominate beauty parlors.) Other lightly capitalized lines that attracted large numbers of Greeks were candy stores and confectioneries of various kinds. In addition, there were in the Greek American community, as in any immigrant community of any size, numbers of businesses that catered to the ethnic group—grocery stores (stores that not only stocked large containers of olive oil and resinated wine, but where the shopper could do business in Greek), ticket brokers, and similar enterprises. Most of these were quite small enterprises and the failure rate in immigrant businesses was probably even higher than the general small business failure rate. Few of their owners achieved anything more than a modest competence and many did not even get that.

One line of enterprise in which some Greeks pioneered that did grow into a giant industry was the showing of motion pictures and other cheap amusements for the working public. The first movie houses in immigrant neighborhoods were simply stores in which a few benches, a screen, and a projector were the only equipment. The motion picture industry came to be dominated by immigrants, mostly Jews, but a significant number of Greeks, notably Alexander Pantages (died 1936), who developed one of the largest theater chains showing both films and vaudeville acts, and Spyros Skouras (1893–1971) and his brothers who controlled the 20th Century–Fox studios.

The first Greek immigrants sometimes found existing Orthodox churches—organized Orthodoxy came to North America in 1794 when Russian monks began missionary work among the Aleuts and Tlingit Indians of Alaska—but most Greeks wanted not just Orthodox churches but ones that were Greek in language and culture. The first Greek Orthodox Church in the United States predated the major groups of immigrants by decades: It was founded by Greek cotton merchants in Yankee-occupied New Orleans in 1864. No other churches were established until the 1890s—in New York, Chicago and Lowell, Massachusetts—and by 1909 they were scattered across the country as far as Portland, Oregon; San Francisco; and Los Angeles.

There were more in the textile centers of Massachusetts than anywhere else.

Church officials in Greece did not send missionary priests to the United States, since it was technically under the jurisdiction of the Russian Orthodox Church. Thus lay initiative was crucial in the founding of Greek American parishes. Typically, the laypeople would raise money or pledges sufficient to start a parish and then ask the ecumenical patriarchate or the Synod of Greece to send a priest or priests to operate the parish. In such parishes clashes between laity and clergy were all but inevitable. In addition, the governance of all the Greek American churches was quite complex. Initially they were under Russian jurisdiction, as noted. Between 1908 and 1922 they were under the governance of the Church of Greece. But since that church was a state church, and since Greece was torn between monarchist and republican factions, this created confusion and, at times, schisms. At one time, in the early 1920s, there were in the United States the deposed metropolitan of Athens, and two bishops, one appointed by him, one by his successor, each claiming authority over and seeking the loyalty of all Greek American churches. The result was chaotic, and lawsuits and secessions of parishes seemed to be the rule rather than the exception until a general settlement was made in 1930.

These divisions were not over essentially religious matters but over Greek politics and reflected chiefly the disputes between King Constantine I (1868–1923), his supporters, and Eleuthérios Venizélos (1864–1936), head of the Greek Liberal party and his supporters. Greece's two victories in the Balkan Wars of 1912–13, during which some forty-two thousand Greek Americans returned to fight for their mother country, created a wave of Greek patriotism and visions of a new Greek empire with headquarters in Constantinople. But in the early 1920s Greek irredentism suffered a series of disasters at the hands of the troops of Kemal Atatürk. The liberal/royalist split in Greek politics was mirrored by rival post–World War I Greek American organizations—the American Hellenic Educational Progressive Association (AHEPA) and the Greek American Progressive Association (GAPA). Their dispute was not just over Greek politics but over what loyalties Greek Americans should have: issues that arise in every immigrant population in one way or another.

AHEPA preached doctrines of Americanization and nonsectarianism and was accommodationist and aggressively middle class. Its rival accused it of forcing Greek Americans to deny their roots, and many were outraged that it used only English in its proceedings (at least officially), that it sanctioned mixed marriages with non-Orthodox, and that its members needed only to affirm a belief in a Supreme Being, not be members of the Greek Orthodox Church.

GAPA, although it, too, insisted that it was "progressive," was stubbornly anti-assimilationist. GAPA generally thought that all Greek Americans should learn and use Greek, cultivate the Greek heritage, and, above all, adhere to the Orthodox church. AHEPA won the battle hands down. Second- and third-generation Greek Americans are no more likely to have a functional knowledge of the tongue of their ancestors than is any other American ethnic group. And, it should be noted, despite GAPA accusations, AHEPA did stress pride in Greek cultural achievements, particularly those of its classical civilization, about which almost all Greek Americans seem to know a great deal.

The late arrival of most Greek immigrants, and their numbers and relative dispersion, meant that there would be few Greeks of the immigrant generation in American politics, and no Greek Americans played any role on the national political stage in the period under consideration. The most famous Greek in America was clearly Jim Londos (1895–1975), the "golden Greek" who was a world champion wrestler when wrestling in America was still a sport and not an exhibition of sadomasochistic fakery.

Arabs

Although most Americans equate the ethnic term *Arab* with Muslims, the fact of the matter is that until the very recent past almost all Arabs who immigrated to the United States were Christians of several Eastern Rite churches. They came from what was, in the later Turkish Empire, the autonomous administrative district of Mount Lebanon, essentially the area served by the two ports of Beirut and Tripoli. Christianity in that part of the Levant, of course, dates from the beginnings of the Christian Era, centuries before both the rise of Islam and the Christianization of most of Europe. Most of the earlier writing about these immigrants, about one hundred thousand of whom came before World War II, called them Syrians, while many of the immigrants called themselves Lebanese, particularly after the foundation of the republic of Lebanon in 1946. In parts of West Africa and Latin America, where immigrants from this region have also settled, they are known by various names, including Turks, since they were, until after World War I, subjects of the Ottoman Empire. In recent years the umbrella term *Arab Americans* has come to be used to describe a variety of national and religious groups—most of them Muslim—and that usage will be applied retroactively here.

Alixa Naff, the leading student of Arab Americans, believes that the first Arabs "to discover the economic opportunities of the United States" were a group of Christian tradesmen who came under Ottoman sponsorship to exhibit various Syrian wares at the Philadelphia World's Fair celebrating the centennial of American independence in 1876.[9] The enthusiastic reports of these merchants, the activities of recruiting agents, economic pressures at home, and, after 1908, compulsory military service, all combined to create a chain migration from the slopes of Mount Lebanon to the United States.

Although the Arab immigration was initially one of bachelors, women soon began to come, almost always as members of families. Fragmentary records indicate that perhaps 47 percent of the immigration before 1924 was female. The overwhelming majority of early Arabs—as many as 90 percent before 1914—began earning their living in the United States as peddlers. Like other groups this was a specialization they adopted after emigrating: In their homeland peddling was largely confined to minority groups—Greeks, Armenians and Jews. The Arab peddlers covered the entire nation, some of them working a distant territory or route for weeks and even months before returning to their base. Others were city peddlers and slept at home every night. Initially most peddlers sold rosaries, costume jewelry, and notions—the sort of things that would fit into a small suitcase or pack. As they quickly acquired English from dealing with their customers, many learned to stress the "fact" that their religious goods came from "the Holy Land." Later, when they had more capital, some peddlers made their rounds on horseback, with wagons, and eventually automobiles, and were able to vend larger wares including imported rugs and linens.

The supplier, often himself a former peddler who had moved up a rung on the economic ladder, performed the same kinds of functions as did labor contractors and *padroni* for other ethnic groups. He encouraged fellow villagers to emigrate and sometimes financed their passage; he supplied them with goods, usually on credit, assigned routes, banked their money, tried to smooth their conflicts with officials (peddlers were always running afoul of some local ordinance or other), and generally served as mentor during their period of acculturation. As it had been for Jews earlier, peddling was most often a transitory occupation for individuals. The peddler had to acculturate, and acculturate quickly. Unlike the immigrant who spent his life in ethnic enclaves and worked at ethnic job sites, the peddler had an economic stake in learning the language and the folkways of other Americans. Once he—or she, since there were a few women peddlers—had learned the ropes, the economic rewards were greater than those for most contemporary immigrants. Naff, perhaps oversanguine, feels that "relatively few" were failures and that peddlers commonly calculated their annual earnings in thousands. Since in 1907, for example, the average annual wage for all nonfarm employees was $595—and most immigrants made less than the average—the Arab peddlers did quite well. Most peddlers who did well eventually went into business—typically in retail shops—and many of them, of course, became suppliers for other peddlers—often relatives and usually fellow villagers, whom they sponsored.

Not surprisingly New York was the initial center of Arab immigrants, and most of the early suppliers were importers who had settled there. But the expansion of Arab peddling territory meant that suppliers had to move—largely south and west—as well. By 1930 the census—not at its best in identifying small ethnic groups—noted just over seven thousand five hundred "Syrians" in New York City, while Detroit was becoming the Arab American second city, with more than five thousand. By 1970 the motor city would be the undoubted Arab American capital, with more than seventy thousand persons, Christian and Muslim, so identifying themselves. New York was also, in the years before 1914, home to a small group of Arabic-speaking and -writing intellectuals who had little contact with the larger Arab American community. The best known of this group, and the only name known to non-Arabic readers, was the translated poet Kahlil Gibran (1883–1931).

Religious institutions, as among the Greeks, were established by lay initiative after settlement and a certain amount of financial success. Between 1890 and 1895 New York Arabs founded three churches, one for each of the three major sects represented among the immigrants—Melkite, Maronite, and Orthodox—and imported priests to serve them. In the next three decades more than seventy additional Arab American churches were founded, with masses conducted in Arabic or Syriac. As more than half of these were in the East, this meant that many people in the far-flung settlements were not served by any ethnic church. Many of these families, and other Arab Christians, attended "American" churches, both Roman Catholic and Protestant.

For the small Muslim minority the problems were even more difficult. It is hard to be a Muslim in a non-Muslim society. Although Muslims, like Jews, have no priesthood and can pray almost anywhere, the imam and the mosque have become as central to Muslim worship as the rabbi and the synagogue have become to the Jewish. Naff illustrates this by talking about one of the few early Muslim settlements, near Ross, North Dakota, established around the turn of the century. There was no mosque until the 1920s, and prayer and ritual were conducted in private homes. The small community, which received no reinforcement, soon lost the use of Arabic. Many adopted Christian names, married non-Muslims, and others moved away. The ethnically conscious community shrank, and the mosque was abandoned by 1948. Only two other mosques are known to have been built anywhere in the United States before the 1930s.

Armenians

Unlike the other groups discussed in this chapter, Armenians are not a Mediterranean people but originated in northeast Asia Minor. Converted to Christianity early in the Christian Era, the Armenians became in the Later Middle Ages one of the subject peoples of the Ottoman Empire. In the process many Armenians reestablished an Armenian kingdom in Cilicia, in southern Mediterranean Turkey just north of present-day Syria. This, too, was overrun in 1385. While some Armenians achieved high economic status under the Ottomans, becoming bankers, skilled artisans, and at times even advisors to the sultans, most were farmers, and many emigrated in late medieval and early modern times to such places as the Crimea and Poland. A few even came to America: The records show that one Martin the Armenian came to Jamestown in either 1618 or 1619, but nothing more is known of him. Handfuls of other Armenians found their way to America and other parts of the New World in the next two and a half centuries, but it was only in the late 1880s that statistically significant migration began. As is the case with other subject nationalities, it is difficult to differentiate Armenians in the American immigration records. Probably around one hundred thousand came to the United States between the late 1880s and the virtual closing of immigration for them in 1924. The period was punctuated by an awful event few Armenians can forget: the massacres—some would say genocidal massacres—of Armenians by Turks in 1915, which had been preceded by waves of persecution and killings in 1894–96 and again in 1909. Today the Armenian SSR contains more than two million Armenians, while another million live elsewhere in the USSR. The largest non-Soviet Armenian community is in the United States, where perhaps half a million Armenian Americans now live.[10]

The fifty thousand Armenians who came in the years before World

War I arrived mainly by two distinct routes to establish two very different kinds of communities. Initially most Armenian immigrants came through New York and settled in eastern cities and worked in factories much as did other immigrants of that era. Most of the prewar Armenian migration was of Turkish origin, and, after the 1909 atrocities, it increased rapidly: In 1913 nearly ten thousand came.

Migration from Russian Armenia was much smaller. Only after members of two pacifist sects who lived in and near Russian Armenia—Dukhobors and Molokans—began to migrate to Canada's Prairie Provinces in 1898 did some Russian Armenians follow them. Most did not find Canada to their liking and eventually made their homes in and around Fresno, California, which became and remains the Armenian American capital. Perhaps two thousand five hundred had settled in California by 1914. They were one of the few immigrant groups of this era to settle on the soil, although many worked in factories to save the money necessary to buy land.

Armenian immigrants of the prewar era present a profile both similar to and different from those of other immigrants of that time. They were highly male and had a moderate return migration rate: Archdeacon's table shows 71.3 percent and 18.1 percent. Almost all males over fourteen were literate, about two-fifths had been town dwellers, and there were many artisans, businessmen, and professionals among them, particularly among the Armenians from Turkey. These latter had some of the characteristics of later refugees from the Third Reich. The pre–World War I immigrants, except for a few who had been subsidized by Canada, used their own or community resources to come to America. Most of those in the postwar era were aided by refugee organizations. This immigration presents a vastly different profile: More than thirty thousand persons were involved, just over half of them women and a fifth children.

The Armenians of the Fresno region, made famous by William Saroyan (1908–81), the most celebrated Armenian American author, have a unique history. Their history in this rich farming region, where some thirty thousand foreign-born Armenians lived in 1930 (New York City's seven thousand ranked second and Detroit's three thousand five hundred third) is a story of successful response to special challenge. Only here were there enough Armenians to establish an all-Armenian town—Yettem, which means Eden in Armenian—and only here were there enough Armenians for special discriminatory measures to be used against them. To be sure, all foreigners—especially those who looked and sounded "foreign"—suffered at one time or another in America—

but it takes a certain critical mass relative to the general population in an area for special, pointed discrimination to take place. In Fresno the most enduring form of discrimination was the restrictive covenant that kept them, along with Asians, Mexicans, and blacks, from buying homes in Fresno's better neighborhoods. (Elsewhere the most frequent other targets of restrictive covenants were Jews; in some parts of upstate New York, Italians were the target group.) They were also barred from most of Fresno's social organizations, including the YMCA and veterans' groups, and were even expelled from a Congregational Church some of them had helped to fund. More threatening were attempts to have them declared aliens ineligible to citizenship, along with Asians. The federal government began denying Armenian petitions for naturalization in 1909, but a federal court ruling of that year, reinforced by an appellate court decision of 1924 in the *Tateos Cartozian* case prevented such denials. Thus the California Alien Land Acts of 1913 and 1920 affected only Asians (for details see chapter 9). Restrictive covenants, in Fresno and elsewhere, remained legal until struck down by the U.S. Supreme Court in *Shelley* v. *Kraemer* (1948).

The Armenian Apostolic church in America has had a conflict-ridden history based, as was true for the Greek Orthodox, on Old World problems. The first parish was established in Worcester, Massachusetts, in 1891, and the poor and relatively small community was supporting ten churches and seventeen imported priests by 1916. (Only in 1962 was an institution to train Armenian rite priests opened in the U.S.) The Russian Revolution caused a severe split in the Armenian church: The crux was not theological but political. Should Armenian Americans recognize the church in Soviet Armenia, which had made its peace, more or less, with the Soviet authorities? One faction, Ramgavars, said yes; the other, Tashnags, said no. Full schism came when Archbishop Levon Tourain was assassinated while he was celebrating mass in New York City on Christmas Eve 1933. Nine members of the Tashnag faction were convicted of the crime. A small minority of Armenian Americans, perhaps 5 percent, are Protestants. There is an even smaller number of Armenian American Roman Catholics, whose few American churches are under the authority of the Armenian Patriarchate in Beirut and use an Eastern-rite mass in Armenian.

8

Eastern Europeans: Poles, Jews, and Hungarians

According to the *HEAEG*, members of at least twenty-six European ethnic groups emigrated to the United States from the area north of Greece and east of Germany.* Most of these groups cannot be treated in a work of this kind, just as certain Western European groups were not treated and as many Asian and Latin American groups will not be treated. (From a demographic point of view it can be argued that the most serious omission has been the failure to talk to any length about those whom Charlotte Erickson has called "invisible immigrants," non-Irish emigrants from the British Isles, persons who could all but disappear into the mass of the American people almost as soon as they arrived.)[1] More than three million such immigrants arrived during our century of immigration from the end of the War of 1812 to 1924. The failure to talk about an ethnic group here should *not* be regarded as dismissive of its importance. For example, five times as many Slovaks came to the United States as did Armenians, yet the former will not be discussed in detail here. Those interested in Slovaks, and in other groups about whom modern histories have been written, can discover appropriate titles in the bibliography or in the *HEAEG* essays. Omission is not a slight. This book, as previously noted, is intended to be illustrative, not encyclopedic. Some groups, like Poles and Eastern European Jews, were simply too large to be ignored. Among the others, choice was arbitrary and was, in part, dictated by the quality of available secondary works.

Most members of these groups who came to the United States began

*Albanians, Belorussians, Bosnian Muslims, Bulgarians, Carpatho-Rusyns, Cossacks, Croats, Czechs, Estonians, Finns, Georgians, Gypsies, Hungarians, Jews, Latvians, Lithuanians, Macedonians, North Caucasians, Poles, Romanians, Russians, Serbs, Slovaks, Slovenes, Wends, Ukrainians.

Prospective Italian emigrants headed toward the Emigrant Aid Office (Opera Assistenza Emigranti) in Naples, about 1900. (*Touring Club Italiano, Milan*)

Emigrants waiting to be transported to vessels in the port of Naples, about 1900. (*Touring Club Italiano, Milan*)

A view of Castle Garden (about 1888), looking out toward the newly erected Statue of Liberty in New York harbor. (*National Park Service: Statue of Liberty National Monument*)

Emigrants in Patras, Greece, embarking in small boats for their steamer bound for America, about 1910. (*Library of Congress*)

Two young Italian immigrants "just off the boat" (the British liner *Canopic*), Boston, March 3, 1920. (*Michigan Historic Collections, Bentley Historical Library, University of Michigan*)

Prospero Frazzini and his brothers were obviously among those Italian immigrants to the American West whom Andrew Rolle has called "the upraised." As this photo, taken about 1900, shows, their substantial Denver establishment included wholesale and retail businesses, a bank catering to immigrants, and a saloon. (*Center for Migration Studies*)

Rose Halleck Boosalis counting the day's receipts in her Greek American family business, the Olympia Confectionary in Minneapolis, probably in 1903. (*Minnesota Historical Society*)

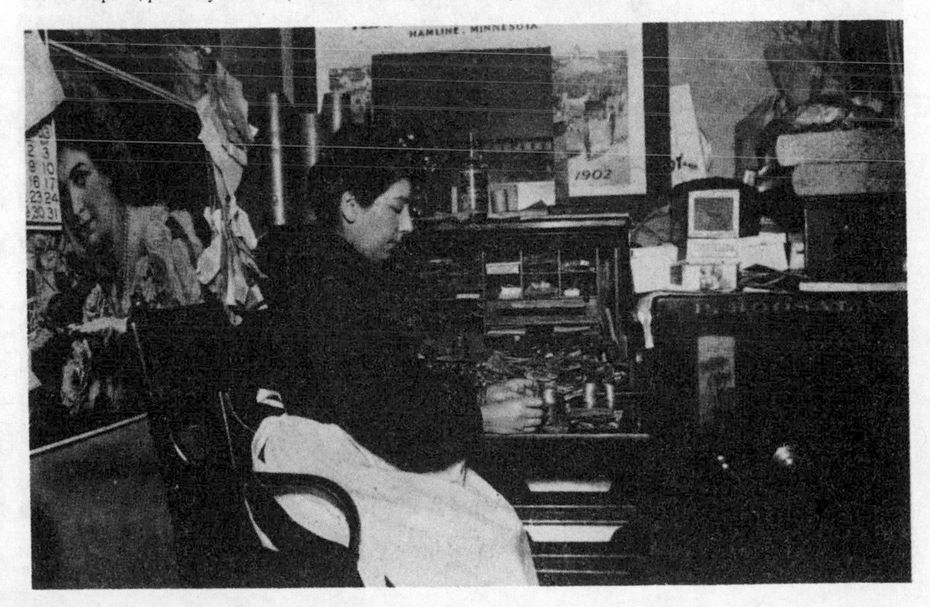

A Utica, New York, Arab American amateur theatrical troupe posed in costume, about 1910. These immigrants were from Baskirta, Lebanon. (*Smithsonian Institution*)

An Armenian American couple being married by Archpriest Moushegh der Kaloustian in St. Illuminator's Armenian Cathedral, New York City, 1978. (*Archpriest Moushegh der Kaloustian, Pastor, St. Illuminator's Armenian Cathedral*)

A street in the Jewish quarter of Chicago, 1890. (*Photography Archive, Carpenter Center for the Visual Arts, Harvard University*)

Some of the first sponsored Jewish immigrants to arrive in Galveston, Texas, on July 14, 1907. The aim of this project, which brought more than ten thousand Jewish immigrants there between 1907 and 1914, was to disperse Jewish immigrants more widely, particularly in the trans-Mississippi West. (*Congregation of B'nai Israel, Galveston, Texas*)

Of the many ethnic dramatic groups that flourished in America, none was better patronized than the Yiddish theater. This scene, from Oscar Carter's play *The Three Little Businessmen*, took place in the Nora Bayes theater on Broadway in 1923. The actor with upraised arm is the famous Boris Thomashefsky. The others, in the usual order, are Rudolph Schildkraut, Ludwig Satz, and Mme. Zuckerberg. (*Museum of the City of New York*)

These immigrants, most of them from Eastern Europe, were posed on the deck of the German liner SS Amerika in 1907, the all-time peak year of immigration to the United States. (*Library of Congress*)

Although many states, like California, established agencies for the protection of immigrants, they were not particularly effective. This poster, from about 1910, is in English, Spanish, Hungarian, Italian, Portuguese, Greek, Russian, Polish, Czech, Yiddish, Croatian, and French. The English text reads: "The State of California Commission on Immigration and Housing is created to protect and aid immigrants in California. Immigrants who feel that they have been wronged, abused or defrauded or who wish information are asked to come in person or write to the office of State Commission, Underwood Building, 525 Market Street, San Francisco. The Commission will furnish information and will aid all in obtaining justice. We speak and write all languages." Note that neither Chinese nor Japanese is included. (*Weidenfeld & Nicolson, Ltd.*)

Many Ukrainians settled on the northern Great Plains, in both the United States and Canada. This congregation is posed after Sunday services in front of St. Josaphat Ukrainian Catholic Church in Gorham, North Dakota, in the late 1890s. (*Ukrainian Cultural Institute, Dickinson State University, North Dakota*)

Lithuanian American Stanley P. Balzekas (in white coat) behind the counter of a grocery/meat market in Chicago at about the time of World War I. (*Chicago Historical Society*)

A very high proportion of the first generation of Finnish Americans embraced radical politics in the United States. This gathering of "Red Finns" and their families (probably on a Sunday or holiday) poses outside a meeting hall in Glassport, Pennsylvania, early in this century. (*Tuomi Family Photographs, Balch Institute for Ethnic Studies Library*)

The diagram and five photographs that follow show some aspects of emigrating from Hamburg and feature the immigrant depot, or *Auswandererhallen*, set up early in this century by the Hamburg-Amerika steamship line. Similar depots in Bremen and Liverpool (set up by the North German Lloyd and Cunard lines, respectively) were part of the migration experience of millions of European emigrants in the peak immigration years before World War I. (*Staatsarchiv Hamburg*)

The Hamburg depot was primarily for emigrants from Eastern Europe traveling steerage. German and Scandinavian emigrants could opt for private facilities such as this large establishment, which was also located at the eastern end of the harbor.

The *Auswandererhallen* were located within the customs harbor (*Zollhafen*) and the emigrants, like the goods in the adjoining customs warehouses, were sealed away from the outside world. This isolation was primarily a health quarantine measure, but it also afforded the immigrants protection from those ripoff artists who preyed on the gullible. Most of the buildings shown are dormitory and eating facilities (the latter included a kosher kitchen). The open area at the right end of the complex was for outdoor concerts and included what we would call a snack bar. The other services of the depot were included in the price of the ticket. Also in the open area was a Catholic church, while the synagogue was relegated to a less prominent location at the top center of the enclosure.

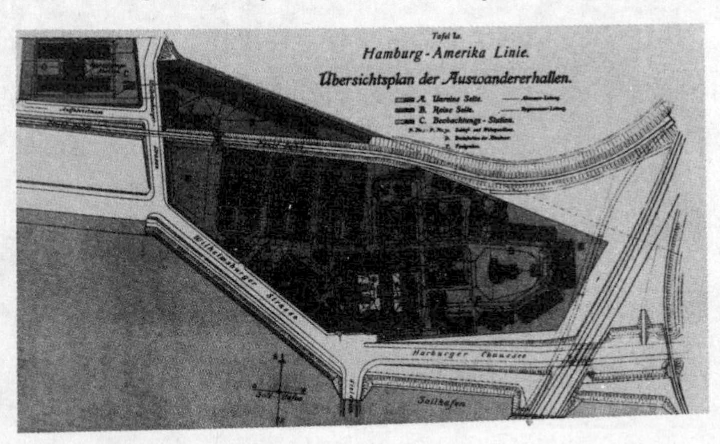

These two photographs may have been taken at the same time—note the same uniformed Hamburg-Amerika line employee in each picture. The first, obviously posed, was used in the steamship line's publicity; the second, more realistic, seems never to have been so used.

As many as five thousand emigrants at a time could be accommodated in the Hamburg *Auswandererhallen* in these double-decked dormitory rooms, which were clean and, although cramped, provided more space than the emigrants would have in steerage.

Another Hamburg-Amerika publicity shot showing the *Abschied von Auswanderern* ("Emigrants' farewell"). These people are waving good-bye to German passengers, probably first or second class, from Hamburg's *Amerikakai* (immigrant pier). The emigrants from the east used the same facility, but there was almost never anyone to see them off. Their leave-takings, if any, were elsewhere.

The great American photographer Lewis W. Hine captioned this study "A Slavic boarding house in factory district near New York City—1912." (*Lewis W. Hine Collection, United States History, Local History & Genealogy Division, The New York Public Library, Astor, Lenox and Tilden Foundations*)

These Polish immigrants, just off the liner SS *Touraine* on September 23, 1920, were part of the postwar immigrant stream that triggered the restrictive acts of 1921–24. (*Michigan Historical Collections, Bentley Historical Library, University of Michigan*)

Chinese American construction workers building the "Secrettown Trestle" in the Sierras, 1877. (*National Park Service: Statue of Liberty National Monument*)

Interior of the Ching Chong Wong provision store, Clifton, Arizona, about the turn of the century. (*Arizona Historical Society*)

A Chinese American nuclear family at home, about 1900. The father was almost certainly a treaty merchant. (*Museum of the City of New York*)

Japanese immigrants being vaccinated on shipboard while en route to Hawaii, 1904. (*Library of Congress*)

Asian immigrants arriving at the quarantine station at Angel Island, San Francisco Bay, about 1911. (*National Archives*)

French Canadian Americans in the mill yard, Amoskeag Manufacturing Company, Manchester, New Hampshire, about 1885. (*Manchester Historic Association*)

Members of the Lessard and Arseneault families, French Canadian Americans, in the filling bay, fifth-floor spinning room, Appleton Mill, Lowell, Massachusetts, late nineteenth century. (*Center for Lowell History*)

French Canadian American potato picker, Caribou, Maine, 1940. (*National Park Service: Statue of Liberty National Monument*)

to arrive in the last years of the nineteenth century and the initial ones of the twentieth. They settled predominantly in the cities of the northeastern and north-central states, cities known today as the "rust belt" for the demolished or obsolete and rusting factories that once made the region the greatest manufacturing center in the world. It was these and other immigrants who, in the main, made these factories go and provided the human raw material that transformed the United States into a great industrial power. The distinguished American historian, Carl Wittke, when he wrote a general book about European immigration, called it *We Who Built America.* If we use the verb *built* literally instead of figuratively, as Wittke did, it would apply to no groups more appropriately than to the European ethnics of this period. To be sure, so-called Anglo-Americans tended to own and operate the factories, and members of better-established or more easily assimilated ethnic groups, tended to be the managers, superintendents, and even foremen. But the brawn that fed the blast furnaces, laid rails, and actually built things was predominantly that of the later arrivals. They came to a country whose frontier, in a traditional sense, was closing. In a report accompanying the census of 1890, the superintendent of the census demonstrated that the frontier, defined as a large contiguous region with fewer than one person per square mile, had ceased to exist in the United States. (Alaskans today amend this to say "in the lower forty-eight.") And it was in 1893 that a young historian named Frederick Jackson Turner first propounded what has become known as the Turner thesis, one of the two or three most influential notions ever put forth about American history.

> Up to our own day American history has been in a large degree the history of the Great West. The existence of an area of free land, its continuous recession, and the advance of American settlement westward explain American development.[2]

Turner, and the generation or two of historians who followed him, all but ignored the role of the cities and the immigrants who were filling them even as he spoke. And Turner spoke, ironically, in the city of Chicago, where the American Historical Association happened to be meeting, a city that was becoming home to hundreds of thousands of immigrants and their children.

Although they came to American cities, these immigrants did not, generally, come from European cities. They were, overwhelmingly, peasants from the less-developed regions of Europe, who lived mainly in villages and small towns. Had there been land available, and had they been able to buy it, many of them would surely have become farmers in America. Indeed, for many, *the* reason for coming to American was to earn enough money to buy land back home, and many sojourners in America were able to save enough from their meager earnings to achieve that goal. And, as Eva Morwaska has pointed out, for many Eastern Europeans in American industry, gardening was a source both of dietary and income supplementation.[3]

The two most important historical surveys of immigration to focus on these immigrants each used a titular organic metaphor to describe them: Oscar Handlin's book, perhaps the most influential and widely read historical work on immigration, was called *The Uprooted;* John Bodnar's was titled, somewhat mimetically, *The Transplanted.* Each metaphor, whether or not its author realized it, tends to reduce the immigrants to objects rather than subjects of history, makes them seem persons to whom things happened rather than persons who caused things to happen. While no person can control completely her or his destiny, and since poor people—and most immigrants were poor—generally have fewer options than do better-off individuals, immigrants were often buffeted by forces beyond their control. But immigrants, or more properly the decision makers in immigrant families, all made at least one crucial decision: They chose to come to America. They were thus, in this sense, movers rather than the moved. Metaphors, however striking or dramatic, that suggest otherwise are misleading and contribute to the stereotype of the "dumb" immigrant. To English-speaking Americans, of course, these immigrants did indeed seem literally "dumb": They could not talk in a language appropriate to "God's country." From this it was easy to assume that immigrants were "dumb" in another sense, that is, not smart. And the statistical evidence shows clearly that most immigrants of this period had very little formal education and were thus ignorant of many things. Many, perhaps most, came with gross misconceptions about what kind of place an industrializing America was. But that is not to say that immigrants were stupid, or at least not more stupid than the general run of humanity. And in general, it seems to me, many of them came to understand the real nature of America much better than did most of those who despised them.

Poles

As indicated earlier, it is difficult if not impossible to tease out of the American immigration statistics accurate data about the immigration of Poles. Between the third partition of Poland in 1795 and the creation of a new Poland by the Treaty of Versailles in 1919, no Polish state existed. Interwar Poland lasted just twenty years; a new one was created by the victorious allies after World War II. During the century of immigration the Poles who came to America lived mainly in one of the three great European empires—the German, the Austro-Hungarian, and the Russian. Other Poles, to be sure, came from elsewhere, including France, to which large numbers of Polish workers traditionally migrated. Poland was also the home of large numbers of non-Polish peoples, including Germans, Jews, and Lithuanians. Almost all the ethnic Poles who came to America shared two common characteristics: They came speaking Polish as a mother tongue, and they were Roman Catholics.[5]

And it is from mother tongue that we obtain our best clues from American statistics as to the number of Poles who immigrated here. In the 1910 census, for example, persons were asked about their mother tongue. Of the more than thirteen million foreign-born white persons counted that year, more than nine hundred thousand (7 percent) told the census taker that Polish was their native tongue; and of 32 million persons of what the Bureau of the Census called foreign white stock—that is, second-generation Americans, the children of immigrants—some 1.7 million (5 percent) were the children of persons whose mother tongue was Polish. Table 8.1 summarizes the census findings for Poles.

As the table shows, almost 45 percent of the Polish immigrants came from Russia, 35 percent from Austria-Hungary, and 20 percent from Germany. The greater apparent rate of increase of second-generation German Poles—2.7 times foreign born vis-à-vis second-generation Russian and Austrian Poles—1.5 times foreign born—represents, not higher fertility on the part of German Poles but the fact that they came earlier and thus (1) more had completed their child production; and (2) more of the Germans would have died by then. What this and similar tables cannot show is immigrants who had come and returned before the census was taken. Archdeacon's return rate for Poles is a fairly high 33 percent, so the probable discrepancy is quite high. Other mother tongues from Eastern Europe that show significant percentages (10 percent or more) coming from different states are Romanians, more than 50 percent from Romania, more than 33.3 percent from Hungary, and 9 percent from Austria (in this table, Austria means the Austro-Hungarian Empire minus Hungary); Slovaks, almost 66.6 percent from Hungary, the rest from Austria; Russian, almost 70 percent, not surprisingly, from Russia, but most of the rest from Austria; Serbo-Croat

Table 8.1

Country of Origin, Polish Mother Tongue, 1910 Census

COUNTRY OF ORIGIN	FOREIGN BORN	FOREIGN STOCK (2ND GENERATION)
Russia	418,370	655,733
Austria	329,418	494,629
Germany	190,096	513,466
Hungary	2,637	4,005
England	1,484	1,848
Canada	471	636
France	143	188
Romania	137	188
At sea	124	162
South America	101	112
Other categories[a]	800	1,327
Mixed foreign[b]	—	35,366
Total	943,781	1,707,640

[a]There are twenty-nine of these, ranging from "Country not specified," 234 and 441, and "Europe not specified," 78 and 179, to "Montenegro," 2 and 2, and "Atlantic Islands," 1 and 2.

[b]*Mixed foreign* means, according to the census, native whites whose parents were born in different foreign countries. In this table, one or sometimes both parents spoke Polish as a mother tongue. Children of one foreign-born Polish speaker and a native-born American are not recorded here.

speakers, 75 percent from Austria, more than 12 percent from Hungary, and 10 percent from various Balkan states; and Bulgarian, almost 60 percent from Bulgaria, 30 percent from Turkey, and nearly 9 percent from Hungary. The distribution of Yiddish and Hebrew speakers will be discussed later.

Table 8.2 looks at the mother tongue data from the 1910 census in a different way: It examines the languages spoken by immigrants and their children from the three European empires and Hungary. Although the mother-tongue enumeration is certainly not exact, nothing else in the American census gives so relatively clear a picture of the complexity of ethnic and linguistic stocks in Eastern and Southern Europe. More than seventy years later, in the 1980s, ethnic unrest would be making headlines from the Baltic to the Soviet Republics of Georgia and Armenia, and other ethnic conflicts attracted attention all over Europe, from the Basque provinces of Spain, to the Italian Tirol, to the frontier area between Romania and Hungary, and that between Bulgaria and Turkey. Unlike many of these groups, the stateless Poles

Table 8.2

Foreign-Born White and Foreign White Stock, by Mother Tongue, from Russia, Austria, Germany, and Hungary, 1910

MOTHER TONGUE	FOREIGN BORN	FOREIGN STOCK (2ND GENERATION)
RUSSIA		
Yiddish and Hebrew	838,193	1,317,157
Polish	418,370	655,733
Lithuanian and Lettish	137,046	204,070
German	121,638	245,155
Russian	40,542	65,612
Finnish	5,865	8,861
Ruthenian	3,402	4,798
Other[a]	37,646	66,149
Total	1,602,752	2,567,535
AUSTRIA		
Polish	329,418	494,629
Bohemian and Moravian	219,214	515,183
German	157,917	275,002
Yiddish and Hebrew	124,588	197,153
Slovenian	117,740	174,943
Croatian	64,295	81,094
Slovak	55,766	110,829
Ruthenian	17,199	23,793
Russian	13,781	23,622
Serbian	11,618	13,304
Slavic, not specified	11,196	21,821
Italian	10,774	17,182
Other[b]	41,448	73,305
Total	1,174,924	2,021,860
GERMANY		
German	2,260,256	7,725,598
Polish	190,096	513,446
Yiddish and Hebrew	7,910	15,510
Dutch and Frisian	6,510	21,580
Bohemian and Moravian	6,263	17,382
Danish	5,232	9,766
French	3,131	8,271
Lithuanian and Lettish	1,486	3,840
Other[c]	20,297	115,073
Total	2,501,181	8,430,466

Table 8.2 (continued)
HUNGARY

MOTHER TONGUE	FOREIGN BORN	FOREIGN STOCK (2ND GENERATION)
Magyar	227,742	318,596
Slovak	107,954	168,636
German	73,338	99,412
Yiddish and Hebrew	19,896	32,539
Romanian	15,679	16,613
Croatian	9,034	11,140
Slavic, not specified	6,837	9,367
Slovenian	5,510	7,919
Servian	5,018	5,613
Ruthenian	4,465	6,616
Polish	2,637	4,005
Other[d]	17,490	26,698
Total	495,600	707,154

[a]Includes 29,330 and 52,943 unknown.

[b]Includes 30,672 and 59,021 unknown.

[c]Includes 16,864 and 109,374 unknown.

[d]Includes 12,374 and 18,687 unknown.

brought to America a relatively high degree of national consciousness, a national consciousness nurtured by Poland's long history as a sovereign nation prior to its partitions. Some other groups had a less developed self-consciousness, identifying, as did many Italians, with village or province rather than nation or ethnic group. As one immigrant spokesman told a meeting in Brooklyn just after World War I: "Us Slovaks didn't know we were Slovaks until we came to America and they told us!"

As was the case for so many other groups, a few Poles came to colonial America, including a handful in the very first colony at Jamestown and a group of Polish Protestants who came to participate in William Penn's "Holy Experiment" in Pennsylvania in the first half of the eighteenth century. The first real impact of a Polish presence came during the American Revolution. About a hundred Poles with military experience came to participate in a fight for freedom that inspired many European liberals and revolutionaries. The two best known, both recruited by Benjamin Franklin in Paris, were Count Casimir Pulaski

(1748–79), a cavalryman killed during the Battle of Savannah, and Tadeusz Kósciuszko (1746–1817), a military engineer who survived to lead a liberal insurrection against the Czarist forces in 1794. Other early Poles to come to America were refugees from later failed rebellions or insurrections in 1830 and 1863.

The first Polish American parish in America, in Texas southeast of San Antonio, was established by German Poles from Silesia, in 1854 and named Panna Maria after the Virgin. They and other Poles who came to Texas to found similar villages in subsequent years, were part of the 10 percent minority of Polish immigrants who settled on the land. Most of them were not landowners but agricultural laborers. German Poles also pioneered rural villages and enclaves in at least eleven other states, most of them in the north-central region.

These, and the majority of the Poles who settled in American cities, were largely impelled by economic motives, *za chlebem* (for bread), as the Poles said, which Eva Morawska has modified to "for bread, with butter." As early as the 1850s some Poles referred to America as the land of the "golden mountains"; in the same decade the Chinese characters developed to stand for California may be translated in the same way. The mass movements of Poles can be seen as west-to-east movement within the partitioned areas, with first German Poles, then Austrian, and finally Russian Poles joining the exodus. As we have seen German Poles began to come in the 1850s and, all told, more than four hundred thousand landed, with perhaps seven-eighths of them arriving in the nineteenth century. The eight hundred thousand Poles from Austrian Galicia almost all arrived between 1890 and 1914—about half in the 1890s, the rest in the twentieth century. The emigration from Russian Poland was predominantly a twentieth-century phenomenon. More than three quarters of the eight hundred thousand from that territory arrived after 1900. Thus we are talking about from two to two and a half million ethnic Polish immigrants, of whom perhaps one-third returned to Poland. A Polish folk song, "When I Journeyed from America," describes the return of a foundry worker; successive stanzas tell of his work, his departure from New York, arrival in Hamburg, passage across Europe, and arrival in Krakow:

There my wife was waiting for me.
And my children did not know me,
For they fled from me, a stranger.
"My dear children, I'm your papa;
Three long years I have not seen you."[6]

Most Poles, of course, stayed, and settled primarily in the "rust belt." Victor Greene, the leading scholar of Polish America, has estimated first and second generation Polish American urban settlement as follows:

Table 8.3

Seven Leading Polish American Population Centers, 1905 and 1920

CITY	1905	1920
Chicago	250,000	400,000
New York	150,000	200,000
Pittsburgh	70,000	125,000
Buffalo	70,000	100,000
Milwaukee	65,000	100,000
Detroit	50,000	100,000
Cleveland	30,000	50,000

While the primacy of Chicago's Polonia was early established and remains unchallenged, the incidence of Poles in some smaller centers, such as Buffalo, was much higher, and in a few smaller cities, such as Hamtramck, Michigan, Polish Americans were the predominant ethnic group.

Polish immigrants often had the lowest or next to the lowest jobs. One anonymous letter from Brooklyn in a collection of contemporary letters sent back to Poland complained bitterly about conditions:

> What people from America write to Poland is all bluster; there is not a word of truth. For in America Poles work like cattle. Where a dog does not want to sit, there the Pole is made to sit, and the poor wretch works because he wants to eat.

Yet, as the writer complained, this was not the majority report. Most Poles were induced to come to America by other Poles: priests, relatives, and friends. Greene reports that in 1908, for example, "virtually all" incoming Poles told immigration officials that they were joining family or friends.

As noted, the words *Polish* and *Catholic* are all but inseparable in describing the American ethnic group, but Polish American Catholicism was different. Poles often respected and followed their clergy—the beginnings of the Silesian settlement in Texas had been prompted by the Franciscan Leopold Mozygemba—but they also fought fiercely with them about lay control and other matters. One unfortunate priest,

Joseph Dabroski, was the intended victim of an assassination plot in rural Wisconsin in 1870—disgruntled parishioners put hollowed-out logs filled with gunpowder in his woodpile—and was physically thrown out of a Detroit church by women parishioners when he tried to replace a popular priest. But Poles—and some Polish clergy—had a more fundamental quarrel with the American Church hierarchy—what many of them called, bitterly, "One Holy, Irish, Apostolic Church"—which did not, they felt, represent their interests. One Polish-born priest, Rev. Wenceslaus Kruska of Ripon, Wisconsin, put it nicely in a famous (or infamous, depending on one's point of view) article he wrote just after the turn of the century, "Polyglot Bishops for a Polyglot Church." He argued, "If a diocese is polyglot, the bishop must be polyglot, too." He went so far as to insist that a bishop without language skills who took his chair in such a diocese was committing a mortal sin. What Kruska and many other Poles wanted was *equal* treatment within the ecclesiastical hierarchy, or, more pointedly, the appointment of Polish priests as American bishops. Kruska stayed within the church, but his statements seemed even more menacing to the established order because in the previous decade dissident Poles in Pennsylvania had begun the secession which became the only full-fledged schism to develop in American Catholicism: the formation in 1904 of the Polish National Catholic Church.

Although the standard history of American Catholicism writes off the schismatic church as arising out of a "quarrel over control of church property and ecclesiastical jurisdiction" the root of the matter was clearly Polish American nationalism and the feeling that American Poles were putting more into the church than they were getting out.[7] The man who led the schism, Rev. Francis Hodur of Nanticoke in the Pennsylvania coal region, insisted:

> The Polish people [should] control . . . all churches built and maintained by them; . . . the Polish people [should] administer their own church property, through a committee chosen by their own parishioners; . . . the Polish people [should] choose their own pastors.

By 1916 Hodur's church had more than thirty parishes and perhaps thirty thousand communicants and half a century later had a quarter of a million adherents. These were, of course, but a fraction of the five million Polish Catholics then in the United States. And, it should be noted, the American hierarchy did react, without ever admitting that it had done so. In 1908, just four years after the schism was formalized, a Chicago priest, Paul Rhode, was appointed the first Polish American Catholic bishop.

The devotion of most Polish Americans to the church was demonstrated by their strong support for parochial schools. By 1921 the 511 schools in the 762 parishes that constituted Polish America taught 219,711 pupils, about two-thirds of Polish Americans' schoolchildren. John Bukowczyk suggests that even this figure understates the influence of the parochial schools, as some immigrant children attended church schools for part of their education, especially while preparing for confirmation, and argues that, since classes were conducted largely in Polish, the system was "a veritable bulwark against assimilation."

Nationalism also was the crucial issue among Polish secular organizations which were long oriented more to Polish politics than to American. And Poles in America were joiners. It has been estimated that by 1910 some seven thousand Polish ethnic organizations existed in America and that two-thirds of Polish Americans belonged to at least one of them. The two largest, the Polish National Alliance (PNA), founded in 1880, and the Polish Roman Catholic Union (PRCU), founded in 1873, had some 220,000 and 188,000 members respectively in the mid-1920s. For a long time the two organizations were bitter rivals, a rivalry that focused on what kind of Poland each envisaged. The PNA leaders tended to favor a middle-class, secular Poland; those of the PRCU envisaged a Poland under clerical leadership or tutelage. Their opposition was not polar and both groups lobbied—and lobbied effectively—with Woodrow Wilson, who called for the establishment of "an independent Polish state" after World War I, a goal all the elements in Polish America could and did support. But beneath the religious and the secular conflict lay a deeper question: Were the immigrants who stayed in America essentially Roman Catholic Poles or Polish Roman Catholics? Eventually, they became Polish Americans, but only in the 1930s would that term become part of the name of an important ethnic organization in the United States.

Although, as we have seen, most Poles were working people, they were generally in industries which were not organized until the drives of the CIO in the mid-1930s and after. Poles were not hostile to trade unions as such. In mining, the one unionized industry in which they were represented in any numbers, Poles were enthusiastic and loyal trade unionists, but the leadership of their unions remained almost completely British American until after World War II. Industrial employers had learned to play ethnic groups off against one another, hiring one national group for one kind of job, a second for another kind of job within the same plant. And the ethnic slurs—Polak, Hunky—of earlier immigrants and their children did nothing to promote solidarity among the multiethnic work forces. Similarly, in politics, Polish Americans had far less impact than their numbers would suggest. Even in Chicago, where every eighth person was a Polish American, no person of Polish birth or ancestry was elected to Congress until 1920. There were a number of reasons for this, including a low level of participation in American politics, a preoccupation with the politics of Poland on the part of many activists, the fact that many were not citizens in an age when alien suffrage was rapidly disappearing, and the desire to hold on to power by ethnic groups already established. All these reasons contributed to the relative lack of Polish clout measured, in the traditional American way, by success in gaining elective office and patronage positions. But Polish American nationalists might well reply that since their influence contributed to the existence of a new Poland they had achieved much in American politics.

Eastern European Jews

In 1880 there were perhaps two hundred fifty thousand Jews in the United States, a few of them descended from Jews who had come during the colonial period, and most of the rest German Jews and their descendants.[8] Fewer than fifty thousand were from Eastern Europe. When, in 1924, Congress cut immigration from Eastern Europe down to almost nothing, there were perhaps four million Jews in the United States, more than three million of them Eastern Europeans and their children and grandchildren. Since neither the immigration records nor the census recorded religion, counting Jews is even more difficult than counting subject nationalities. Almost all Poles spoke Polish as a mother tongue, but Jews might record Yiddish—as most of the Jewish immigrants of those years surely did—or any one of a number of national languages: German, Russian, Romanian, and so on. Table 8.4 recapitulates the 1910 data on Yiddish and Hebrew as a mother tongue.

The Jews of Eastern Europe were impelled to migrate for two basic reasons: Like other contemporary Eastern European migrants they wanted to improve their standard of living; in addition, they fled from religious persecution that became more pronounced after 1881. The number of Eastern European Jews had increased dramatically during the nineteenth century, from perhaps 1.5 million to nearly 7 million, a rate of increase roughly twice that of the population generally. Their

Table 8.4

Country of Origin, Yiddish or Hebrew "Mother Tongue," 1910 Census

COUNTRY OF ORIGIN	FOREIGN BORN	FOREIGN STOCK (2ND GENERATION)
Russia	838,193	1,317,157
Austria	124,588	197,153
Romania	41,342	56,524
Hungary	19,896	32,539
England	13,699	15,100
Germany	7,910	15,510
Canada	1,434	1,541
33 other categories[a]	4,705	5,870
Mixed foreign[b]	—	35,368
Total	1,051,767	1,676,762

[a]This includes five of each generation from Japan.

[b]*Mixed foreign* means, according to the census, native whites whose parents were born in different foreign countries. In this table, one or both parents spoke either Yiddish or Hebrew as a mother tongue. Children of one foreign-born Yiddish or Hebrew speaker and a native-born American are not recorded here.

increased incidence was a factor in their persecution. Although few of these Jews were really well off, most were seen by their neighbors as having a better standard of living, and this made the poison of religious prejudice more effective. Some of that prejudice was fanned by churches and governments. The Russian government, particularly after the assassination of Czar Alexander II in 1881, became even more repressive, sponsoring pogroms and passing laws which restricted Jewish residential, occupational, and educational mobility. Decrees banished 20,000 Jews from Moscow in 1891, and later, more Jews from Saint Petersburg and Kharkov. In 1905 the outbreak of the Russo-Japanese War, the abortive revolution of that year plus a series of pogroms caused Jewish emigration from Russia to peak in 1906 at perhaps 150,000 persons. (In that same year about 275,000 Italians came; those two groups made up almost 40 percent of the 1.1 million immigrants of that year.)

Migration was physically difficult for the Eastern European Jews. Not only were distances to the ports of embarkation long, but serious legal difficulties often presented themselves. One example will have to suffice. Jews from the Ukraine might simply have gone to the Black Sea port of Odessa and embarked for the United States from there. However, most traveled west, often through Austria-Hungary—avoiding Romania where the border guards had a deserved reputation for brutality—to German ports. They did so because the sea voyage was longer and more expensive from Odessa than from Hamburg or Bremen, and Russian passport control was unpleasant, especially for young men of military age. Most Russian Jews who emigrated in these years probably did not have the necessary papers, so that most of them were crossing borders illegally. However, for a determined emigrant crossing a border rarely presents an insuperable problem. In addition, in this instance, Jewish emigration was good business for German shipowners, so the government often winked at illegal border crossing. The immigrants traveled across Germany on sealed trains as they were not welcome there. Albert Ballin's Hamburg-America Line alone was making an annual profit of ten million marks on the emigrant trade by the 1890s, and much of that trade was Eastern European Jews.[9]

The Jews who left Russia were not representative of the population there; as is usually true of emigrants they were young, with some seven out of ten between the ages of fourteen and forty. Those with industrial skills, usually in the needle trades, were more likely to emigrate, those in mercantile pursuits less likely. Arthur Goren has suggested that nearly a third of employed Russian Jews were mercantile, but only about a twentieth of all Russian Jewish emigrants were. Conversely only two-fifths of the Russian Jewish population was skilled while nearly two-thirds of Russian Jewish immigrants with occupations were skilled.

Apart from their religion, what most clearly set Jews off from other contemporary migrants was their great propensity to stay in the United States. There is a general consensus among scholars that about one Jewish emigrant in twenty returned to Europe, and some of those were sure to emigrate again later. Of all the other ethnic groups of the period for whom return rates have been calculated, only the Irish have a rate below 10 percent. Among other Eastern Europeans the remigration rates were higher: 20 percent for Lithuanians; 36 percent for Slovaks; 66 percent for Romanians; and a high of some 87 percent for several Balkan nationalities. A corollary of the permanence of Jewish migration was the large number of females who came, although fathers and/or older sons would often lead the way. Females have been calculated at 45 percent of Jewish migration; only the Irish, whose later migration contained large numbers of single women, had a higher rate, 55 percent. About 33 percent of Lithuanians were female, about 35 percent of Slovaks, 16 percent of Romanians, and just under 10 percent of several Balkan groups.

Unlike the relatively dispersed German Jews of midcentury, the

Eastern European Jewish immigrants settled overwhelmingly in New York and other cities in the Northeast and Midwest. Most came through Ellis Island and perhaps seven out of ten stayed in New York City. It has been estimated that a quarter of American Jews lived in New York in 1860, about a third lived there in 1880 and close to half in 1920. In the latter year about three-fifths of Jewish Americans lived in the region from Boston to Baltimore; in terms of incidence, Jews were about a quarter of New York's population, a tenth or more of the populations of Cleveland, Newark, Boston, Baltimore, and Philadelphia, and just under a tenth for Chicago. Most were members of the working poor and crowded into ethnic enclaves that were breeding grounds for disease and crime. In 1910 more than five hundred thousand Jews were wedged into tenements in the 1.5 square miles of New York's Lower East Side. The five- and six-story walk-up tenements, often without hot water and with only one toilet on a floor, were divided into three- and four-room apartments that were dark and often served as workplace as well as living quarters for families with four or more children and boarders as well. One 1908 survey showed that in a quarter of the apartments two people slept in each room, in half three or four slept in each room, and in a quarter there were five or more. Similar conditions prevailed in Jewish neighborhoods elsewhere and in most contemporary large city immigrant neighborhoods.[10]

That generation of Eastern Europe Jews lived in a Jewish world in which Yiddish was the medium of communication and most contacts were with other Jews. Goren quotes a 1905 observer who noted:

> Almost every newly arrived Russian Jewish laborer comes into contact with a Russian Jewish employer, almost every Russian tenement dweller must pay his exorbitant rent to a Jewish landlord.

This was a different pattern than that experienced by most other contemporary immigrants, who might have a *padrone* or straw boss of their own ethnicity but did not generally toil for fellow ethnic entrepreneurs. The vertical socioeconomic structure of the American Jewish community was heightened by the Jewish concentration in the garment industry. In New York City alone before World War I, perhaps three-quarters of the more than three hundred thousand garment workers were Eastern European Jews. Likewise, most of the industry's sixteen thousand factories, or shops as they were called, belonged to Eastern European Jews. One of the great industrial tragedies of the era, the

Triangle Shirt Waist fire of 1911, in which 146 workers, almost all young women, were burned to death or died leaping from high windows in the sight of horrified crowds, took place in one of those Jewish shops.

These conditions, and the fact that many of the urbanized skilled workers who came had already had exposure to socialist ideas, led to the relatively early organization of garment workers roughly a generation before other mass-production industries were organized. These garment worker unions were largely Jewish unions with Jewish leaders. Gary Fink's *Biographical Directory of American Labor Leaders* (1980) lists no fewer than thirty-nine foreign-born Russian Jews, four of them women; by comparison, only five German-born persons are listed, none of them Jews. The other great economic concentration of these Jewish immigrants was in retail trade, which accounted for perhaps a fifth of the gainfully employed. These retailers ranged from ragged pushcart vendors and peddlers to proprietors of substantial stores. From the earliest days of their migration there was thus a significant entrepreneurial element among the Eastern Europeans, but most were, in Mike Gold's memorable phrase, "Jews without money."

The culture of the Eastern European Jews, both religious and secular, was quite different from that of the Jews of Germany and other Western European countries and even more different from that which had been developed by the highly acculturated American Jewish community with its largely German and Iberian roots. The newcomers from Eastern Europe were poor, had a communal tradition that had been nurtured in the *shtetls*—largely self-contained rural communities—and their religious observances often had a messianic fervor foreign to the more staid American Jews, whether of the Orthodox, Reform, or emerging Conservative persuasion. In addition, the ideology of many of the newcomers, particularly those who had come from the cities of western Russia or Russian Poland, was socialist or Zionist or both. American Jewry, on the other hand, was essentially bourgeois and anti-Zionist. A great deal has been written, much of it by the grandchildren of the Eastern Europeans, about the conflicts between the "uptown" or German-American-Jewish leadership and the masses of "downtown" Eastern European Jews and their leadership. That these conflicts were real and deeply felt is beyond any question. That most of the German-American-Jewish leaders patronized the newcomers and were embarrassed by their squalor and their enthusiasms—religious and political—and sneered at their language—Yiddish—as a "jargon," and that the Eastern European Jews knew and resented this, are among the basic facts of the communal history of American Jewry.

But it is often forgotten that the existence of an established American Jewish community, no matter how condescending and patronizing it might be, was of great advantage to the Eastern European newcomers—an advantage most other contemporary immigrant groups did not have. The "uptowners" gave financial assistance right from the start, just as Jews in Germany had formed organizations like the Hamburg Jewish Committee for the Support of Destitute Jewish Emigrants. Even though some of the initial distaste of the uptowners for their coreligionists was stimulated at least in part by fears that resentment of the newcomers would heighten American anti-Semitism, both groups joined to fight a growing anti-Semitism in twentieth-century America. One important uptown leader, Louis Marshall (1856–1929), became in the process the first civil rights lawyer in American history, combating discrimination in the courts in cases that involved not only the rights of Jews but of Catholics, Asians, and blacks. This collaboration produced, in 1906, the American Jewish Committee, the first national organization that claimed—and still claims—to speak for all American Jews. In 1915, a rival group, the American Jewish Congress, was founded by American Zionists and drew its greatest support from the Eastern European Jewish community. Its leaders, however, were long established American Jews such as the native-born Louis D. Brandeis (1856–1941) and the Hungarian-born, American-educated Rabbi Stephen S. Wise (1874–1949). And, despite continuing differences, American Jews were unified in helping Jews overseas, particularly in Russia and Eastern Europe. The American Jewish Joint Distribution Committee raised and disbursed more than sixty million dollars for their relief during and immediately after World War I.

But most downtown communal activity was micro- rather than macro-organizational. Most characteristic were *landsmanschaftn,* organizations of persons who came from the same town *(landsleit).* These organizations were both sacred and secular. Most of the synagogues on the Lower East Side—one 1907 survey found 326—were basically for *landsleit,* some of them so small that they had difficulty in maintaining a *minyan,* the ten "adult" Jewish males necessary to hold formal religious services. Both these and secular *landsmanschaftn* performed the kinds of communal services as had the associational groupings of earlier immigrants: They helped greenhorns find jobs and learn the ropes, provided interest-free loans and sick and death benefits, established cemetaries or burial plots, and, not least, provided a familiar setting in an alien land.

The establishment of relatively large and contiguous urban enclaves of Jews transformed the face of American Jewry. While Jews had participated in American politics with some individual successes, they did not generally represent ethnic constituencies. Certainly none of the six Jews who sat in the United States Senate before World War I—David Levy Yulee, who represented Florida (1845–51; 1855–61); Judah Philip Benjamin and Benjamin Franklin Jones who represented Louisiana (1853–61 and 1879–85); Joseph Simon of Oregon (1898–1903); Maryland's Isidor Rayner (1905–12); and Simon Guggenheim from Colorado (1907–13)—was a surrogate for any large body of Jewish voters. This began to change in the twentieth century: The Congress that declared war in 1917, for example, contained six Jews in the House of Representatives. Four of them clearly represented ethnic communities in New York, New Jersey, and Chicago; one of them, the Socialist Meyer London, was the darling of the Lower East Side.

Many have written—and some still write—as if all Eastern Europe Jews were devoted adherents of their religion. Such was not the case with Jews any more than for any other large immigrant group. Many of the earliest immigrants were the most secularized, the least traditional, persons in their communities who had only what the great Berlin rabbi, Leo Baeck, called *Milieu-Frömmigkeit,* piety by association. A distinct minority were agnostics or atheists, who scandalized most of their neighbors with such goings-on as Yom Kippur balls. Rabbinical authorities in Eastern Europe did their best to discourage emigration to what one called the *trefa* (ritually impure) land. Another rabbi, who actually came to America to see for himself, went back after writing a tract designed to inhibit immigration, which Jonathan Sarna has recently translated as *People Walk on Their Heads.* A recent writer in the *American Jewish Yearbook* has noted that the immigrants conspicuously neglected Jewish education (while eagerly embracing the American public school system) and usually failed to provide adequate *mikvoth,* ritual baths for women. Joseph Blau's conclusion on this matter is judicious: "Accounts of the extreme religiosity of the immigrants of the 1880–1914 wave [have been] greatly exaggerated."

The enthusiasm with which most Jews participated in American education is one of the hallmarks of American Jewish history and a key to their relatively rapid upward social mobility. As a people of the book, Jews had traditionally respected the learned man in ways that some other cultures do not. In addition, the bar mitzvah, the traditional rite of passage by which young Jewish men were publicly proclaimed adults, required the candidate to read aloud a passage from the Hebrew scriptures, so that most Jewish men achieved at least marginal literacy. Although the great flowering of Jewish intellectual activity in the United States would be largely a product of later generations, even by the 1920s Jewish academic success was so pronounced that Harvard, Yale, and other elite institutions were establishing quotas to keep the number of Jewish students below a certain level. Nevertheless, Jews were already overrepresented in these colleges, as they soon would be in certain of the learned professions, notably law and medicine.

One Jewish American writer, Abraham Cahan (1860–1951), a graduate of a Russian teacher's college before he came to America in 1882, was influential in both Yiddish and English at the very height of the Eastern European migrations. His novel *The Rise of David Levinsky* (1917), which was first published in Yiddish, is one of the most important immigrant novels. He is, however, best known for the *Jewish Daily Forward,* a Yiddish newspaper, which he founded in 1897 and edited for the rest of his life, save for an early four-year interval. Its peak circulation of 175,000 made it the largest foreign language daily of its time—during World War I there were four other Yiddish dailies in New York City—and its influence on the Jewish masses is hard to overestimate. Anthologies today almost always print examples from its delightful letters-to-the-editor column, called a *Bintel Brief* (a bundle of letters), which include answers that are a guide to life in America. (That some of the letters were fabricated in the *Forward* office is beside the point.) What is too often ignored is the fact that the *Forward* was a socialist paper. While the following twenty-fifth birthday tribute to the *Forward,* by its longtime managing editor, B. Charney Vladeck (1886–1938), is obviously biased, it nevertheless gives some notion of the role the newspaper played in the community.

> The *Forward* steadily grew to be a great leader of the Jewish masses. Conditions in New York shops fifteen and twenty years ago were appalling—long hours, unsanitary lofts, brutal treatment on the part of the employers and foremen and utter lack of self-respect and self-confidence. Under the leadership of the *Forward* the scattered and demoralized masses of Jewish immigrants in New York and elsewhere began to acquire spirit and to organize. . . . The *Forward* led every strike. It served the purpose of a large trumpet which warned of danger and summoned help. It was with the working immigrant in his shop, on the picket line, at his home. It collected money for strikers, and it created for them a favorable public opinion. It lifted the Jewish immigrant from the position of

> a slave and competitor to the American working man, to the position of leader and forerunner in the American Labor Movement.

The *Forward* survives today only as a pale weekly shadow of its former self. Its multistoried building on the Lower East Side has long since been sold and, symbolic of the ethnic succession among New York's immigrants, its exposed side is now emblazoned with a large sign that says, in Chinese characters, "Jesus Saves": The building is headquarters for a group seeking to Christianize Chinese immigrants.[11]

Many Jewish immigrants, however, never worked for others, and many more quickly left labor's ranks. Large numbers were petty tradesmen, and, in the garment industry, others soon became substantial entrepreneurs. Young immigrant Jews were among the first to show motion pictures in "theaters" hastily set up in vacant stores, and others soon began to make films. Most of the major studios in Hollywood were founded, or were soon controlled, by immigrant Jewish magnates such as Samuel Goldwyn, Carl Laemmle, and Harry Warner. There was no particular reason for Jews to become moviemakers: It was simply an economic niche, an opportunity, that developed just at the time when many ambitious Jews were, so to speak, "on the make." (The same could be said for the other prominent immigrants in the business, Greeks like the Skouras brothers or the theater magnate Alexander Pantages.)

Jews also attained prominence on the performing side of the entertainment world. There was an important Yiddish theater both in Europe and in New York, the latter noted for imperious stars such as Jacob Adler, Maurice Schwartz, and Molly Picon, who played in original Yiddish works as well as a repertoire in translation ranging from Shakespeare, Chekov, and Ibsen to turgid melodrama. Few of its personnel ever made the transition to the English language stage and many did not want to. But from Broadway and Tin Pan Alley to Hollywood and Vine, Jewish creative talent was very important. In the musical theater, for example, immigrants such as composer Irving Berlin and the singer Sophie Tucker were important pioneers, to be followed by second-generation figures such as Jerome Kern and Fanny Brice. And a list of important Jewish American comedians, many of whom started in vaudeville, would have to begin with the Marx Brothers and go on almost indefinitely.

One other area of particular Jewish American influence must be noted: the burgeoning field of social work and social reform. Here were joined together middle-class children of German Jewish immigrants, such as Julia Richman (1855–1912) and Lillian Wald (1867–1940), and immigrant Eastern Europeans such as Boris Bogen (1869–1929) and Jacob Billikopf (1883–1951). In no other professional field were Jewish American women so important. While some were clearly in the "lady bountiful" tradition of Anglo-American social work and none of those mentioned here had the professional training that began to become available around the turn of the century, their role as mediators between the immigrant masses and middle-class society cannot be overestimated. Belle Moscowitz (1877–1933), today best remembered as a "brain truster" for the Irish American politician Alfred E. Smith, is one such person. She had previously been a worker for the Jewish Educational Alliance and an investigator of Lower East Side dancehalls. She also became an expert on factory legislation and was employed for a time as an official of the famous "protocol" that brought peace—or at least an armed truce—between employers and employees in the New York garment industry.[12]

Hungarian Americans

Hungarian—Magyar—immigrants are, in one sense, more representative of emigration from Eastern Europe than are either Poles or Jews. Males were highly predominant among the Magyars (about two-thirds), and nearly half of them returned to Hungary, some of them, to be sure, to return again to America. Unlike most Eastern European nationalities, Magyars are found in significant numbers in only one nation, Hungary. The data from the 1910 census show, in Tables 8.1 and 8.2, that more than 99 percent of those immigrants claiming Magyar as a mother tongue emigrated from Hungary and that, conversely, Magyars were a minority (46 percent) of immigrants from Hungary.[13]

There was a small Hungarian presence as early as the American Revolution, when several professional soldiers came, including the cavalryman Mihály Kováts de Fabricy, killed in the battle of Charleston in 1779. Later, there was an occasional entrepreneur such as Agaston Haraszthy (1821–69), who introduced Hungarian Tokay grapes to California and became a pioneer winemaker. But the first Hungarians to arrive in any number were political refugees, followers of Lajos Kossuth after the failed revolutions of 1848. Kossuth himself, after a triumphal tour of the United States, returned to Europe, but hundreds stayed. Many of them, military veterans, were among the roughly eight hundred Hungarians who served in the Union Army during the Civil War.

These men left few traces after the war and established no communities. Hungarian America really began with labor migration in the late nineteenth century.

Worsening economic conditions in Hungary after 1880 and the attraction of relatively high-paying jobs in the United States drew, first of all, members of ethnic minorities in Hungary to the United States. After some of these began to send home remittances, and the worsening economic conditions extended to the better-off Magyars, the latter, too, began to come to America. Hungarian statistics show that in 1899 only one emigrant in four was a Magyar speaker, but by 1903 they comprised a majority. They, too, sent home remittances, and these began to transform the rural economy of their local regions. In one county alone, Veszprém, more than half a million dollars were received from America in 1903. While some of this surely went to buy railway and steamship tickets, some settled overdue taxes, reduced mortgages, bought land, and improved old houses and built new ones—all of which, of course, increased the attractions of America for those who remained.

This Magyar migration was short lived, extending from the late 1890s to the outbreak of World War I. In that time more than 450,000 Hungarians came. Most were under thirty, and the immigration data show that more than 88 percent were literate, about 30 percent higher than the rate for Hungary as a whole. Coming largely without industrial skills, they, like members of most other Eastern European ethnic groups, took dirty, dangerous jobs at wages that were low for America but high for Hungary. Initially they worked long hours, spent little, and saved relatively large amounts to send or take back home with them. Like so many others, most Hungarians clearly came intending to sojourn, and obviously many did. Many others stayed, whatever their original intention, and soon began establishing families and communities. While many Hungarians started in coal mining, more eventually worked in heavy industry in the Northeast and Middle West. Ohio in general and Cleveland in particular became a focal point of Hungarian American settlement, with churches, both Catholic and Protestant, and beneficial organizations in the typical American immigrant pattern we have seen before.

Because of the superb work done by the Hungarian scholar, Julianna Puskás, we can use a Hungarian American extended family case history to illustrate concretely an important and usually neglected aspect of the phenomenon of immigration to America—the complex pattern of relationships and residence that a pattern of sojourning could create. No one knows enough to say precisely how representative the family experience described below was, but it was clearly not a unique instance.

Puskás tells us of one Lajos P., born into a farming family in 1883, the youngest of seven children. In 1903 he left for America (he may have been there before), where he already had a sister. He went to West Virginia and worked in mines there. He also encountered there a young woman from his own village, Lea L., who bore their illegitimate daughter, probably in 1905. By 1908 Lajos P. had abandoned them and returned to Hungary, where he married Hermina A. They lived with his parents on the family farm and had two sons, Jozsef P. (1911) and Lászlo (1912). In 1913 Lajos P. and his wife returned to the United States, but left their sons with the grandparents. Lajos and Hermina first returned to West Virginia, but then moved from job to job. The couple had three more sons, each in a different town: Lajos, Jr., in Placement, New Jersey (1915); Ferenc (1917) in Philadelphia, and István in New Brunswick, New Jersey (1919). Shortly after that the couple and their three American-born sons returned to the native village in Hungary, and the family of seven was reunited. With money saved they built a house and purchased about an acre and a half of land. The middle son, Lajos, the eldest born in America, attended a university and became a Protestant minister. The other four completed only elementary school. In 1938 and 1939 the two youngest boys born in America returned to the United States. As they were native-born citizens, they were able to obtain American passports and were neither reflected in the immigration data nor affected by the quota system. At least one of them served in the U.S. Army during World War II, both married Hungarian American women, and they and their descendants still live in the United States. In the meantime, Lea L., the mother of Lajos P.'s illegitimate daughter, Julianna, had returned to their common village in the early 1920s and brought their daughter with her. As Puskás puts it, "The village knew who the father was." Lea married a man from the village and they had four children. Julianna also married, but the marriage was not a success; in 1937 or 1938 she remarried and, soon after, used an American passport to return to her native country and shortly thereafter her husband was able to join her. As the husband of an American citizen, he was admissible without reference to the quota.[14]

As this complex family tale demonstrates in petto, the notion that people either came or did not come is far too simple. Generalizing about this Hungarian village she studied, Puskás notes that the migration of couples and family from there was unusual. A more typical pattern was for the siblings in a large family to migrate serially at intervals of a year or more; most were either married men or single young men and women. If the single immigrants got married in America, as most of them did, they were more likely to stay there. The married men most often returned to their families in Hungary. If a couple came to America, they often, as Lajos and Hermina P. did, left children with grandparents and returned with their American-born children.

In the subsequent decades very different kinds of immigrants came from Hungary. In the 1920s and 1930s refugees from the Horthy regime and later from Nazism came to the United States. Perhaps fifteen thousand in number, these included some very distinguished intellectuals. Leo Szilard (1898–1964), the physicist who encouraged Albert Einstein to write President Franklin D. Roosevelt about the necessity of the United States developing an atomic bomb, was only one of a number of distinguished scientists from Hungary who aided the United States during World War II and after. Others included two leading atomic physicists, Eugene Wigner (1902–95) and Edward Teller (1908–), and the distinguished mathematician and game theorist, John von Neumann (1903–47). There were also a number of important figures in the world of music and the theater, including one of the twentieth century's greatest composers, Béla Bartók (1881–1945), Hungary's most noted playwright, Ferenc Molnár (1878–1952), and five men who made their reputations as conductors of American symphony orchestras: Fritz Reiner (1888–1963), George Szell (1897–1970), Eugene Ormandy (1899–1985), Antal Doráti (1906–1988), and Georg Solti (1912–97). Many of these, of course, were Hungarian Jews.

In the years immediately after World War II some twenty thousand Hungarians were among the displaced persons and other refugees admitted to the United States, while after the failed Hungarian Revolution of 1956 some thirty-five thousand persons, many of them freedom fighters who had resisted the Soviet occupation forces, also came to the United States. Needless to say, few of these returned to Europe. While some have settled in places where long-standing Hungarian American communities exist, such as Cleveland or New York, many others settled, or were settled by various voluntary agencies, in other parts of the country, including the Sun Belt of the South and West.

Similar capsule histories could be written about most of the other immigrant groups from Eastern Europe mentioned at the beginning of this chapter. I say "most" only because some of them are still awaiting their first historian. And each history would be somewhat different from the others, just as biographies differ. For Finns, for example, one would note their settlement pattern in the Great Lakes region, and how many of them either came here as convinced socialists or were converted by others after they came. Although the stereotype of the Jewish immigrant radical is widespread and has some historical basis, on a per capita basis Finnish American radicalism was much more pronounced. But what these groups have in common may be as important or almost as important as their differences. All came at a time when backbreaking physical labor was the rule in American industry under conditions that present-day American workers can hardly imagine. In the steel industry, for example, well into the 1920s, many workers—mostly immigrant and black workers—worked 12-hour shifts. As the open hearths ran twenty-four hours a day, there were no shut-down days. Workers got every other Sunday off but at a price: a thirty-six-hour shift around that other Sunday.

Nor were immigrant women exempt from toil. Married Eastern European women made important economic contributions, largely by keeping boarders. One study of Johnstown, Pennsylvania, by the federal government before World War I showed that fewer than a third of all immigrant households drew their entire income from the husband's earnings. Nearly half relied on the combined earnings from husbands and boarders, while the rest had contributions from children and other sources. Eva Morawska reports that the typical income derived from boarders in that era was about five dollars a month per boarder, and that many kept five or more boarders. (Twenty-five dollars a month would be between two-thirds and three-fourths of a husband's earnings.) Her research shows that, in Johnstown, 10 percent of boarder-keepers had ten or more boarders, and one woman she interviewed reported that her mother had kept fifty boarders in a three-story house with four bedrooms.

Single immigrant women also scrimped and saved. Of the quarter of Johnstown's immigrant women who were single, Morawska reports:

> They stayed with relatives or boarded with fellow immigrants from the home country. For the single woman, the source of income was housework, cooking, or working in the cigar factory . . . or in the match factory. . . . If they could read and write, young immigrant women could work as clerks and cashiers in the neighborhood stores. Young women working as housemaids and servants received $2.00 to $3.00 a week with board. In the local cigar and match factories, a young foreign woman could make as much as

> $4.00 to $5.00 a week. In a store, if she was not working for a relative who paid her nothing, earnings were also about $4.00 or $5.00 a week. According to an estimate made by the local newspaper in 1913, a single American woman needed an absolute minimum of $10.00 a month to support herself ($2.50 for rent, $5.00 for food, $1.50 for clothes, $.75 for church and other purposes, $.25 for insurance.) A young immigrant woman, spending no more than $2.00 for a "cot with the family" in the foreign colony and about $3.00 for food plus the maximum of $1.00 for clothing, $.25 for insurance, and $.25 for church, needed even less. A thrifty East Central European housemaid could then, have saved $50.00 to $90.00 a year. . . . The cigar and match factories, if they worked steadily for at least ten months a year, allowed their young women employees willing to reduce spending to the bare minimum to save $90.00 to $130.00 annually.[15]

All these groups had their immigration process short-circuited, first by the outbreak of World War I, and second, by the curtailment of the immigration of their groups by the American government in 1921 and 1924. This happened at a time when many of their communities were nascent, while sex ratios were still heavily male, and before the emergence of a large adult second generation. The Great Depression of the 1930s, in any event, would have brought one stage of American economic growth to a halt. These were, until very recently, the last European groups to come to the United States as economically motivated immigrants; most of those after them, such as the latter-day Hungarians, would be refugees.

9

Minorities from Other Regions: Chinese, Japanese, and French Canadians

During the century of immigration more than 90 percent of all immigrants were Europeans, and for many writers the words *European* and *immigrant* were all but interchangeable. Africans, so numerous in the formative years of the American colonies and the new nation, were kept out of the immigrant canon by definition; later, many authors did the same for Chinese, arguing that they were mere "sojourners" and thus not immigrants at all. These arguments have little attractiveness today; few scholars any longer deny the relevance of the Afro-American and Asian American experience for immigration history, and the once-overwhelming Eurocentricity of the field has weakened significantly. This chapter will examine the early experience of the first two Asian immigrant groups to come, Chinese and Japanese, along with the major discrete group from Canada, the French Canadians. Two other groups that might have been treated here, the Mexicans and the Filipinos, will be treated in later chapters because, although their immigration began before 1924, it continued thereafter and was not affected by the quota system—Mexicans because they came from the Western Hemisphere, Filipinos because they were considered American nationals. Also not treated here are English-speaking immigrants from Canada—what contemporary writers call "Anglophones." The 1910 "mother tongue" census showed them more than twice as numerous as French Canadians—781,000 to 385,000—but, as we have seen, many were immigrants who had merely paused in Canada en route to the United States. Even more important, except for clusterings in towns and cities along the border, and later in the National Hockey League, there were no English-speaking Canadian communities until, in recent years, they began to develop in retirement areas of Florida. Even more than Charlotte Erickson's English, these Canadians were invisible immigrants.

Chinese

If we do not count the ancestors of the Amerindians, who presumably crossed what is now the Bering Strait in prehistoric times, Chinese are the first immigrants from Asia. Although a few Chinese were present in Mexico in the seventeenth century, presumably having arrived on the annual Manila galleons, and others came to eastern U.S. ports in the late eighteenth and early nineteenth centuries, meaningful Chinese immigration to the United States begins roughly with the California gold rush of 1849. This identification of California—and America—with gold was so prevalent in the Chinese mind that the characters that came to stand for California in the Chinese language may also be read as "gold mountain." When, in the next decade, another gold rush took Chinese, some of them veteran miners from California, to Australia, the characters for that country could be read as "new gold mountain." It is beyond dispute that these Chinese and their immediate successors came, like so many Europeans, with the intention of sojourning and returning with a nest egg. Yet despite the similarity between European and Chinese sojourners, some scholars do not like to call these Chinese immigrants. What still bothers some scholars about nineteenth-century Chinese Americans is the false and essentially racist notion that they—and perhaps other Asians in that period—were, somehow, different from the other immigrants. It is a variant of the notion that while involuntary white persons brought to America were immigrants, involuntary black persons so brought were not. Such notions are simply no longer tenable. By the definition used in this book, which has gained growing acceptance among scholars in the field, it is now beyond dispute that Chinese were also immigrants. If any nineteenth-century migration qualified as a "change of residence involving the crossing of an international boundary," the ocean journey from Canton or Hong Kong to San Francisco or other western ports certainly did. And it is the basic principle of this book that all immigrants to the New World in historic times have faced the same kinds of challenges and that the things that various immigrant groups have in common are as important as those that differentiate them.[1]

Between the beginnings of Chinese migration in 1848 and the passage of the Chinese Exclusion Act in 1882, perhaps three hundred thousand Chinese entered the United States. As was the case with European sojourners, there was much coming and going, so that, in all probability, the Chinese American population hit an intercensal peak of perhaps one hundred twenty-five thousand in the early 1880s. The census data, along with some indication of Chinese American distribution within the United States, are given in Table 9.1.

The Chinese had a migratory tradition long before any came to the United States, chiefly involving migration to Southeast Asia—the region the Chinese call *Nanyang* or South Seas. At the beginning of the nineteenth century Western entrepreneurs, with the help of Chinese middlemen, began importing unfree Chinese labor to various parts of the plantation world, largely as surrogates for African slaves. This "coolie trade," as it came to be called, first brought Chinese to Trinidad in the Caribbean in 1808. The major New World destinations were Cuba and Peru, but few nations or colonies in the Caribbean and Latin America were untouched by it. It was a brutal and infamous system that in some ways was worse than slavery, in that some employers literally worked their coolies to death before their indentures ran out. This was particularly true in the guano islands off the coast of Peru, where that was the fate of the overwhelming majority.[2]

Table 9.1

Chinese in the Contiguous United States, 1870–1930

YEAR	UNITED STATES	CALIFORNIA	PERCENTAGE	OTHER WEST[a]	PERCENT	REST OF U.S.	PERCENTAGE
1870	63,199	49,277	78.0	13,554	21.4	368	0.6
1880	105,465	75,132	71.2	26,970	25.6	3,363	3.2
1890	107,488	72,472	67.4	24,372	22.7	10,644	9.9
1900	89,863	45,753	51.5	21,976	24.4	22,134	24.6
1910	71,531	36,248	50.7	15,686	21.9	19,597	27.4
1920	61,639	28,812	46.7	9,792	15.9	23,035	37.4
1930	74,954	37,361	50.1	7,522	10.0	30,071	40.1

NOTE: *Chinese* in the U.S. Census was a racial definition and included both immigrants and their descendants.

[a]*Other West* here means the states or territories of Oregon, Washington, Idaho, Montana, Wyoming, Colorado, Utah, Nevada, Arizona, and New Mexico.

There is no evidence that any coolies were ever brought to the United States and, after the Civil War, coolie contracts would have been unenforceable at law. In the 1850s American consular officials in China explained to Washington the differences between immigrants and coolies. Based on this and other information, a report of the U.S. House of Representatives in 1860 distinguished between

> the "Chinese coolie trade" . . . a servitude in no respect practically different from . . . the . . . African slave trade [and the flow to]

> California, a Chinese emigration which has been voluntary and profitable to the contracting parties. The discovery of gold in Australia divided the migration, which hastened to both places at the option of the immigrants themselves.[3]

The financing of emigration is a crucial problem. Well into the nineteenth century, as we have seen, Western Europeans indentured themselves to American employers in order to have their passage paid. Chinese, who in the early days of the migration expected to work for themselves in the "diggings," borrowed from Chinese moneylenders. A British official in China in the early 1850s reported to London that some Chinese were borrowing seventy dollars (fifty dollars for the ticket and twenty dollars for expenses) and obligating themselves to pay back two hundred dollars. Some would have been able to draw on family resources, while others would utilize various kinds of rotating credit mechanisms then prevalent in South China. In any event, hundreds of thousands of Chinese got the money to come, and, since the credit ticket and other devices described above persisted into the twentieth century, we must assume that most of the money owed was paid, although we are not clear about the methods employed, kinds of security accepted, and so on. Clearly there were defaulters, just as some indentured servants ran away and just as today some people don't pay their credit card bills, but shrewd lenders always try to build a certain margin for loss into their interest rates. If all the above seems relatively simple—and I think it does—it did not seem so to anti-Chinese American officials in the 1870s and 1880s or to generations of later historians who made the system—or rather a caricature of it—seem a sinister plot to undermine American standards of living and even the republic itself.

Emigration from China to America was, for almost a century, not so much Chinese as Cantonese emigration. Well over 90 percent of the immigrants of that era were not only from Canton in South China but from a very few counties centered on the Pearl River Delta there. And, although male migration was characteristic of many groups, among no large group of immigrants to nineteenth-century America was the sex ratio as skewed as it was among the Chinese. By the 1880s, according to the censuses of 1880 and 1890, Chinese males outnumbered females by more than twenty to one. (In Australia the imbalance was incredible: In Victoria in 1857 there were 25,421 Chinese males and just 3 females!)

Within the United States, as table 9.1 shows, Chinese were concentrated in the West in general and in California in particular, although the percentage in the rest of the United States grew steadily after 1870.

Within California there was concentration as well. Initially centered in the mining districts of the Sierras and their foothills, Chinese soon made San Francisco, the port of entry for most of them, the *dai fou,* or big city, with more than a fifth of all Chinese Americans recorded as living there by the 1880 and 1890 censuses. These probably significantly understate the Chinese population, as many were migratory workers who lived in San Francisco when not on the road. Wherever they lived, Chinese Americans were increasingly urbanites, and big city urbanites at that. In 1880 just over one in five lived in large cities, those with more than one hundred thousand population. By 1910 almost half of Chinese Americans did, and by 1940, more than seven out of ten.

In those large cities they lived almost exclusively in ethnic enclaves known as Chinatowns. Unlike the enclaves of European immigrants, the populations of Chinatowns of any size were almost totally Chinese. "Little Italies," by contrast, had much lower concentrations of Italians: In Chicago, for example, only a few blocks had a concentration of as high as 50 percent. Only the black neighborhoods of twentieth-century American cities have the kinds of concentration found in the big-city Chinatowns of the nineteenth century.

San Francisco's Chinatown was the first and most important: It was replicated in large cities across the United States as far away as Boston and, even though today there are more Chinese in New York than in San Francisco, the latter retains its cultural primacy. One of the remarkable things about San Francisco's Chinatown has been its geographical stability; in the 1850s an immigrant community was formed in the area centering on the intersection of Dupont and Stockton Streets, and for almost a century and a half of growth, earthquake, fire, and urban renewal it has remained in that neighborhood with only slight variation, mostly expansion. While Chinatowns have become tourist attractions, that is largely a twentieth century phenomenon, although some whites went to nineteenth-century Chinatowns for "thrills," including prostitutes and opium. Chinatowns were primarily places where Chinese Americans lived, worked, shopped, and socialized. They were overcrowded slum areas, but as such were not too different from other immigrant enclaves except that the well-to-do urban Chinese lived there too. The classic pattern of ethnic succession, by which one group moved out of the slums while another group or groups moved in, did not work for "colored" immigrants. And despite the restrictiveness of the larger society that kept Chinese confined, as it were, within the enclave, there were for Chinese, as for other ethnic groups, positive aspects to life in an ethnic community. In San Francisco and the other urban Chinatowns there were shops, services, communal organizations, and entertainment—all provided by and for Chinese.

Economically the Chinese were at first largely occupied in mining; in California alone, about a fifth of Chinese workers were so engaged as late as 1880. Another fifth in that year were laborers; a seventh, in agriculture; another seventh were in manufacturing, mostly of shoes and clothing; yet another seventh were domestic servants, and a tenth were laundry workers. The thirty thousand Chinese workers outside of California in 1880 were concentrated in mining, common labor, and the service trades. In the 1860s as many as ten thousand Chinese workers were engaged in building the western leg of the Central Pacific Railroad. Most Chinese on the railroad payroll seem to have received thirty-five dollars a month. Since food costs were estimated at fifteen to eighteen dollars a month, and the railroad provided shelter, such as it was, a frugal workman could net close to twenty dollars a month. Not all Chinese were laborers. Sucheng Chan has shown that hundreds of Chinese owned or operated farms and that they played a vital entrepreneurial role in California agriculture, introducing new crops and pioneering distribution systems.[4] As early as 1870, one Chinese farmer in Sacramento County produced a crop worth nine thousand five hundred dollars. Most Chinese, in agriculture or anywhere else, made much less. Most of those in manufacturing, for example, made a dollar a working day or less, but 1880 census data indicate that 8 percent of Chinese cigar workers in San Francisco made four hundred dollars a year or more, well above the sweatshop level. At the very top of the Chinese American economic pyramid were the merchants who became the power elite of the community. We know very little and are never likely to know much about them. Many came with some capital and presumably had ties with Canton mercantile houses. Merchants made money in a number of ways: by normal international trade and the importation of the exotic goods desired by the Chinese American community; many also had a "piece" of the credit ticket system and served as labor contractors. A merchant who provided labor to the Central Pacific or other employers of large numbers of Chinese often was able to make an additional profit by selling rice and other foodstuffs and opium to the employer or to the laborers themselves. I suspect, but cannot demonstrate, that there were proportionately more individuals of real affluence within the Chinese American community in the late nineteenth century than there were in any other contemporary immigrant group.

The communal life of Chinese America was distinctly different from that of most other immigrant groups in that churches were not major organizations. This was because the focus of most Chinese religion was the family. (There were, to be sure, a small but growing number of Chinese who came here as Christians or who were converted after they arrived.) For most immigrant Chinese the family association or clan was the most important organization. These associations united all those who had a common last name—all the Lees or Lis, for example—and thus presumably a common ancestor. In small Chinese villages the clans tended to be village or intervillage associations based on real as opposed to theoretical kinship, but by migrating—either to a large Chinese city or to America—Chinese broke the original village relationship. All Chinese in America were also—in theory—members of a district association based on the regions of Kwangtung Province from which almost all of them came. Without tempting to trace the evolution of these groups, suffice it to say that they were soon governed by a new institution made in America, a still-existing umbrella group called the Chinese Consolidated Benevolent Association, popularly known as the "Chinese Six Companies." These were clearly *landsmanschaftn* type of organizations, but with Chinese American variations. They were run by the Chinese American power elite, the merchants, and they became the spokesmen for the entire Chinese American community and its intermediary to the white establishment. When, for example, a congressional investigating committee came to San Francisco in 1876 to look into Chinese immigration, it was the Six Companies that hired Caucasian attorneys to conduct a "defense" of the community. And the word *benevolent* in its title was not mere window dressing: In an age when government assumed few of the obligations of what is now called the welfare state, the Six Companies, like other associations all over ethnic America, assumed them; it helped new arrivals find jobs and housing, fed the hungry, nursed the sick, buried the dead, and—a uniquely Chinese function—arranged for their bones eventually to be sent back to China for burial in the appropriate ancestral cemetery.

And, as is the case with most benevolent associations as well as with the contemporary welfare state, the Six Companies served to exercise some forms of what sociologists call "social control": That is, they encouraged their members to conform to certain community norms, with the implied loss of benefits or protection for those who deviated from them. The merchant-run association naturally encouraged Chinese to pay their debts and to pay their dues to the association, which helped defray the costs of the welfare system. It tried to see that every returning Chinese was checked at the dock to make sure that he was debt free and had paid his "tax": it was not a fail-safe system, but, since steamship companies often cooperated, it was at least partially effective. Since most Chinese planned to return to China, this presumed check on returnees was a real threat to the sojourner's security. The association also settled disputes between individuals and groups within the community and otherwise assumed roles that, in the larger community, were assumed by government. This led to the charge, not without truth, that the Six Companies were an "invisible government." The adjective *invisible* was plain silly: Its headquarters was one of the most prominent buildings in Chinatown. But anti-Chinese Caucasians gave the word a sinister spin that was inappropriate. Actually, they should have welcomed the association's role as it insured more order in the community than would have existed without it.

In addition to the establishment and public family associations and the Six Companies, there were the antiestablishment and private organizations known as tongs. Tongs became notorious as criminal organizations, with links to Chinese American crime—prostitution, gambling, and drugs. They used thugs known as "hatchet men," who soon became Americanized and used guns. But the tongs, about which we have very little reliable evidence, had other functions as well. The tongs parallel, and may have had direct relations with, the Triad Society, an anti-Manchu, antiforeign secret society that flourished in South China. Like so many revolutionary organizations elsewhere, it also had a criminal side. While the Chinese American tongs seem to have had some political connections—funds for Sun Yat-sen and other anti-Manchu leaders were probably funneled through them—they were primarily criminal.

In all these matters the structure of the Chinese American community was, in essence, a variation on the American ethnic pattern, showing differences in degree but not in kind. What makes the Chinese experience unique in American ethnic history was not what they did but what was done to them. What was done to them includes both discrimination and extralegal violence and a whole series of discriminatory ordinances and statutes from the municipal to the federal level. The details of this discrimination will be treated in the next chapter, but two specific federal statutes must be briefly noted here: the Naturalization Act of 1870 and the Chinese Exclusion Act of 1882. The first, which limited naturalization to "white persons and persons of African descent," meant that Chinese immigrants were in a separate class: They were aliens ineligible for citizenship and would remain so until 1943. (When other Asian groups came, they too were in this category, some until 1952.) The second, which was the first significant inhibition on free immigration in American history, made the Chinese, for a time, the only ethnic group in the world that could not freely immigrate to the United States.

The Exclusion Act, in effect, froze the Chinese community in its 1882 configuration, a configuration that included, as has been noted, a highly imbalanced sex ratio, which was characteristic of most American immigrant communities in the later nineteenth century. As Table 9.1 shows, the Chinese population of the United States went into a long decline that ended only in the 1920s. The act ossified the gender structure of the Chinese community for more than half a century, making it an essentially bachelor society and one in which old men always outnumbered young men. The resourceful immigrant community devised ways to replenish itself, however. Not only was there a significant but incalculable amount of illegal border crossing, but Chinese also created the elaborate system of immigration fraud the community called "paper sons," exploiting the combination of an anomaly in American law and a natural disaster. Although Chinese could not become naturalized citizens, the Fourteenth Amendment to the Constitution, which had been passed in 1868 to protect the rights of newly freed blacks, made "all persons born . . . in the United States" citizens. Chinese born here were thus citizens, although there were not very many of them. Such citizens could travel to China, marry, and have children. These children—but not their mothers—could come to the United States because they were children of an American citizen. The great San Francisco earthquake and fire of 1906 destroyed birth records. This enabled many Chinese to make fraudulent but successful claims of American citizenship. Some of these citizens traveled to China and brought back their own children, almost always sons. Others, however, stayed there and sold the "slots" their trips created to other Chinese who then brought over their own relatives. Victor and Brett de Barry Nee, in their brilliant book *Longtime Californ'*, interviewed some paper sons in the 1960s as part of their oral history of San Francisco's Chinatown. One told them:

> In the beginning my father came in as a laborer. But the 1906 earthquake came along and destroyed all those immigration things. So that was a big chance for a lot of Chinese. They forged themselves certificates saying they were born in this country, and when the time came they could go back to China and bring back four or five sons just like that! They might make a little money off

it, not much, but the main thing was to bring a son or a nephew or a cousin in.[5]

How the Chinese American community would have evolved if there had been no Exclusion Act is impossible to say, but there is reason to believe that it would have come to resemble other immigrant groups, in that those men who decided to stay and had some degree of success would have sent for a wife or gone back to China to get married. And, although we speak of Chinese America before World War II as essentially a bachelor society, large numbers of the "bachelors" were married men: Their wives, however, lived in China. One of the arguments for the Exclusion Act, an argument that would be used later against other Asian immigrant groups and against many European groups as well, was that they were "unassimilable" and/or did not wish to assimilate to American life and American standards. Since family formation was a major vehicle of acculturation and Americanization, it is obvious that American law helped to retard this process. But some nineteenth-century Chinese immigrants did both acculturate and assimilate into American life. Two examples follow, one from the lower social levels of the immigrant generation, the other from the very highest.

The first, Sing On, was born about 1860 somewhere in China. In 1873 the teenager somehow showed up in Montana Territory, where he supported himself and attended school. He then moved to Chouteau and then to Teton County, where he farmed 480 acres. In 1879, at his request, the Montana Territorial Assembly passed an act changing his name to George Taylor, which he later embellished to George Washington Taylor. In 1890 he married Lena Bloom, also an immigrant, from Sweden, and they had seven children, four boys and three girls. In 1917, when the Taylor family appears fleetingly in the national historical record, their eldest son, Albert Henry Taylor, was serving with American troops on the Mexican border, part of General Pershing's "punitive expedition" seeking Pancho Villa. The Taylor family history up to that time may be found in a formal petition of the Montana legislature asking that Congress grant citizenship to the elder Taylor, described as an "honorable . . . and upright man . . . opposed to anarchy and polygamy." Congress did not grant the petition, and I know nothing more of the Taylor saga.

The second was the most famous Chinese to live in nineteenth-century America and the first to publish an autobiography in English, which I had reprinted some years ago. Yung Wing (1828–1912) was the first Chinese to graduate from an American college and thus the forerunner of tens of thousands of immigrant students. Born near Macao, he came under the tutelage of Christian missionaries and attended one of their schools in Hong Kong. He arrived in America in 1847, entered Yale College in 1850, and received a bachelor's degree four years later. In the meantime he had joined a Christian church and, in 1852, became a naturalized American citizen. (This was contrary to the law, which then restricted naturalization to "free white persons," but naturalization procedures were chaotic until reform took place in Theodore Roosevelt's administration.) After graduation he was involved in large business transactions on both sides of the Pacific and at one time had the Gatling gun concession for all of China. For eighteen years, 1863–81, he performed missions for the Chinese government abroad, including investigating the coolie trade in Latin America. His most celebrated task was serving as codirector of the Chinese Educational Mission of 1872–81, which brought 120 men to Connecticut for a Western education, some of whom were able to follow in his footsteps at Yale. For the last six years of that period he was also assistant minister of China to the United States. In 1875 he married a native-born American citizen, Mary L. Kellogg, in a Christian ceremony. After 1881 Yung was out of favor with the Chinese government, which had closed down the educational mission, but remained active in transpacific commercial activity. In 1898, again in China, he applied to the American minister for assistance and, as a result, had his citizenship cancelled by the American secretary of state, John Sherman, who admitted that, since Yung had been a citizen for twenty-three years, it "would on its face seem unjust and without warrant" to cancel his citizenship, but that is exactly what he ordered done. This meant that Yung could not legally return to the United States. Nevertheless he did so in 1902—we don't know how—and lived in Connecticut until his death.[6]

George Washington Taylor and Yung Wing were clearly exceptional persons, but so are all the nineteenth century immigrants we know anything about: Most immigrants are simply represented statistically in what Abraham Lincoln called the short and simple annals of the poor. Some native-born Chinese Americans also demonstrated that the American environment had had its way with them. In 1895, for example, a group of Californians formed an American-style rather than a Chinese-style benevolent association and called it Native Sons of the Golden State, a title deliberately mimetic of an established and virulently anti-Chinese organization, the Native Sons of the Golden West.

The desire to be "American" can be seen in this clause from its initial constitution:

> It is imperative that no members shall have sectional, clannish, Tong or party prejudices against each other or to use such influences to oppress fellow members. Whoever violates this provision shall be expelled.[7]

Immigrant women, too, show the influence of America. One Chinese woman, the wife of a merchant, remembered, years later:

> When I came to America as a bride, I never knew I would be coming to a prison. Until the [1911] Revolution, I was allowed out of the house but once a year. That was during New Year's when families exchanged . . . calls and visits. . . . After the Revolution . . . I heard that women there were free to go out. When the father of my children cut his queue [Chinese men were required to wear a queue by the Manchu dynasty] he adopted new habits; I discarded my Chinese clothes and began to wear American clothes. By that time my children were going to American schools, could speak English, and they helped me buy what I needed. Gradually the other women followed my example. We began to go out more frequently and since then I go out all the time.[8]

While other Chinese were and remained classic sojourners—working and scrimping only to be able to improve their lot in China, as did countless thousands of European immigrants, such as the Hungarian couple Lajos and Hermina P., many thousands of Chinese remained in America either by choice or by necessity. All of them, no matter how deeply embedded they were in Chinese enclaves, which some of them hardly ever left, were affected by the American environment, and many soon began to relate to American as well as Chinese cultural patterns. That more of them did not do so was at least as much the fault of the American society that rejected them as it was due to the deep hold that Chinese culture had on most of its members, even the emigrants. American society seemed so closed to Chinese that even ardent Americanizers among its first generation leadership, such as Ng Poon Chew (1866–1931), often despaired of an American future for the American-born second generation. Toward the end of his life this pioneer Chinese American editor could write that perhaps "our American-born Chinese will have to look to China for their life work" as there were simply no appropriate jobs for educated Asians in white America between the world wars.[9]

Japanese

Although Chinese and Japanese were linked in the American mind as Oriental immigrants, and each group eventually suffered exclusion on the grounds of its race, their history and immigration experience are at least as different from one another as those of Germans and Poles. Unlike China, Japan had no long emigrant tradition; by the time Japanese began to immigrate to the United States in significant numbers in the 1890s, Japan was a nascent imperial power with aspirations to the leadership of East Asia, while China was a victim of imperialism, some of it Japanese.

The first group of Japanese to come were political refugees in 1869, who founded a short-lived agricultural colony near Sacramento. In the same year about 150 Japanese were brought to Hawaii to work on sugar plantations, where large numbers of Chinese were already employed. The Japanese immigration to Hawaii was renewed in 1884, and about thirty thousand Japanese were brought there under contract to plantation owners. After 1898, when the United States annexed Hawaii, many of these workers emigrated to the American West Coast. In the meantime, beginning in the 1890s, small but significant numbers of Japanese arrived directly from Japan at North American Pacific ports such as San Francisco, Seattle, and Vancouver. In the years prior to 1924 fewer than three hundred thousand Japanese came to the continental United States; many returned and many others made more than one trip. Table 9.2 gives the census data, along with an indication of Japanese American geographical distribution. As that table shows, the concentration

Table 9.2

Japanese in the Contiguous United States, 1900–1930

YEAR	UNITED STATES	PACIFIC COAST[a]	PERCENT	CALIFORNIA	PERCENT
1900	24,326	18,269	75.1	10,151	41.7
1910	72,157	57,703	80.0	41,356	57.3
1920	111,010	94,490	85.1	73,912	66.6
1930	138,834	119,892	86.4	97,456	70.2

NOTE: *Japanese* in the U.S. census was a "racial" definition, and included both immigrants and their descendants.

[a]*Pacific Coast* here means California, Oregon, and Washington.

of Japanese on the Pacific Coast and in California increased with each census, and would continue to do so until the United States government forcibly removed them to interior concentration camps in 1942. The Chinese, as we have seen, were more dispersed with each census.[10]

But the most important differences in the demography of the two communities do not appear in those tables: They were gender and age distribution, which are illustrated in charts 9.1 and 9.2.

These charts show that, as of 1920, the Chinese American community was not only still predominantly male (87.4 percent) half a century after significant numbers were recorded in the census, but also that it was still a bachelor society with the two largest cohorts being men in their fifties. The small female population (12.6 percent), conversely, was relatively young, with the two largest cohorts being girls under ten years of age. In the same census, by contrast, the Japanese American

Chart 9.1
Age and Sex Distribution of Chinese Americans, 1920

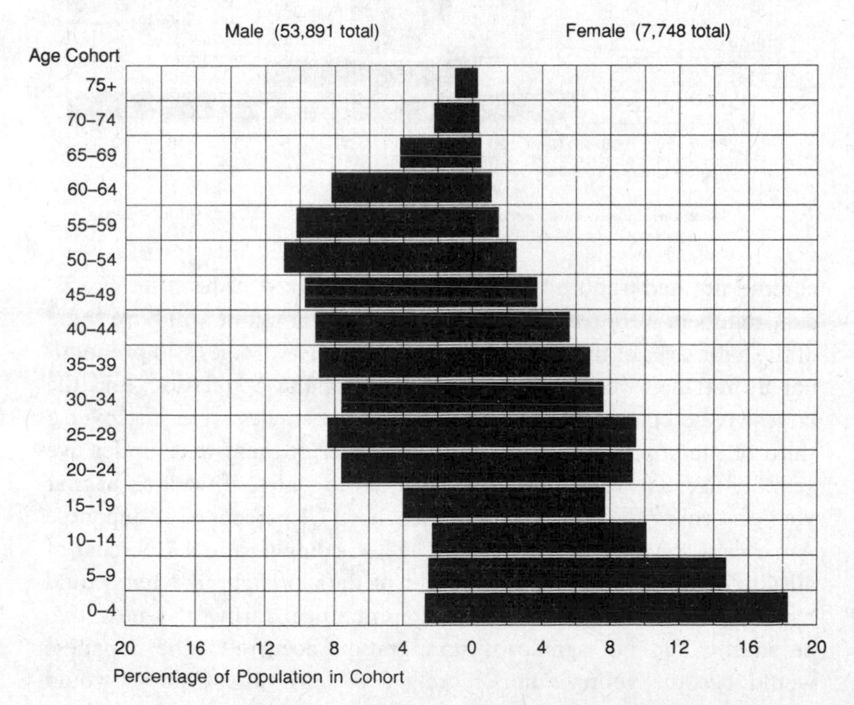

Source: United States Department of Commerce, Bureau of the Census, *1920 Census of Population* I, tables 4, 5,and 10, pp.157, 166–67.

Chart 9.2
Age and Sex Distribution of Japanese Americans, 1920

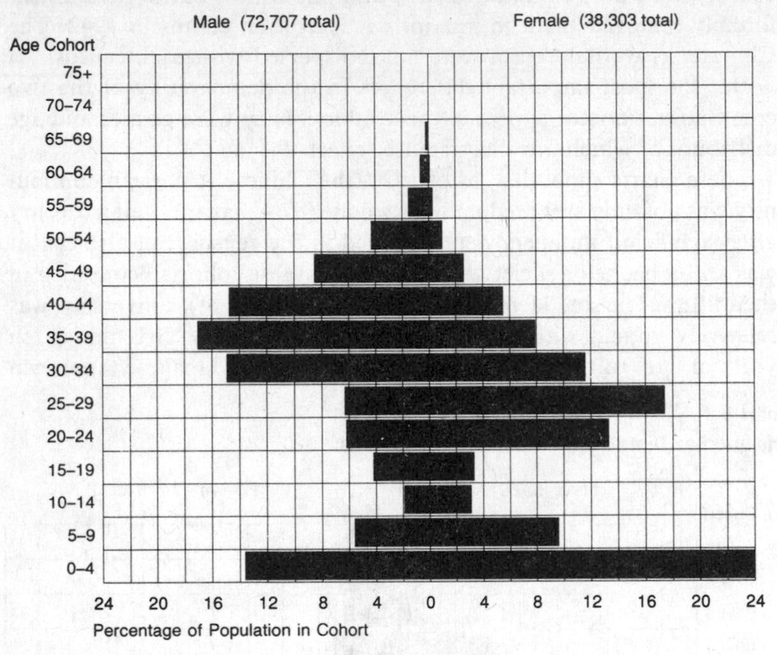

Source: United States Department of Commerce, Bureau of the Census, *1920 Census of Population* I, tables 4, 5, and 10, pp. 157, 166–67.

community had a more "normal" look, just two decades after significant numbers were recorded in the census, although it still bore some of the hallmarks of the immigrant bachelor society. Males still predominated, but they were "only" 65.5 percent of the population, and the largest male cohort was in its late thirties. Females were just over a third of the population, and almost a quarter of them were under five years of age, as were almost a seventh of the males. If we look at that youngest cohort comparatively, we see that 17.1 percent of all Japanese Americans were under five years of age, as compared to 4.7 percent of all Chinese Americans. This last set of data prefigured what would happen demographically in both communities during the next two decades, when no significant immigration occurred: The Japanese would become younger and more native born; the Chinese would change much more slowly. As of 1940 more than two-thirds of the 125,000 Japanese Americans were native-born citizens, as against a bare majority of the 75,000 Chinese Americans. Furthermore significant numbers of that majority of Chinese Americans—those with forged birth certificates and the "paper sons" those certificates brought in—were persons acculturated in China rather than America. How and why these differences came about, largely as a result of American law, is a fascinating story.

The early years of Japanese immigration to the American mainland were marked by a heavily male immigration. In the 1880s and 1890s many and perhaps a majority of Japanese worked at urban occupations but, by 1900 their economic focus was in agriculture where it remained. In the last decades of the century of immigration they were the only sizable ethnic group to have such a concentration. Initially most Japanese worked as agricultural laborers, in part replacing aging Chinese who were becoming more urban, in part meeting some of the ever-increasing demands of California agriculture for cheap labor. Soon Japanese began to acquire farmland. Large numbers of Japanese immigrants were from farm families who were being squeezed by the industrialization of Japan, just as German and other European peasants had been squeezed in the earlier nineteenth century. And, to be sure, the scarcity of arable land was even more pronounced in Japan than it had been in Europe. By 1904 Japanese were already farming more than fifty thousand acres in California alone; by 1909 it was more than one hundred fifty thousand acres and by 1919 more than four hundred fifty thousand acres. This latter figure represented only about 1 percent of all California agricultural land, but on it Japanese farmers, utilizing a labor-intensive style of agriculture, grossed about sixty-seven million dollars in that year, about a tenth of the total value of all California produce. While most were small family-style operations, one spectacular California entrepreneur, George Shima (1863–1926), ran a huge operation of the type that journalist-historian Carey McWilliams would later style "factories in the field." Born Kinji Ushijima in Fukuoka Prefecture, he came to the United States in 1889 with some capital he later described as less than one thousand dollars. Within twenty years he was the most famous Japanese in America, described in the press as the "Potato King," from the crop he introduced in California on the drowned islands of the Sacramento Delta. In 1913 Shima and his associates controlled twenty-eight thousand acres and through marketing agreements with other Japanese farmers sold the produce grown on other thousands of acres. His work force in that year was more than five hundred persons, including agronomists, boat captains, engineers, and common labor and its supervisors. When Shima died his pallbearers included David Starr Jordan, the chancellor of Stanford University and James Rolph, Jr., the mayor of San Francisco. Shima's success, and the early successes of most of his compatriots, came in Northern California, but soon Los Angeles became the quintessential city of Japanese America. By 1930 more than 35,000 Japanese lived there—more than a quarter of the nation's Japanese population.

John Modell, the historian of Japanese Los Angeles, has described the unique ethnic economy that developed there and demonstrated that "agriculture was the foundation of much of the enterprise and prosperity" of that ethnic community, and this holds true for Japanese communities all up and down the Pacific Coast and as far east as Colorado, and even to a small community in Florida. In Los Angeles, Japanese dominated the production of fresh green vegetables and some fruit crops, particularly strawberries. They not only grew the produce but organized the wholesale marketing of it for local consumption, while white wholesalers controlled most of the produce shipped elsewhere for sale. In the City Market of Los Angeles a few of the large Japanese-owned operations grossed a million dollars or more, but most were small stall operations. By the late 1930s there were radio programs in Japanese that reported market prices. Although initially almost all labor on Japanese agricultural enterprises in Los Angeles was Japanese, by the 1930s there were more Mexicans and Mexican Americans in the local agricultural labor force than there were Japanese, and Japanese growers' attitudes toward labor were similar to those of the entrepreneurial classes everywhere.[11]

Obviously, by American standards, Japanese immigrants and their children were highly successful. They made a significant contribution to the growth of California in particular and much of the West in general. But Japanese, whatever their other virtues, were not white people, and Californians and other Westerners lobbied vigorously and well-nigh unanimously for their exclusion beginning in the early years of the twentieth century. Although the details of the anti-Japanese movement will be treated in the next chapter, it is important to note here that if Japan had been a weak nation such as China, it is clear that something very much like the Chinese Exclusion Act would have been directed against Japanese in about 1906 or 1907. But Japan was not weak and few Americans were more aware of her military might than President Theodore Roosevelt, who raged, in a private letter, about the "foolish offensiveness" of the "idiots" of the California legislature, and who publicly pointed out that the "mob of a single city [he had the anti-Japanese mobs of San Francisco in mind] may at any time perform acts of lawless violence which would plunge us into war." Roosevelt successfully headed off anti-Japanese federal legislation and negotiated an agreement with Japan—the so-called Gentlemen's Agreement of 1907–8—which ended the immigration of Japanese laborers to the United States by having the Japanese government refuse to issue passports to such persons. The Gentlemen's Agreement did provide for family unification through a provision that allowed passports to be issued to "laborers who have already been in America and to the parents, wives, and children of laborers resident there."

The Gentlemen's Agreement forced Japanese immigration into an essentially female mode: Between its adoption and its abrogation by Congress in 1924, some twenty thousand adult Japanese women migrated to the United States. A few were wives whom Japanese immigrants had left at home, but most were newly married women. Some married men who went back to Japan for the ceremony but most were married in Japan by proxy to men they saw for the first time only after they landed in America. Often there would be a second ceremony in the United States. This custom, called "picture bride marriage," was consonant with Japanese tradition and was not unknown among European migrants to the United States, but it seemed to anti-Japanese Californians as a plot to flood the Golden State with Japanese. But, because of it, the Japanese American gender ratio was much less distorted than it was among Chinese Americans or among certain other heavily male ethnic groups.

The Japanese American community thus developed along sharply differentiated generational lines. The immigrant generation, the Issei (literally first generation) came in two echelons: most males between the 1890s and 1908; most females between 1918 and 1924. Their children, the Nisei (literally, second generation) were born into families in which the father was usually a decade older than the mother. When, in 1942, the government incarcerated the West Coast Japanese and got a detailed count, in the typical Japanese family children had been born in the years 1918–22, and the surviving male parents were in their late fifties, the females in their late forties.

Economically the population was heavily engaged in agriculture and other outdoor activities. In 1940 slightly more than half of all employed Japanese males and a third of working Japanese women worked in agriculture, forestry, and fishing. In the Pacific Coast states where most of them lived those sectors employed only one worker in eight. Unlike Chinese Americans, who by 1940 were more than 90 percent urbanized, just a little over half Japanese Americans were urbanized (54.9 percent), which was slightly below the national figure. Most of the rest of Japanese Americans were in wholesale and retail trade (about a quarter) and personal service (more than a sixth).

The cultural organizations of the immigrant generation were unique in that the most important of them were sponsored by the Japanese government. As part of its responsibilities under the Gentlemen's Agreement the Japanese government caused the Japanese Association of America to be founded, with headquarters in San Francisco, the cultural capital of both Japanese and Chinese America, with local and regional associations developing wherever significant numbers of Japanese settled. Tokyo delegated to this organization and its branches the issuing of documents necessary for immigrants who had continuing relations with Japan. The most important of these were the documents needed to get a passport for an existing wife or bride. It may seem that the Japanese government took what appears to be an inordinate interest in the lives of its citizens in America until one realizes two things. First of all, like Chinese and other Asians, they were aliens ineligible for citizenship in the United States, so that if Japan didn't look out for them, no government would or could. Second—and even more important from the point of view of Tokyo, which was not much given to worrying about the welfare of its peasantry—was the fact that it was convinced that its prestige as a nation was, in part, dependent on the respect given to Japanese abroad. The nightmare, for Japanese diplomats and officials from the 1890s on, was that there might eventually be a "Japanese Exclusion Act," along the lines of the Chinese Exclusion Act. When Japanese exclusion finally came, in a different form, in 1924, there were riots and even a suicide or two as part of Tokyo's deep resentment. A few Chinese diplomats, conversely, made the odd protest, but it was never a cardinal point of Chinese diplomacy. The redoubtable Wu Ting-fang (1842–1922), the Chinese minister in Washington around the turn of the century, for example, complained of the constant abuse Chinese Americans received in the press: "Why can't you be fair? Would you talk like that if mine was not a weak nation? Would you say it if Chinese had votes?"[12]

The Japanese government, through the Japanese associations, encouraged Japanese to acculturate: to adopt Western dress and, above all, to educate their children. One of the first of the diplomatic crises over the immigration issue between the United States and Japan was the so-called San Francisco School Board Affair. In 1906 the local authorities tried to force the Japanese students in San Francisco—then fewer than a hundred—to attend the already established segregated school for Chinese, which was quite proper under California law and the American Constitution as then interpreted. Only intervention by President Theodore Roosevelt got the San Francisco authorities to rescind their order. This stress on acculturation and education from the top at a time when most contemporary immigrant groups were indifferent—or worse—to anything more than a rudimentary education for most of their children, was an important influence within the community and was surely one of the factors that led, within a generation, to native-born Japanese having more years of education than the average American.

Religious organization among the Japanese Americans was diverse. A majority of the first generation continued to practice Buddhism (the church in America, and in other countries where there were overseas Japanese, was subsidized by Japan), but a large minority were or became Christians. Japanese attitudes toward religion tended to be more plastic than those of many immigrant groups whose loyalty to the religion of their fathers has been pronounced. Not atypical were these remarks of an Issei resident of Seattle:

> I told my children that it didn't matter whether they went to a Christian church or a Buddhist church but that they should go to some kind of church. Since their friends were going to the Methodist Church, they went there, but after I joined the Congregational Church, I transferred them to the latter.[13]

Eventually, the majority of the second generation became Protestant Christians; in Brazil, a majority of its Nisei became Roman Catholics; and in Utah a significant percentage of Japanese living there have become Mormons.

However successful the acculturational strategies pursued by most Japanese Americans eventually were, as far as Tokyo was concerned they were a failure. In the final analysis, by 1924, Chinese, whose acculturation was slow, and Japanese, whose acculturation was quite rapid, were dumped into the same ignominious category: They were not only "aliens ineligible for citizenship" but also, as such, inadmissible to the United States as immigrants. The two-decade delay in Japanese exclusion was in no way due to any qualities demonstrated by the Japanese American people but rather to the respect inspired by Japan's military power. A contemporary analogy can be seen in the way that South Africa treats Asians: Most—and all regular residents there—are nonwhites: Japanese businessmen, however, are "honorary" white people.

French Canadians

In 1881 Carroll D. Wright (1846–1909), then Massachusetts commissioner for labor statistics and a major figure in American reform movements, launched a diatribe at one immigrant group:

> The Canadian French are the Chinese of the Eastern States. They care nothing for our institutions, civil, political, or educational. They do not come to make a home among us, to dwell with us as citizens, and so become a part of us; but their purpose is merely to sojourn a few years as aliens, touching us only at a single point, that of work, and, when they have gathered out of us what will satisfy their ends, to get them from whence they came, and bestow it there. They are a horde of industrial invaders, not a stream of stable settlers. Voting, with all that implies, they care nothing about. Rarely does one of them become naturalized. They will not send their children to school if they can help it, but endeavor to crowd them into the mills at the earliest possible age.[14]

The comparison with Chinese seemed particularly ominous to French Canadian leaders—Congress was in the process of enacting Chinese exclusion—and, some months later, a delegation of editors, priests, and other community leaders met with Wright in an attempt to show him that their people were not as he had said. To his credit, when confronted with evidence, Wright withdrew some of his remarks, or at least tempered them. (Not surprisingly, no one attempted to speak up for the Chinese: Rather the thrust was that the French Canadians were not at all like the Chinese.)

What was it that caused an intelligent man like Wright to link the French Canadians and the Chinese, two groups that in most characteristics were quite dissimilar? To answer that question it is necessary to examine the history of French Canadian migration to the United States, chiefly to New England. It was, like most migrations to the United States economically motivated, and it was unique in just one respect: The French Canadians are the only ethnic group whose migration was chiefly accomplished by rail.

The French-speaking population of Quebec, which numbered only about sixty thousand as late as 1763, had multiplied by 1871 to more than a million persons. This growth was entirely due to natural increase: There was no significant French-speaking immigration to Quebec, and there had been steady outmigration—to other parts of Canada and to the United States—from the time of conquest. Although Quebec is a huge province, nearly six hundred thousand square miles, more than twice the size of Texas or France, the combination of hardscrabble soil and a short growing season made agriculture difficult. The same factors that caused millions of Norwegians and Swedes to migrate to the United States and some eight hundred thousand New Englanders to migrate to western farmlands in the thirty years after 1790 impelled hundreds of thousands of Quebecois to come to the United States between the end of the Civil War and the turn of the century. It has been pointed out that they did not so much displace native New Englanders as replace them. While a few went from Quebec agriculture to New England agriculture, and thus shifted from one marginal farming area to another, most went into the textile mills and other factories that were the heart of New England's economy. Just how many French Canadians came is all but impossible to determine, as enumeration at the land boundaries was at best erratic and the ebb and flow of individuals and families back and forth across the border only compounds the difficulty.[15]

The census data can give us some approximations of the numbers involved. Table 9.3 shows foreign-born and foreign-stock French Canadians as reported in the census between 1890 and 1920.

We can get an excellent notion of the attractions and realities of the New England mill towns for Quebecois in the 1870s from the first French-Canadian-American novel, *Jeanne la Fileuse (Jeanne the Mill Girl)* published in 1878 by an immigrant journalist, Honoré Beaugrand (1849–1906). Essentially what we would call today a docudrama, it is not great literature but is very useful for the student of history who wants to get a feel for how some contemporaries viewed industrial conditions. Jeanne, a sixteen-year-old orphan and her adopted family,

Table 9.3

French Canadians in the United States, 1890–1920

YEAR	FOREIGN BORN	FOREIGN STOCK	TOTAL
1890	302,496	224,483	526,934
1900	394,461	435,874	830,335
1910	385,083	547,155	932,238
1920	302,675	545,643	848,309

the Dupuis, leave Montreal at four in the afternoon and arrive at Fall River, Massachusetts by train at two the following afternoon. Their tickets cost ten dollars each. They have to change trains twice, at Boston and at Lowell, but the railroad has provided bilingual personnel to help immigrants, and it ships and delivers their baggage as well. One member of the family, a seventeen-year-old son, has already been working in Fall River for a year, and has arranged for jobs for the whole family and a flat in a company tenement. The family arrives with thirty dollars, with which they purchase furniture and pots and pans. Beaugrand provides a capsule history of French Canadians in Fall River; they had been coming there only since 1868 but already number some six thousand persons, about an eighth of the population. There is already an established immigrant community and a newspaper, *L'Echo du Canada,* which the elder Dupuis reads, and French Canadian tradesmen. Since the Dupuis are in company housing, their rent is deducted from their wages. Their assured jobs mean they can get instant credit from the butcher, the baker, and the grocer until payday at the end of every month.

The father and the older children, including Jeanne, begin work in the mills. The three youngest children, aged eight, ten, and twelve, must, under Massachusetts law, attend school a minimum of twenty weeks a year, but, as soon as that minimum has been met, they too go to work in the mills. Everyone arrives at the mill at 6:30 A.M. When the mills are running at capacity a sixty-hour week is the rule. Mill workers were paid about $1.22 a day; children, depending on age and ability, get from twenty-eight cents to a dollar a day. By the third monthly payday—even before the three youngest children are able to work—the Dupuis family is beginning to put money in the bank, money they believe will buy them a farm in Quebec.

In other hands the Dupuis saga could have been a savage exposé of the conditions of exploitive capitalism, but to Beaugrand the Dupuis are an American success story. To be sure, he admits that the work is not only drudgery but also that the conditions are somewhat like slavery, with the workers suffering from domination by foreigners, regimentation, and strict supervision. Beaugrand, however, looks at the positive side, as most French Canadian mill workers seem to have done at first as they fiercely resisted unionization. He sums up his treatise on conditions by insisting: "One is very unhappy the first weeks but when payday arrives, this unhappiness generally changes into satisfaction at the prospect of receiving regular wages, which is only natural."

Beaugrand shows that the immigrants lived, worked, and socialized almost exclusively among their own kind. They attend a Catholic church, Saint Anne's, with a Quebecois priest; the father not only reads a French-language newspaper but belongs to a French Canadian sodality, the Société Saint-Jean-Baptiste, while the already established eldest son is a member of the Cercle Montcalm, a local cultural organization. The novel, originally serialized in a French Canadian newspaper, was read and apparently approbated by the very population it describes.

An essentially family migration, like that of the Dupuis, moved hundreds of thousands of French Canadians from Quebec to New England between the 1860s and the 1920s and, in the process, helped to change the face of New England. Arriving between the onset of the Irish and the coming of Southern and Eastern Europeans, the French Canadians experienced constant cultural reinforcement, both from continued migration and remigration and from being able to visit their homeland almost at will. Thus their acculturation has, in some respects, been slower than that of other groups.

The French of New England's French-Canadian-American second and subsequent generations has persisted much more significantly, for example, than the Italian of New England's Italian Americans. As late as 1970 more than half of those nearly 2.6 million Americans who reported French as a mother tongue were described by the census as "native born of native parentage"; in the same census fewer than 15 percent of those reporting Italian as a mother tongue were so described. In the same census nearly a million New Englanders, about a tenth of the population, reported French as a mother tongue. Part of this language persistence must be attributed to the constant reinforcement from Quebec, but part of it is also due to the fierce determination of the Quebecois, after their conquest by the British, to keep their language alive.

> "Let us worship in peace and in our own tongue," they said. "Let us read and write in our own tongue. All else may disappear, but this must remain our badge."

With a century of this tradition behind them before they came to the United States, it would not be easily shed after a short train trip.

All over New England, but particularly in the mill towns where "les petits Canadas" were established, French-speaking priests were to be found, almost all of them missionary priests from Quebec who treated New England as other religious groups treated Africa or China. Not surprisingly the French Canadians clashed with the Irish-dominated hierarchy, as Germans had done before them and Poles would do after them. The key issues over which they struggled were, first, the appointment of non-French-speaking priests; second, Irish as opposed to Quebecois forms of worship; and, third, relative parish autonomy, traditional in Quebec but frowned upon in the Irish American church. Although both the French and Irish were seen in New England as "papist interlopers" by nativists, they rarely united to face their common enemies. French Canadian Catholics worshiped in three different kinds of parishes: the national parish, in which they were the dominant group and had a Quebecois priest, with the main or even all services in French; the mixed parish, with a large number (often a majority) of French Canadians, bilingual services, but no Quebecois priest; and parishes in which all services were in English despite a substantial French Canadian membership.

The worst conflicts were, classically, in the small- to medium-size mill towns and cities. In Fall River, for example, the very stronghold that Beaugrand chose to write about, the Quebecois missionary priest died in 1884 and the Irish bishop appointed an Irish successor despite the fact that French Canadians were some 85 percent of the parish. The Quebecois withdrew from the parish in a bloc and the bishop placed them all under an interdict. Rome, however, intervened, and urged the bishop to appoint a French Canadian. Similar struggles took place elsewhere. In Maine, during the incumbency of Louis S. Walsh, appointed bishop of Portland in 1906, the struggle was over attempts of the laity to control church property and again resulted in an interdict being pronounced against those who resisted his authority. On this issue Rome did not intervene and, although the Quebec hierarchy tried to, it was to no avail. The climax of these struggles came after the appointment of William Hickey as bishop of Providence, Rhode Island, in 1921. Here the struggle was over lay control of church property and, more ominously for most Quebecois, the Americanization of their parochial schools, forcing them to shift from French to English as the language of instruction. Around the turn of the century four out of ten parochial schools in New England were taught in French and had more than fifty thousand pupils. One French-Canadian-American newspaper went so far as to argue that:

> A parish without a church is preferable to a parish without a Catholic school for the excellent reason that where the second is lacking, the first often becomes useless.

Bishop Hickey's drive to Americanize was thus seen as an attack on the vital center of Quebecois culture in the United States. The cudgels were taken up by two important French Canadian national organizations: The Union St. Jean-Baptiste d'Amérique (USJB), the largest of them all, took the side of the bishop, while the Association Canado-Américaine (ACA) supported his rebellious parishioners. The last straw came when the bishop began to levy funds from unwilling parishes to support English-language parochial schools and other of his pet programs. His opponents called for more national parishes, bilingual schools—which meant, for them, French as the main teaching language—and the inviolability of parish funds. After an appeal to Rome failed, the most determined of the bishop's opponents tried unsuccessfully to get the Rhode Island courts to block the bishop's levies. After he won in the civil courts, the bishop hauled out the big gun: excommunication of the rebellious leaders. Despite extensive support from Quebec and the community, virtually all the leaders, including Elphège Daignault, president of the ACA, capitulated by early 1929.

The relatively slow acculturation of the French Canadians may also be seen in the political arena, where, for a long time, their potential influence, as reflected in the number of French Canadians elected to political office, was decidedly lower than either their numbers or degree of concentration would have predicted. Elliott R. Barkan, a leading authority on ethnicity and naturalization, has pointed out that French Canadians had one of the lowest naturalization rates of any American ethnic group. In 1910, for example, only 45 percent of all French Canadian males over twenty-one were naturalized and only 37 percent in the core area of New England, substantially lower than among other Canadians, Irish, English, and Scandinavian adult males. Such disparities continued for decades, and by 1930 a majority of French Canadians were still unnaturalized and their rates were well behind Italians and Poles, most of whom arrived later, as well as most other immigrant groups. In addition, the relatively slow increase in the number of native-born French Canadians, as reflected in Table 9.3, reflects the high emigration rate of French Canadians born here. The Canadian census of 1931 showed more than fifty thousand persons of French Canadian descent who were born in the United States living in Canada. This meant that the political impact of eligible French Canadian voters was lessened.

Not surprisingly, the rate of intermarriage by French Canadians was very low: Even the thought of such a thing was to many French Canadian leaders "a crime against God and a national abomination." (One suspects that for some the second was more important than the first.) In 1880, for example, 66 percent of second-generation French Canadians had both parents born in Quebec, and about 7.5 percent had either a foreign-born mother or father of a different ethnic group. The rest had one Quebecois and one U.S.-born parent, most of whom were certainly second-generation French Canadians. One 1926 study showed that in Woonsocket, Rhode Island, the "Quebec of New England," only 11 percent had married out, and that most of those spouses were Irish and presumably Catholic. Rates for Fall River, an early settlement area, were considerably higher: In 1880 the rate was 14 percent; it was 30 percent by 1912, 50 percent by 1937, and 80 percent by 1961, one-fifth of which were unions with non-Catholics. As Barkan notes, "even in the self-styled 'third French city in America,' " after Montreal and Quebec City, the process of acculturation was all but irresistible.

10

The Triumph of Nativism

During most of the century of immigration, immigration to the United States remained largely unfettered by government regulation. Most Americans understood that it was necessary to fill up the country and welcomed most of the foreigners who came. That welcome was not unalloyed, and three discreet phases of anti-immigrant activity, or nativism, can be ascertained. Each was a response to a specific aspect of immigration to the United States. The first phase, anti-Catholic, was aimed at Irish Catholic and to a lesser extent German Catholic immigration and flourished from the late 1830s to the mid-1850s. In some respects, it never completely died out. The second phase, anti-Asian, much more specific, was triggered by Chinese immigration and flourished from the early 1870s until the passage of the Chinese Exclusion Act of 1882. Successor movements would be directed against Japanese in the period from about 1905 to 1924 and against Filipinos in the 1920s and 1930s; this anti-Asian component of American nativism, like anti-Catholicism, has never entirely disappeared. Finally, the third phase, anti-*all* immigrants, began in the mid-1880s, when a movement for general restriction of immigration gained popularity, and finally triumphed in the Immigration Act of 1924, which dominated American immigration policy for the next forty years. There was never a time when nativist attitudes were not present in American society. They existed in the colonial period and are enjoying a revival today. And while nativists have always been able to point to some specific danger, real or imagined—Franklin's fear of the German language and culture taking over Pennsylvania, for example, or the Federalist fears of Irish and French political subversion—successful nativist movements have almost always been linked to more general fears or uneasiness in American society. When most Americans are generally united and feel confident about their future, they seem to be more willing to share that future with foreigners; conversely, when they are divided and lack confidence in the future, nativism is more likely to triumph.

In the years immediately after the War of 1812, capped as it was by Andrew Jackson's glorious and seemingly providential victory at New Orleans, self-confidence—this period was sometimes called the Era of Good Feeling—characterized the national mood. During the war itself Congress enacted a law prohibiting the naturalization of any Briton who had not formally declared the intention of becoming a citizen before the war started, but it was probably not effective and was repealed soon after hostilities ended. Encouragement of migration to the West and of immigration from Europe again became the order of the day. Congress did pass, in 1819, a law requiring that immigrants be enumerated at the various points of entry, but no bureaucracy was created to do the counting. That was left to customs and other port officials. Naturalization proceeded rapidly, and many states, especially the newer western ones, gave immigrants the right to vote and hold office even before they became citizens. Legal alien suffrage, in fact, would continue at various places in the United States until 1926 when the last state to allow it, Arkansas, changed its laws. Many states treated the declaration of intention to become a citizen as if it were citizenship, and the whole process of naturalization was haphazard: Federal courts, state courts, territorial courts, and even justices of the peace issued certificates of citizenship or, sometimes, simply declared persons to be citizens. There was also much fraud in naturalization, most notoriously in New York City, where the Democratic political organization, Tammany Hall, would routinely arrange mass naturalization "ceremonies," if they can be called that, before tame local judges, often in close proximity to election day.

In politics the Era of Good Feeling did not survive the controversial election of 1824, which Andrew Jackson's supporters insisted had been stolen by a corrupt bargain between aristocrats John Quincy Adams and Henry Clay. It was in just that period that the anti-Catholicism that had been endemic in American life almost from its beginning became what its historian, Ray Allen Billington, called the "Protestant crusade."[1] During most of the eighteenth century it had been the external threat from Catholic French Canada with its Indian allies that had stimulated the ever-present Protestant animus against Catholicism. And even though the American Revolution had been supported by Catholic France, for reasons of state, of course, not ideology, during the early republic the association of Catholicism and the pope with monarchy and reaction was a given of political debate. What was different about the 1820s and subsequent decades was that, with growing Irish and German Catholic immigration, Catholics and Catholicism could be seen as an internal threat, as subversive not only of republican principles but of the republic itself. In the 1820s anti-Catholicism took a back seat to antimasonry in American politics. The Masons, like the Catholics, were seen as secretive, monarchial, and conspiratorial, and there were, to be sure, many more Freemasons than Roman Catholics in America. The anti-Masonic party, which held a national convention and nominated a presidential candidate in 1832, was the first American third party. But antimasonry, as a political force, did not survive the 1830s, while anti-Catholicism was important at least into the 1960s.

When relatively large numbers of Irish and German Catholic immigrants, many of them desperately poor, began to arrive in the late 1820s and early 1830s, what had been a largely rhetorical anti-Catholicism became a major social and political force in American life. Not surprisingly, it was in eastern cities, particularly Boston, where anti-Catholicism turned violent, and much of the violence was directed against convents and churches. Beginning with the burning down of the Ursuline Convent just outside Boston by a mob on August 11, 1834, well into the 1850s violence against Catholic institutions was so prevalent that insurance companies all but refused to insure them. Much of this violence was stirred up by Protestant divines, ranging from eminent church leaders such as Lyman Beecher (1775–1863) to anonymous self-appointed street preachers. Billington notes that:

> Frequently crowds of excited Protestants, whipped to angry resentment by the exhortations of some wandering orator, rushed directly to a Catholic church, bent on its destruction. A dozen churches were burned during the middle 1850s; countless more were attacked, their crosses stolen, their alters violated, and their windows broken. At Sidney, Ohio, and at Dorchester, Massachusetts, Catholic houses of worship were blown to pieces with gunpowder. . . . In New York City a mob laid siege to the prominent cathedral of St. Peter and St. Paul, and only the arrival of the police saved the building. In Maine Catholics who had had one church destroyed were prevented from laying the cornerstone of a new one by hostile Protestants, and statues of priests were torn down or desecrated.

Nor were the priests themselves safe from public assault. The abuse in public was all but constant. At least two were badly beaten while on their way to administer last rites. In 1854 one Portland, Maine, priest described his ordeal:

> Since the 4th of July I have not considered myself safe to walk the streets after sunset. Twice within the past month I have been stoned by young men. If I chance to be abroad when the public schools are dismissed, I am hissed and insulted with vile language; and those repeated from children have been encouraged by the smiles and silence of passers by. The windows of the church have frequently been broken—the panels of the church door stove in, and last week a large rock entered my chamber unceremoniously about 11 o'clock at night.

If convents, churches, and priests were seen as something to attack, nuns were seen essentially as victims, first of the church's authoritarianism, and later as the targets of sexual abuse and worse by priests and bishops. There was a spate of "confessions" of former nuns—or in most instances of persons who claimed, falsely, to have been nuns. The first of these of any significance, Rebecca T. Reed's *Six Months in a Convent* (1835), was relatively mild, described nothing either illegal or immoral, and was concerned mostly with the penances she was allegedly forced to endure. But it quickly sold hundreds of thousands of copies and served as an encouragement for further confessions, which were soon numbered in the dozens. Far and away the most important were Maria Monk's *Awful Disclosures of the Hotel Dieu Nunnery of Montreal* (1836) and its inevitable successor, *Further Disclosures* . . . (1837). Although the first and more influential book was execrably written, it has been called, with good reason, the *Uncle Tom's Cabin* of nineteenth-century anti-Catholicism. Maria, or her ghostwriter, told a lurid and preposterous tale of secret passageways leading from a nearby priests' residence to the convent so that the fathers could exercise their carnal lust on the nuns, and of babies born to nuns there being strangled regularly by the mother superior. Maria herself, according to the tale, was seduced by a priest and made pregnant. Not wishing to see her child murdered, she fled the convent, was rescued, and taken to a hospital, and was eventually saved by a Protestant clergyman who brought her to the United States, where her story was written and published. That Maria was unmarried and pregnant was true. All the rest was fantasy, perhaps psychotic fantasy. Maria had never been a nun or even been inside the Hotel Dieu convent, and eventually even many of her supporters came to disbelieve her stories, especially after she again became pregnant. Her managers took most of the enormous profits from the books; she spent the rest; and in 1849 poor Maria was arrested for picking the pocket of a customer in a New York whorehouse and died in prison shortly thereafter.

It was against this background of religiously inspired anti-Catholicism, that the political and economic anti-immigrant attitudes of the pre–Civil War decades take on their full meaning. Many of the immigrants, as we have seen, were poor, others utterly destitute. The costs of maintaining the poor were mounting and were borne solely by the port cities and their states. In an effort to regain these costs, some eastern states passed modest head taxes—New York charged $1.50 for cabin passengers, Massachusetts a simple $2.00 a head—to be paid by the owners of the immigrant vessels. Not in themselves a great deterrent to immigration, they led the United States Supreme Court to lay down an important principle. In the *Passenger Cases* (1849) the court declared these state laws unconstitutional, holding that the right to regulate immigration under the commerce clause of the Constitution—Article I, Section 9, gives Congress the power "to regulate Commerce with foreign nations, and among the several states, and with the Indian Tribes"—was prescriptive. Thus even though Congress had passed no legislation concerning immigration, individual states could not tax it for any purpose, since, as John Marshall had put it earlier, the power to tax was the power to destroy. For the time being the court left the police powers of the states unimpaired: A state could, for example, quarantine a ship on which smallpox or cholera was raging.

This ruling only added supporters to an anti-immigrant bloc that was already flourishing in the country. As early as 1837 a nativist-Whig coalition was able to elect a mayor and council in New York City, and in Germantown, Pennsylvania, a Native American Association was formed that opposed foreign-born officeholders and voters. In New Orleans a similarly named organization denounced the immigration to the United States of "the outcast and offal of society, the vagrant and the convict—transported in myriads to our shores, reeking with the accumulated crimes of the whole civilized world." The major strategies of these movements, which coalesced in the 1840s and early 1850s in the American, or Know-Nothing, party, were to call for a change in the naturalization laws. The most common proposal was to require a twenty-one-year period for naturalization and bar the foreign born from holding any but minor local offices. Other measures proposed in

Congress including forbidding the immigration into the United States of paupers, criminals, idiots, lunatics, insane persons, and the blind. Although such proposals had much support on both ideological and economic grounds, they never had enough to force a vote on them in either house of Congress. At the same time the new Free-Soil party, which would eventually be absorbed into the Republican Party, was advocating a program of continued immigration and land for the landless. The Republican party platform of 1864 stated well the ideological attitude of most Americans toward immigration. A specific immigration plank of that year—echoed in later years—read:

> Foreign immigration which in the past has added so much to the wealth, resources, and increase of power to this nation—the asylum of the oppressed of all nations—should be fostered and encouraged by a liberal and just policy.

Nativism grew in the pre–Civil War years for a variety of reasons, including a growing uncertainty about the future of the nation. Much of the direction that future would take was decided by the Civil War. The truly dangerous subversive forces, it suddenly became clear, were not foreigners but Southern white Americans; those with a penchant for seeing a conspiracy in every threat no longer had to worry about the pope, the Jesuits, or the crowned heads of Europe: They had instead a homegrown slave power conspiracy to worry about. In addition, immigrants and foreigners had been of great assistance to the Union forces. Whole ethnic regiments, chiefly Irish and German, sustained the Union cause, and the Civil War draft worked even more against the poor—including immigrants—than have subsequent drafts. A drafted upper-class or middle-class individual could, if he wished, legally hire a substitute to go in his place, usually by providing a cash bounty of three hundred dollars or more. A future president, Grover Cleveland, chose this method of avoiding military service, as did the father of Theodore Roosevelt and thousands of other persons, almost all of them native-born Americans. Immigrants, it should be noted, fought in the Confederate armies as well.

As a result of the Civil War, Congress did change the Constitution and the naturalization statute, but not in the way that Know-Nothings and their allies had imagined. The Fourteenth Amendment, ratified in 1868, for the first time established a uniform national citizenship and provided that "all persons born or naturalized in the United States . . . are citizens of the United States and of the State wherein they reside." Intended to protect the rights of the former slaves, it would serve, in the twentieth century, to protect the rights of second-generation Asians. In addition, the abolition of slavery made the phrase *free white persons* in the naturalization statute redundant, and in 1870 Congress made the first significant change in that law since Jefferson's time. A few Radical Republicans, led by Senator Charles Sumner of Massachusetts, sought to make the statute color-blind and refer simply to "persons." In this Sumner and his allies were almost a century ahead of their time: Congress chose instead to broaden the law to allow the naturalization of "white persons and persons of African descent." Asians were pointedly excluded, and in the brief debates it was clear that a desire to exclude Chinese from citizenship was, for the majority, the main point. While the courts would later haggle about what the phrase "white persons" really meant, the intent of Congress was clear: Whites and blacks could be naturalized, yellows could not. This meant that the thousands of Chinese already in the United States and the hundreds of thousands of other Asians who would come in the following eight decades were in a new category: "aliens ineligible to citizenship" by federal law.

Twelve years later, in 1882, a bipartisan majority in Congress passed overwhelmingly the Chinese Exclusion Act. Mistakenly treated by many scholars as a regrettable but relatively unimportant event, the Chinese Exclusion Act was the hinge on which American immigration policy turned, a hinge on which Emma Lazarus's "golden door" swung almost completely shut. Few national figures of any prominence had anything good to say for the Chinese: One who did, Republican Senator George Frisbie Hoar of Massachusetts, insisted correctly that Chinese exclusion represented nothing less than the legalization of racial discrimination. Chinese exclusion passed for a number of reasons. The economic interests of white workingmen in California and elsewhere in the West were surely important factors and influenced the stand that organized labor would later take toward all immigration. But there was also, as Senator Hoar had noted, the important factor of racial prejudice.[2]

The act was a complex one, largely because of the desire not to interfere with American trade with China. Although called an "exclusion" act for political reasons, it actually only "suspended" the immigration of Chinese laborers for ten years. "Merchants" however, were admissible. In addition, the law recognized that many Chinese workers in America went back and forth to China, and so allowed those already in the country to get a federal certificate before departing which would allow them to come back in. In 1888 Congress cancelled all the outstanding certificates and ended the practice of them. Thousands of Chinese who had left the country in good faith were barred from returning. As he signed the bill, President Cleveland, in a message reeking of election-year politics, said that the

> experiment of blending the social habits and mutual race idiosyncracies of the Chinese laboring classes with those of the great body of the people of the United States . . . [has been] proved . . . in every sense unwise, impolitic, and injurious to both nations.

In 1892 the act was extended for another ten years, and in 1902 Congress made it "permanent," or so it thought. Also significant was the fact that, in a whole series of decisions, the Supreme Court ruled that the exclusion of a particular "class" of immigrant was constitutional, thus paving the way for other restrictions.

In the meantime the Court, imbued with the nationalism engendered by the Union victory in the Civil War, reversed a string of previous decisions, going back to 1837, which had allowed the states to exercise "police power" over incoming immigrants. In *Henderson* v. *Mayor of New York* (1875) it ruled:

> It is equally clear that the matter of these statutes may be, and ought to be, the subject of a uniform system or plan. The laws which govern the right to land passengers in the United States from other countries ought to be the same in New York, Boston, New Orleans and San Francisco. . . . We are of the opinion that this whole subject has been confided to Congress by the Constitution [and Congress should deal with it].

This meant that the administration of immigration, which had been essentially a matter of laissez-faire, or, in a major immigrant port like New York, the concern of state and local government, now became a federal problem. Since there was no existing federal bureaucracy, the federal government, for a few years, simply subsidized New York to continue to operate its massive immigrant depot at Castle Garden, near the southern tip of Manhattan. To finance this, the federal government, then reluctant to spend much money on anything except Civil War pensions, resorted to a head tax, which began at a modest fifty cents in 1882 and would reach eight dollars by 1917. By 1892 the federal government was able to open its new immigration reception center on Ellis Island, formerly the site of a naval arsenal, in New York Harbor.

By 1932, when it stopped functioning as a reception center for steerage-class immigrants—cabin passengers did not generally have to go there—some twelve million immigrants had passed through it on the way to America.

Most immigrants stayed only a short time on Ellis Island and did not even spend the night there: For them it was truly an "island of hope." For others, especially the tiny minority who were refused admittance, it was an "island of tears." After 1932 it was a detention center for persons who were either refused admittance or were being deported, and during World War II it was used as a temporary internment center for enemy aliens. In 1990, as a part of the National Park Service's Statue of Liberty National Monument, a magnificent museum of immigration was opened in its refurbished main building, although much of the rest of the island was still in a state of disrepair and neglect.

The typical steerage passengers were brought to Ellis Island by lighter or ferry from the ships on which they came. The routine went like this: A physical examination—cursory for most—an examination of documents, checked against shipping manifests, a brief questioning hardly worthy in most instances of the designation *interview,* a gathering up of baggage (which was sometimes brought separately), a visit to the railroad ticket office to purchase tickets or confirm prepaid ones, and the passengers were ready to enter America proper, again by ferry, either to the Jersey shore or the foot of Manhattan. For the minority who had to stay overnight or longer, there were dining, bathing, and sleeping facilities. If anyone was or became ill, there was a hospital on the island with isolation wards for those with contagious diseases and special facilities for those judged to be mentally deranged.

At peak times the island could be chaotic, with thousands of persons being processed in a single day. Although there were recurring scandals about its administration, Ellis Island was, all things considered, a relatively benign institution. While it could be terrifying for the awed newcomer, there were almost always persons on each ship who had been to America before and most knew enough about America to know where they were going. The federal staff tended to be polyglot, as all the languages of Europe had to be handled at one time or another. Fiorello La Guardia, who became New York's greatest mayor, worked as an interpreter on Ellis Island early in his career and left a vivid account of what it was like:

> [Many immigrants] were found to be suffering from trachoma, and their exclusion was mandatory. It was harrowing to see families separated. . . . Sometimes, if it was a young child who suffered from

> trachoma, one of the parents had to return to the native country with the rejected member of the family. When they learned their fate, they were stunned. They had never felt ill. They could see all right, and they had no homes to return to. . . . [A] large proportion [of immigrants] were excluded for medical reasons [many of them for mental reasons]. I felt then, and I feel the same today, that over fifty per cent of the deportations for alleged mental disease were unjustified. Many of those classified as mental cases were so classified because of ignorance on the part of the immigrants or the doctors and the inability of the doctors to understand the particular immigrant's norm, or standard.[3]

While creating a new immigration bureaucracy, the government began to make rules and regulations for immigrants. Most of these were minor restrictive changes that affected few persons but that cumulatively changed the once-free immigration policy of the United States. In both 1885 and 1887, as a sop to organized labor, Congress enacted laws prohibiting contract labor, but these statutes were never enforced to any meaningful degree. An 1891 statute showed, for the first time, a concern for both the physical and mental condition of prospective immigrants. It barred the immigration of "all idiots, insane persons, paupers or persons likely to become a public charge, persons suffering from a loathsome or contagious disease, persons who have been convicted of a felony or other infamous crime or misdemeanor involving moral turpitude," and "polygamists," this later bar being aimed at Mormons. The statute's laundry list of exclusion is strikingly similar to that of the Know-Nothings, omitting only the blind, although most of those would be excluded under the "likely to become a public charge" (LPC) rubric. Yet, despite the growing number of excluded classes, relatively few immigrants were either excluded or deported. In 1905, for example, the first single year in which a million immigrants arrived, deportations and exclusions combined also reached a new high—12,724 persons—which represented barely more than 1 percent of the total. This figure does not take into account the number of those who were stopped or dissuaded from coming. Steamship companies, which had to bear the expense of returning rejected immigrants, instituted pre-embarkation checks, and many immigrants, fearing rejection, simply did not try to come.

As we have seen, both the volume and source of immigration began to change at the end of the nineteenth century. By 1920, with a total population of 105 million, nearly 14 million were foreign born, and another 22 million had at least one foreign-born parent. Thus the 36 million immigrants and their children constituted more than a third of the entire population. While we know that these people contributed greatly to our national existence and are the ancestors of many of us, many Americans at the turn of the century felt that their way of life was threatened by what they called the "immigrant invasion." American Protestant leaders regarded Roman Catholic, Greek Orthodox, and Jewish immigrants with alarm. Some Americans perceived the immigrants as contributing disproportionately to crime and, even worse, dangerous radicalism.

There were also objections to immigrants on economic grounds: The trade union movement saw the seemingly inexhaustible supply of European workers, willing to work for almost any wage, as a threat to the standard of living of American workers. "We keep out pauper-made goods, why not keep out the pauper?" ran a standard AFL argument that made an analogy between the protective tariff and proposals to limit immigration. Such sentiments were not confined to American-born workers. During the depression of the 1890s, for example, surveys taken by the Michigan Bureau of Labor indicated that the foreign-born worker was as "emphatic in condemning immigration as his American brother." The depression also seems to have helped the growth of a new wave of anti-Catholicism. The most prominent anti-Catholic organization was the American Protective Association, which was founded in 1887 and had perhaps half a million members by 1893–94. Strongest in the Middle West it appealed mainly to middle-class whites and revived many of the Know-Nothing proposals, with even less effect. These seemingly rational aspects of restrictionist thought—the one motivated by perceived economic disadvantage, the other by perceived religious disadvantage—had continuing importance in restrictionist sentiment. But lurking behind and sometimes overshadowing these objections to continued immigration was a growing and pervasive racism, a racism directed not against non-white races, but against presumed inferior peoples of European origin. Lines by the genteel poet and novelist Thomas Bailey Aldrich (1836–1907) caught the spirit and the fears of many middle and lower middle-class Americans. In "The Unguarded Gates," published in the *Atlantic Monthly* in 1882, Aldrich complained:

> *Wide open and unguarded stand our gates,*
> *And through them passes a wild motley throng,*
> *Men from the Volga and Tartar steppes.*

Featureless figures from the Hoang-Ho,
Malayan, Scythian, Teuton, Kelt and Slav,
Flying the Old World's poverty and scorn;
These bringing with them unknown gods and rites,
Those tiger passions here to stretch their claws,
In street and alley what strange tongues are these,
Accents of menace in our ear,
Voices that once the Tower of Babel knew.

However curious it may seem today, by the late nineteenth century many of the "best and the brightest" minds in America had become convinced that of all the many "races" (we would say "ethnic groups") of Europe one alone—variously called Anglo-Saxon, Aryan, Teutonic, or Nordic—had superior innate characteristics. Often using a crude misapplication of Darwinian evolution, which substituted these various "races" for Darwin's species, historians, political scientists, economists, and, later, eugenicists discovered that democratic political institutions had developed and could thrive only among Anglo-Saxon peoples. It was axiomatic, therefore, that these intellectuals and others in the grip of what Barbara Miller Solomon has called the "Anglo-Saxon complex" should view the immigration of that era with alarm and organize to raise the bars against it. The census of 1890, which, as the historian Frederick Jackson Turner (1861–1932) announced, signaled the end of the frontier in America, demonstrated to these elite leaders that immigration was changing, and changing for the worse, the composition of the country. In 1894 a group of young Harvard graduates formed the Immigration Restriction League, which became the most influential single pressure group arguing for a fundamental change in American immigration policy. According to one of its founders, Prescott F. Hall (1868–1921), the question for Americans to decide was whether they wanted their country "to be peopled by British, German and Scandinavian stock, historically free, energetic, progressive, or by Slav, Latin and Asiatic races [this latter referred to Jews rather than Chinese or Japanese] historically down-trodden, atavistic and stagnant."[4]

The league and its chief political spokesman, Henry Cabot Lodge (1850–1924), the scholar in politics, one of the first to receive a Ph.D. in history from an American university and who represented Massachusetts in Congress from 1887 until his death, chose to work for a literacy test as the best way to improve the quality of the incoming immigrants. The twenty-two-year crusade for a literacy test, first introduced in Congress in 1895, is instructive. It passed the House in 1895,

1897, 1913, 1915, and 1917, and was passed by the Senate on all but the first of those occasions. But it was vetoed by presidents as diverse as Grover Cleveland, William Howard Taft, and Woodrow Wilson.

The various veto messages are also instructive. Cleveland, in 1897, attacked it as "a radical departure" from established policy and, echoing what the Republican party had said in the post–Civil War years, argued that the "stupendous growth" of the nation had been "largely due to the assimilation and thrift of millions of sturdy and patriotic adopted citizens." He pointed out that "the time is quite within recent memory when . . . immigrants who, with their descendants, are now numbered among our best citizens" were also branded as "undesirable." Cleveland was not impressed with literacy as a barrier:

> It is infinitely more safe to admit a hundred thousand immigrants who, though unable to read and write, seek among us only a home and an opportunity to work than to admit one of those unruly agitators and enemies of governmental control who can not only read and write, but delights in arousing by unruly speech the illiterate and peacefully inclined to discontent and tumult.

In general, according to the president, the bill was "illiberal, narrow, and un-American."

The bill had passed Congress in 1897 in part due to the anxieties caused by the depression of the 1890s, still rated as the second worst in American history. Its passage in 1913, 1915, and 1917, represented convinced majorities in both houses and probably a majority of the electorate. Taft's veto was essentially economic, although it did make the point that illiteracy resulted more from a lack of opportunity than from lack of ability. The bulk of his message consisted of the formal opinion of his secretary of commerce and labor that the United States needed labor and that "the natives are not willing to do the work which the aliens come over to do."

Two years later the bill passed again, and this time Woodrow Wilson vetoed it. His message ignored the economic arguments. It stressed, as was Wilson's wont, ethical principles. Immigrants, Wilson insisted, came seeking opportunity, and the bill would reject them "unless they have already had one of the chief of the opportunities they seek, the opportunity of education." His veto was sustained in the House by a mere four votes.

Two years later, in February 1917, the bill passed for the fourth time under entirely changed circumstances. World War I had greatly reduced immigration, especially that from Europe. The figures for the period July 1, 1915–June 30, 1916, showed total immigration below 300,000, with fewer than half the total from Europe. In addition, more than 125,000 left the country during the year, so net immigration was just over 150,000, as opposed to 900,000 for the last prewar year. The pressure of continuing immigration obviously had little effect in 1917.

What did have an effect was the heightened sense of American nationalism engendered by Wilson's preparedness program and the imminent break with Germany. This nationalism was not merely positive; it was also clearly antiforeign. As such, it cut across the lines established in the continuing debate over America's involvement in Europe's war. Interventionists and noninterventionists alike tended to polarize the differences between "good" America and "bad" Europe. In addition, some of the most powerful voices against restriction, such as the German-American Alliance, found their influence largely negated by a national atmosphere which tended to equate "hyphenated Americanism" with disloyalty and subversion. In these circumstances, Wilson's 1917 veto, an echo of his 1915 message, had little effect. The House voted to override, 287 to 106, while in the Senate only nineteen senators supported the president while sixty-two went against him. That a strong and still-popular president should influence so few senators just three months after his reelection is a good indication of how the war fever had strengthened the nativist climate of opinion.

The 1917 legislation was the first significant general restriction of immigration ever passed; in the future all adult immigrants would have to be literate, although, in the case of family immigration, if the husband were literate the wife need not be. The test was a fair one. Unlike the Australian law at that time, in which the examiner could choose the language(s) in which the immigrant was to be tested, literacy was defined as being able to read in any recognized language, including Yiddish and Hebrew. (Extreme nativists had wanted the bill to restrict immigration to those literate in English, but that had little congressional support.) The other major aspect of the 1917 act was the creation of a "barred zone," described in degrees of latitude and longitude, which kept out all Asian immigrants except Japanese and Filipinos.

Ironically the literacy test, a nativist goal for more than two decades, did little to restrict immigration, although, of course, it may have deterred some from attempting to come. During the last full year in which it was the major statutory bar to immigration—July 1920 to June 1921—more than 800,000 immigrants entered the country. About 1.5 percent (13,799 persons) were denied admission on some ground or other, a mere 1,450 of whom were barred by the long-debated test. Rising standards of literacy in Europe had vitiated the impact of the law, which, had it been passed in the 1890s or earlier, would have had much more effect. Despite its ineffectiveness, the passage of the literacy test was an important victory for the forces of immigration restriction. They would reap even greater benefits in the hypernationalism of the postwar era.

While some restrictionists had been concentrating on the literacy test, others had been raising other barriers. The first of these had come in 1903 as part of the reaction to the assassination of President William McKinley by a native-born anarchist with a foreign-sounding name. For the first time Congress demanded an inspection of the political opinions of prospective immigrants. In words that have since become too familiar the law added to the excluded categories "anarchists, or persons who believe in or advocate the overthrow by force and violence of the Government of the United States . . . or the assassination of public officials," words aimed at those "unruly agitators" Cleveland had fulminated against.

Another restriction resulted from a revival of West Coast anti-Asian sentiments, directed at Japanese immigrants who, in the early twentieth century, seemed to pose the same kind of threat that Chinese had in the nineteenth. A succession of chief executives—Roosevelt, Taft, and Wilson—used their influence to suppress popular anti-Japanese immigration measures so as not to offend the increasingly powerful Japanese government and relied on executive action, such as the Gentlemen's Agreement of 1907–08, under which Japan cut off immigration by withholding passports to laborers.[5]

Thus, by 1917 the immigration policy of the United States had been restricted in seven major ways. Admission was denied to Asians (except for Japanese and Filipinos, the latter because they were held to be American nationals); criminals; persons who failed to meet certain moral standards; persons with various diseases; paupers; assorted radicals; and illiterates. After the war was over, the nativist spirit grew, fed by the patent failure of stated American war aims—Europe was clearly not safe for democracy—and a growing hysteria about domestic radicalism, much of it perpetrated by foreigners or persons with "foreign-sounding" names, a hysteria known to historians as the Red Scare. In addition, by late 1920, as Congress reassembled, the nation's press was filled with scare stories about the flood of undesirable immigrants on their way from war-ravaged Europe. The immigration data suggested no such flood: In 1919–20, 430,000 had entered but 288,000 had left.

Even for the year in progress, immigration levels were not even back to prewar norms: Just over 800,000 entered while nearly 250,000 left. Yet such was the panic that in the space of one week in December 1920, a bill was introduced into the House suspending all immigration for a year and, without any hearings being held, it was passed by a vote of 296 to 42. The Senate, partially in response to frenzied protests from the National Association of Manufacturers and other employer groups, shelved that thoughtless bill. It substituted the so-called Dillingham quota bill, introduced by its resident expert on immigration, William P. Dillingham, a Vermont Republican who had headed the United States Immigration Commission of 1909–11. The quota plan, whose original authorship is unclear, was based on the notion that the best way to inhibit immigration was to limit, first of all, the total number to be admitted in any one year, and second, to assign percentages of that total to particular nationalities on the basis of the number of people from that nation already here. Using the 1910 census, the latest available, the Dillingham plan ignored Western Hemisphere immigration: There would be no quotas for Canada, Mexico, or any other New World nation. As long as the existing barriers remained in effect, Asian immigration was already shut out, except for Japanese, who would get a tiny quota, and Filipinos who could not be kept out. Europeans would be limited to 5 percent annually of the number of foreign-born Europeans in the country as of 1910, assigned in proportion to the nationality recorded in the census. This produced a quota of about six hundred thousand slots per year for Europe, the bulk of which would go to British, Germans, and Scandinavians and presumably would not be fully utilized. The Dillingham plan also assumed that, as new census data became available, the quota would be revised accordingly. There was, thus, a certain amount of equity and fair-mindedness in the plan. The only serious objections voiced in the Senate to Dillingham's plan was that it was not restrictive enough. It passed the Senate easily; like the House measure it replaced, it was a one-year "emergency" measure. The House somewhat reluctantly accepted the Dillingham principle of national quotas but insisted on lowering the percentage from 5 to 3 percent, producing a quota of three hundred fifty thousand. The Senate agreed, and the measure was sent to President Wilson just before his second term expired. As one of his final acts, he used the pocket veto, thus avoiding the inevitable override. The delay caused was about sixty days, as President Warren G. Harding called Congress into special session. The bill was reenacted and was so noncontroversial that it passed the House without recorded vote and the Senate by a vote of seventy-eight to one, and Harding signed it. In May 1922, Congress extended the Dillingham plan for two more years, thus setting the stage for a full-scale debate on immigration restriction in the election year of 1924.[6]

Before discussing that, however, one additional change in American law should be noted—a change stemming from the Nineteenth Amendment, which gave women the vote and made them, for the first time, full citizens of the nation. The rights of female aliens had not much concerned the Congress. By some common-law doctrines, the citizenship of the wife followed that of the husband. An 1855 act said that resident alien women who married American citizens were automatically citizens. The so-called Expatriation Act of 1907 provided that an American woman, naturalized or native born, who married a foreigner, lost her citizenship. This angered many women, and in 1922 Congress passed the Cable Act, which ended that discriminatory practice *except* for those female citizens who married "aliens ineligible to citizenship," that is, alien Asians. Most of the women who did so were second-generation Asian Americans. This inequity lasted until 1931 when it was repealed. Other and more enduring provisions of the Cable Act insisted on alien women becoming naturalized separately, although certain requirements did not have to be met if the husband was already a citizen.

A snapshot of the United States in 1924 would have revealed a nation relatively prosperous by contemporary standards but riven by social conflict and confused by social change. John Higham has coined the phrase *tribal twenties* to describe the ethnocultural struggles of those years, struggles between an old-stock, Protestant, smalltown, and rural America and an immigrant-stock, Catholic, and big-city America. The issues over which they fought—apart from immigration restriction—included prohibition, fundamentalism, and the rise of the second Ku Klux Klan. Other aspects of modernization—the emancipation of some young women, greater sexual freedom—were opposed by the majority of both groups. Prohibition, which seemed in the late 1920s a permanent part of American life even to such a sophisticated opponent as Walter Lippmann, endured only to the end of 1933. The triumph of Calvin Coolidge in 1924 over progressives within the Republican party coupled with the nomination of a Wall Street lawyer by the Democrats—who refused to denounce the Klan by name at their 1924 convention—seemed to signal a long era of conservatism. Four years later the election of Herbert Hoover, who represented all the old American values, over the second-generation Irish Catholic American from the sidewalks of New York, Al Smith, was taken as further evidence of the conservative ascendancy. We can now see that, although Smith lost badly, he carried every nonsouthern large city except Los Angeles, the first Democrat ever to do so. Similarly, in Chicago, when longtime Republican Mayor William "Big Bill" Thompson was challenged by second-generation Czech immigrant Anton Cermak, he derided him as "pushcart Tony." Cermak's measured response, that his folks had not come over on the *Mayflower* but had come to America as fast as they could, was effective in polyglot Chicago—Cermak was elected in 1931—but it would not have played in Peoria. And Peoria was still more representative of America than Chicago. The Peorias of America, and smaller towns, rural areas, and many big-city dwellers as well, supported immigration restriction. In retrospect, it is now clear that, insofar as the old order was concerned, restriction of immigration came too late. The coming-of-age of second- and third-generation voters in the 1930s, the impact of the Great Depression, and the political leadership of Franklin Delano Roosevelt would transform American politics. But in the 1920s no such change appeared on the horizon, and it was in the decade's tribalized atmosphere that the old notions about immigrants as national assets would be scrapped and a restrictive immigration policy adopted that would endure, largely unchanged, for almost four decades.

By 1924, when the extended quota law was due to expire, the new system, proposed originally as a one-year emergency, seemed already to be conventional wisdom. The nativist forces in Congress, led by Albert Johnson, a Republican from Washington State who headed the House Committee on Immigration, were not only intent on making the quota system permanent but also wanted to make it more restrictive. The major goal was to cut the total number and, even more important, to move the baseline census back from 1920, where it should have been according to Dillingham's original notion, to 1890. This, of course, as restrictionists openly stated, would make the discrimination against more recent immigrant groups even more pronounced. Johnson calculated that this would cut the annual Italian quota from forty-two thousand to four thousand, that of the Poles from thirty-one thousand to six thousand, and so on. More moderate restrictionists wanted merely to keep the quotas of the 1922 act. This had held down immigration, during the two years it was in effect, to some six hundred thousand persons annually, about half of them from Europe. The moderates lost; the 1924 act set up a two-stage system. Phase one, which was supposed to last until 1927 but actually lasted until 1929, was what Johnson had proposed: a change to the 1890 base and a reduction of the quotas from 3 to 2 percent. This system allowed in some three hundred thousand annually, about half of them from Europe. Phase two—whose effectiveness cannot really be judged because the Great Depression of the 1930s caused major changes in immigration patterns, was intended to be even more restrictive. The quotas were to be based on a scientific study of the origins of the American people going back to the first census of 1790. Even with precise data, which were not (and are not) available, such a study would have been of dubious validity, but its effect was predictable. The switch to national origins further increased the percentage allowed to the British Isles, Germany, and Scandinavia and reduced all the others. One additional change, with little numerical effect, further tightened Asian exclusion. It abrogated the Gentlemen's Agreement with Japan by barring Japanese totally as "aliens ineligible to citizenship." Since the Japanese quota would have been fewer than two hundred a year, this was intended as an international insult and was so taken by the Japanese government and people. The 1924 law also tightened the administrative apparatus and made deportation for a variety of causes much easier.

Congress thus wrote the assumptions of the Immigration Restriction League and other nativists into the nation's statute books. Those assumptions had become, by then, part of the national climate of opinion. President Calvin Coolidge, who signed the bill into law, had published an article when he was vice president entitled "Whose Country Is This?" In it he made clear his adherence not only to the theory of Nordic supremacy but also to the notion that intermarriage between "Nordics" and other groups produced deteriorated offspring. To cite but another example of the prevalent "racism" of the American 1920s, this is how Congressman Johnson, chief author of the 1924 act, justified that legislation three years later:

> Today, instead of a well-knit homogeneous citizenry, we have a body politic made up of all and every diverse element. Today, instead of a nation descended from generations of freemen bred to a knowledge of the principles and practice of self-government, of liberty under law, we have a heterogeneous population no small proportion of which is sprung from races that, throughout the centuries, have known no liberty at all. . . . In other words, our capacity to maintain our cherished institutions stands diluted by a stream of alien blood, with all its inherited misconceptions respecting the relationships of the governing power to the governed.

> . . . It is no wonder, therefore, that the myth of the melting pot has been discredited. . . . The United States is our land. . . . We intend to maintain it so. The day of unalloyed welcome to all peoples, the day of indiscriminate acceptance of all races, has definitely ended.

Whatever one may think of Johnson's racial theories, which in slightly different form became the official ideology of Nazi Germany, most Americans at that time desired the goal he sought—restriction of immigration. And, in retrospect, without in any way endorsing his or others' theories about racial superiority and inferiority, it is easy to see that some kind of limitation of immigration was not only all but inevitable but probably desirable. There were and are limits to the number of immigrants a developed country can absorb. The real tragedy is not that immigration was restricted but that the criteria used to do so were blatantly discriminatory and that the essentially false notions about the dangers of immigration were so firmly fixed in the American consensus that, in the following decade, it seemed politically impossible to adjust the system to save the lives of those fleeing tyranny and death.

PART III

MODERN TIMES

11

Migration in Prosperity, Depression, and War, 1921–1945

The passage of restrictive immigration legislation and the phasing in of the national origins system in the 1920s brought an entire era of American immigration history to an end. The century of immigration was over. Even more effective in curtailing immigration were the Great Depression of the 1920s and World War II. But immigration never ceased. Table 11.1 provides a year-by-year breakdown of immigration and emigration for the years 1920–45.

The nearly five million immigrants of those years are largely ignored in most histories of American immigration, yet they are an important part of the story. Although the average immigration for the quarter century was nearly two hundred thousand a year, that figure was not even approached after 1930. Nearly half (48.8 percent) of all the immigrants during the whole period entered in the four years before the 1924 act took effect. More than a third (36.7 percent) came in the following six years, while the last fifteen years of depression and war saw just over a seventh (14.5 percent). The figures are still more unbalanced if we look at net immigration—that is, immigration minus return migration. During the quarter century there was almost one remigrant for every three immigrants (32.2 percent). Of the 3.25 million net immigration more than half (53.4 percent) came during 1921–24, just over two-fifths (40.6 percent) during 1925–30, and not quite a sixteenth (6 percent) for the 1931–45 period. In fact, under the impact of the worst years of the Great Depression more persons remigrated than immigrated for four consecutive years (1932–35), and in 1936 the positive balance was a mere 516 persons. Of the million and a half remigrants, just under two-fifths (39.1 percent) left during 1921–24, about two-sevenths (28.5 percent) during 1925–30, and nearly a third (32.5 percent) during 1931–45. However, when one looks behind the mere data to the mil-

Table 11.1

Immigration and Emigration, 1921–1945

YEAR	IMMIGRATION	EMIGRATION	NET IMMIGRATION
1921	805,228	247,718	557,510
1922	309,556	198,712	110,844
1923	522,919	81,450	441,469
1924	706,896	76,789	630,107
1925	294,314	92,728	201,586
1926	304,488	76,992	227,496
1927	335,175	73,336	261,839
1928	307,255	77,457	229,798
1929	279,678	69,203	210,475
1930	241,700	50,661	191,039
1931	97,139	61,882	35,257
1932	35,576	103,295	−67,719
1933	23,068	80,081	−57,013
1934	29,470	39,771	−10,301
1935	34,956	38,834	−3,878
1936	36,329	35,817	512
1937	50,244	26,736	23,508
1938	67,895	25,210	42,685
1939	82,998	26,651	56,347
1940	70,756	21,461	49,295
1941	51,776	17,115	34,661
1942	28,781	7,363	21,418
1943	23,725	5,107	18,618
1944	28,551	5,669	22,882
1945	38,119	7,442	30,677
Total	4,806,592	1,547,480	3,259,112
Average	192,264	61,899	130,365

lions of individual decisions they represented, it is clear that there were several distinct kinds of population movements involved. Only by examining the three separate periods—the last years before the Immigration Act of 1924, the first "normal" six years of its operation, and the fifteen-year period of depression and war—will the varying streams and counterstreams of immigration become clear. Tables 11.2 and 11.3 show some of the major trends.

During the first of these periods, 1921–24, conflicting forces were at work. On the one hand, there was a pent-up demand of persons who had wanted to come to the United States and had been prevented by

Table 11.2

Immigration and Emigration, by Period, 1921–1945

PERIOD	NUMBER OF IMMIGRANTS	NUMBER OF EMIGRANTS	NET MIGRATION
1921–24	2,344,599	604,699	1,739,930
Average	586,150	151,168	439,982
1925–30	1,762,610	440,377	1,322,233
Average	293,768	73,396	220,372
1931–45	699,283	502,434	196,849
Average	46,619	33,496	13,123
Total	4,806,492	1,547,480	3,259,012

the war, plus persons, some of them refugees, whose emigration was impelled either by what had happened during the war or by postwar conditions. On the other hand, there was a similarly pent-up demand of persons who wanted to return to Europe and elsewhere. Thus the figures for 1921—unaffected by the Emergency Quota Act of 1921—are inflated. The eight hundred thousand immigrants of that year are comparable to, but not nearly as large as, the numbers coming in just before the war (in the ten years 1905–14 immigration had averaged more than a million a year), while the nearly two hundred fifty thousand returners were also about 80 percent of the prewar average. During the three years that the first quota act was in effect, gross immigration was reduced to about half of the prewar decade's average (about half a million annually), while remigration was at about a third of the prewar level. The numbers for 1922, which saw returners almost two-thirds as numerous as immigrants, were probably affected more by the sharp postwar American depression than by the legislation. Similarly, the figures for 1924, when remigrants were only a tenth of immigrants, are probably skewed the other way by justified fears of a stricter law. However, the notion that the United States was about to be swamped by an unprecedented flood of immigrants is simply not borne out by the data. Nevertheless, the perception was otherwise. Many Americans shared the disgust of Henry James, who, visiting Ellis Island before the war, could only shudder at what he called "the visible act of ingurgitation" and wonder what it could mean to share "the sanctity of his American consciousness . . . with the inconceivable alien."[1]

Europe continued to be the chief source of immigrants, but not quite as heavily as before the war. Then Europeans were about 90 percent of all arrivals, while in the immediate postwar years they were just under two-thirds. Ironically, the nation that most benefited from the new

Table 11.3

Immigration, by Period and by Region, 1921–1945

PERIOD	NUMBER OF IMMIGRANTS	EUROPE	PERCENTAGE	AMERICAS	PERCENTAGE	OTHER	PERCENTAGE
1921–24	2,344,599	1,541,008	65.7	720,393	30.7	83,198	3.6
1925–30	1,762,610	936,845	53.2	796,323	45.2	29,442	1.7
1931–45	669,283	401,355	57.4	269,751	38.8	28,177	4.0
Total	4,806,492	2,879,208	59.9	1,786,467	37.2	140,808	2.9

pattern of immigration was America's former enemy, Germany, which by 1924 was accounting for a fifth of European immigration and a tenth of the total. The prewar figures were less than 4 percent of either. Conversely, as was the intent of the lawmakers, the percentages from Eastern and Southern Europe were slashed. In 1921, the last year before the Emergency Quota Act went into effect, 95,000 Poles and 222,000 Italians came, representing about two-fifths of all entrants. In the next three years about 28,000 Poles and 46,000 Italians came annually: They were about one-seventh of all immigrants.

The greatest increase in postwar immigration, reflecting in part patterns that had been established during the war, were the nations of the New World, particularly Canada and Mexico. In the prewar decade New World immigration had run at about 7 percent of the total and was increasing. During the war years—1915–20—about every fourth immigrant was from the Western Hemisphere, three-quarters of them from Canada and most of the rest from Mexico. During 1921–24 the Western Hemisphere percentage jumped to more than three-tenths of the total (31 percent). More than one hundred thousand Canadians came annually, as did more than fifty thousand Mexicans, while the rest of the hemisphere sent about twenty thousand.

1925–1930: The "Normal" Quota Years

The next six years saw an intensification of the new patterns: Immigration from Europe dropped, absolutely and relatively, while it continued to increase from the New World. Annual immigration, which had run at some 590,000 in the prior period dropped to about 295,000, a cut of 50 percent. If we just look at net emigration, the result is essentially the same: 435,000 annually in 1921–24 as opposed to 220,000 in 1925–30. Within Europe, the effects of the 1924 act could be clearly seen. The number of incoming Poles and Italians was slashed drastically: Only about 8,000 Poles and 15,000 Italians entered annually, representing a mere twelfth of all immigrants. Conversely, German immigration continued at a rate above its prewar level: Almost 45,000 Germans a year entered the country, and they represented more than a quarter of European immigrants and 15 percent of all newcomers. The favored immigrant groups from Northwestern Europe—British, Irish, Scandinavians, and Germans—accounted for 37 percent of all immigration and nearly 70 percent of that from Europe. This was precisely the kind of result that the sponsors of the 1924 act had envisaged.

In the meantime the incidence of immigration from the Americas, still almost exclusively Canadian and Mexican, increased by 50 percent. The nearly half million Canadians and the more than quarter million Mexicans recorded as entering immigrants were almost 95 percent of all immigrants from the New World. These increases had not been envisaged by most restrictionists and there were constant suggestions that the quota system be extended to the New World, but these were resisted in Congress partly because the agricultural interests of the Pacific Coast and Southwest insisted that they needed Mexican labor.

Although the total annual quota was about one hundred fifty thousand a year, actual immigration ran well above that. There were three reasons for this. First of all, as we have seen, the increasingly important immigration from the New World was not subject to any quota. Second, there had been, since 1921, provisions for nonquota immigrants from areas that were subject to quota. These included, after 1924, wives and dependent children of United States citizens—as long as they were not aliens ineligible to citizenship—resident aliens returning from visits abroad, ministers of religion and their immediate families, various professionals, and domestic servants. This introduced two entirely new factors into immigration legislation: family unification and a recognition of certain "skills," although it should be noted that family unification had been recognized in the Gentlemen's Agreement of 1907–8 negotiated between the United States and Japan. This meant that some European nationalities would exceed their quotas in some years, but in fact the total number of Europeans admitted under quota was below the maximum. The quota for Great Britain and Northern Ireland, for example, was some sixty-five thousand annually. It was never filled. And, finally, as the restrictions kept out more and more people, more and more individuals entered illegally, either by fraud, as was the case with paper sons, or by crossing the Mexican or Canadian border without a visa, by jumping ship, or by other means of entrance. One of the more common means of entry was to come on a visitor's visa and simply stay. How many such persons were there? No one can be sure. It is what one expert has called the problem of "counting the uncountable." There is no reason to believe that such persons were a major proportion of immigration in the 1920s and 1930s.

In addition, immigrants learned to cope with the new regulations. One method was to take up residence in a New World country: The law initially allowed anyone who had resided continuously in such a country for a period of one year to enter as a nonquota immigrant, provided, of course, that the prospective immigrant met all the other criteria—was not an Asian or a political subversive, did not have a loathsome

Map 3: Major Immigration Streams from 1941 to 1987

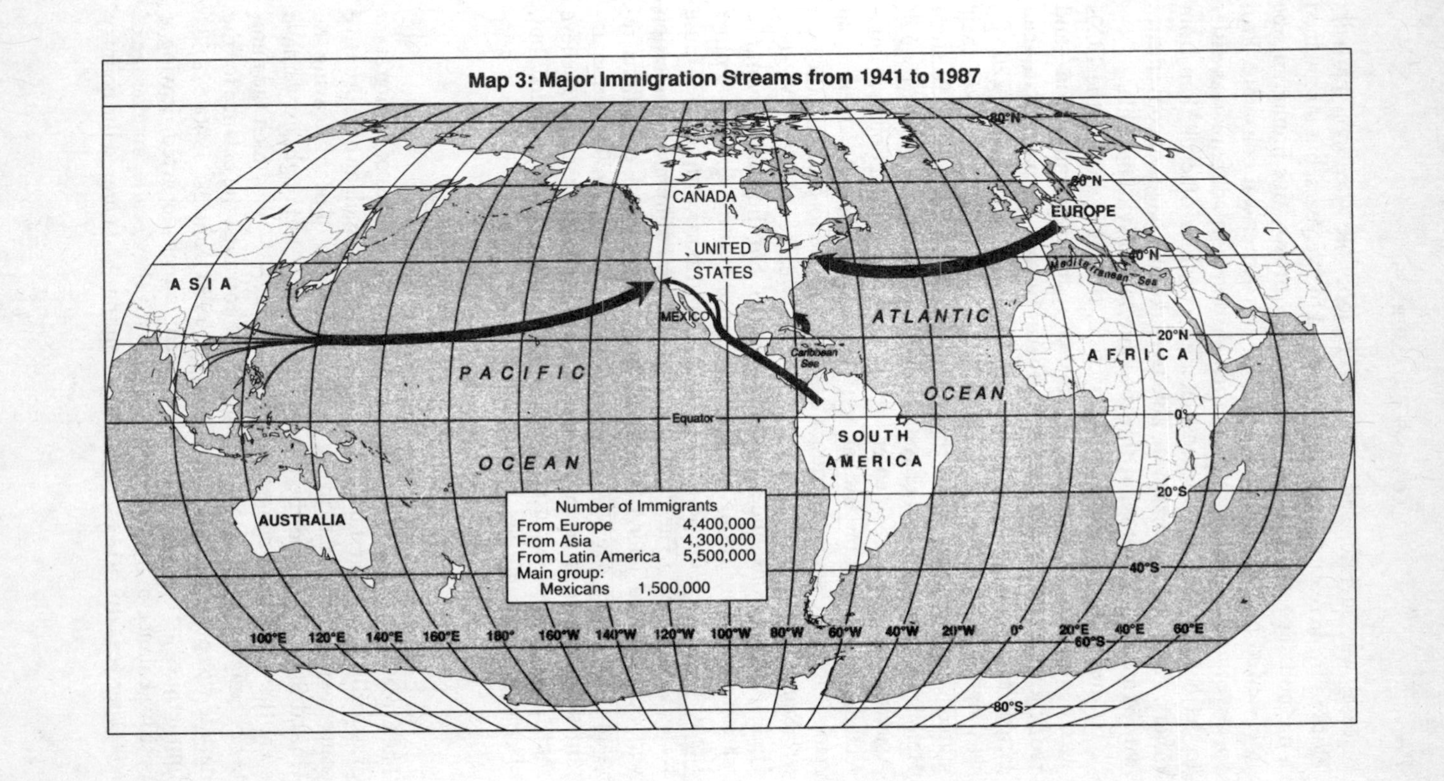

disease, was literate, and so on. This loophole was soon partially blocked by increasing the residence period to five years, but it remained a method. The new law also required that prospective immigrants be interviewed by, and obtain a visa from, an American consulate. Prospective immigrants—or many of them—soon learned that it was easier to get a visa from certain consulates or consular officials than from others. And, finally, some were able to make marriages of convenience with American citizens and thus become instantly admissible.

To the leading advocates of restriction, the legislation of the 1920s was a qualified success. Their goals of reducing immigration—and reducing it from Eastern and Southern Europe in particular—had largely been met, although the average annual net migration of 220,000 was significantly higher than the 150,000 to 165,000 that was their stated goal. Most galling to them was the exemption of the Western Hemisphere from the quota system and the fact that Filipinos, though Asians, could continue to enter the United States as "nationals," unaffected by immigration law, as long as the Philippines remained an American possession. Restrictionists did manage to get a bill through the Senate in 1930 that would have put Mexico, alone of Western Hemisphere nations, under the quota system, but the Hoover administration, which was trying to improve relations with Mexico and other Latin American countries, arranged to keep it bottled up in the House Rules Committee. (The details of Mexican and Filipino immigration will be discussed later.) Had economic and political conditions remained the same, had the relative prosperity of the 1920s been maintained and the world remained at peace, there is every reason to believe that the immigration patterns of 1925–30 would have continued: but, of course, things did not remain the same.

1931–1945: Depression and War

The Great Depression of the 1930s changed the patterns of immigration drastically. It is often stated, even, for example, in such an authoritative source as *HEAEG,* that "in the 1930s the number of people leaving the United States exceeded the number entering."[2] Although, as we have seen, there were four consecutive years—1932–35—in which the number of emigrants *did* exceed the number of entrants, the balance for the decade was positive. In the period July 1, 1930–June 30, 1940, 528,331 immigrants arrived and 459,738 resident aliens departed, leaving a positive balance of nearly 69,000 persons for the decade, an average of 6,900 per year. In 1914 more persons than that had entered the country every two days! Not reflected in these figures are the movements of American citizens, whether incoming persons such as the children of the returned Hungarian immigrants Lajos P. and Lea L., or of naturalized citizens who returned to their mother countries.

The great drop in the number of immigrants from 241,700 in 1930 and 97,139 in 1931 is one of the significant demarcation points in the history of American immigration. Not until 1946 would as many as 100,000 persons again enter the country as immigrants. There were two obvious causes for the decline. In the first place, the impact of the Great Depression was beginning to be felt. Every previous depression in American history had lowered the numbers of immigrants and the Great Depression had similar effects. In addition, the American government changed the rules and did so by executive order. In 1930 President Herbert Hoover issued an order to American consulates to interpret the long-standing LPC clause more rigidly. Previously, only persons who were obviously unable and/or unlikely to be able to support themselves had been kept out by that provision of the law: At one time the possession of twenty-five dollars had been enough. Now consular officials and officials of the INS were given discretionary powers, and it is clear that they used them. It is not now possible even to estimate how many were denied entrance on these grounds as no scholar has yet investigated the visa application files of the various consulates. In time many prospective immigrants had to get affidavits of support from relatives or others already in the country and those providing the affidavits had to be able to demonstrate to an often hostile officialdom that they could, in fact, support the immigrants if necessary. Although Franklin Roosevelt revoked the executive order in 1936, many in the consular and diplomatic services continued to interpret the LPC clause in the new and more restrictive way.

By the depths of the depression, many more immigrants were leaving the country than were entering it. Again, 1930–31 provides the watershed. Net immigration, which had been 191,039, fell to 35,257, and in 1932 began the four-year period when emigration exceeded immigration. In addition, as we shall see, the federal government assisted the departure of many thousands of destitute Mexican Americans who seem not to be counted in the emigration data. As we know, although it is not much talked about, emigration in sizable numbers has been a constant feature of American immigration history since the seventeenth century. Added to ordinary factors causing emigration in the 1930s was the simple fact that it is easier to be poor in a poor country than in a rich one and that many immigrants had support networks in the old country that did not exist in America.

The administrations of Franklin Delano Roosevelt, which changed almost every other aspect of American political, economic, and social life, made no attempt to alter the basic structure of immigration legislation or administration. There was no New Deal for immigration. As we have seen, FDR's government even continued to apply the Hoover interpretation of the LPC after the executive order that promulgated it had been revoked. To be sure, Roosevelt, who attracted the votes of most recent immigrants and their descendants and gave numerous appointments to Catholics and Jews, who formed an important part of the triumphant Roosevelt coalition, did not indulge in nativist rhetoric. To the contrary, he sometimes twitted nativist groups, as in his celebrated remark to the Daughters of the American Revolution in 1938 when he urged them to "remember, remember always that all of us, and you and I especially, are descended from immigrants and revolutionaries." Although no nativist by any criteria, Roosevelt did share with many nativists and other Americans the notion that, since the country was essentially completed—"our industrial plant is already built," he once noted,—immigration was a thing of the past. I know of no evidence to suggest that he ever even considered asking for major changes in the National Origins Act.

This adherence to the status quo by our most innovative president led to one of the major moral blots on the American public record: our essential indifference to the fate of Jewish and other refugees from Hitler's Third Reich. It is hard to improve on the judgment made by Vice President Walter Mondale in 1979 that the United States and other nations that could have offered asylum "failed the test of civilization." The immigration acts of the 1920s made no distinction between "immigrants" and "refugees," and the legislative debates that preceded them show little if any awareness of potential refugee problems.[3] One leading nativist propagandist, Madison Grant, wrote in 1918:[4]

> When the Bolshevists in Russia are overthrown, which is only a matter of time, there will be a great massacre of Jews and I suppose that we will get the overflow unless we can stop it.

Such anticipations were unusual. Without in any way minimizing the widespread anti-Semitism in American life in the 1920s, especially among the elite, almost all of those who participated in the great debates over immigration policy assumed that the future would resemble the past and that most of those who would try to come to the United

States would be attracted for economic reasons, not to save their lives. Thus, refugees from Germany had to surmount barriers that were, in many instances, unsurmountable. And the progressive president in the White House, the political idol of most American Jews, was not willing to swim against the current by attempting to get changes or exceptions made in the law.

From his very first days in office—Hitler had preceded him in power by a little over a month—FDR was made aware of what was happening in Germany and, as a liberal, humane democrat, deplored it. As chief executive, however, he consistently took the advice of his conservative State Department. When, for example, Felix Frankfurter and Raymond Moley urged him to send representatives to a 1936 League of Nations conference on Jewish and non-Jewish refugees and to appoint the outstanding American Jewish leader, Rabbi Stephen S. Wise, to the delegation, Roosevelt instead followed the State Department's advice and sent only a minor functionary as an observer. Secretary of State Cordell Hull told him that: ". . . the status of all aliens is covered by law and there is no latitude left to the Executive to discuss questions concerning the legal status of aliens." FDR, who could usually find a way to do what he wanted to do, in this and other instances involving refugees, meekly accepted the doctrine of executive impotence in the face of congressional will.

Similarly, when Herbert H. Lehman (1878–1963), Roosevelt's successor as governor of New York, wrote him about the difficulty German Jews were having in getting visas from American consulates in Germany, FDR, on two separate occasions in 1935 and 1936, sent replies that had been drafted by the State Department. In each instance, the response assured Lehman of the president's "sympathetic interest," but insisted that "the Department of State and its consular officials abroad are continuing to make every effort to carry out the immigration duties placed upon them in a considerate and humane manner." Lehman was also assured that a visa would be issued in any instance

> when the preponderance of evidence supports a conclusion that the person promising the applicant's support will be likely to take steps to prevent the applicant from becoming a public charge.

Irrefutable evidence exists in a number of places to demonstrate that, despite these assurances at the very highest levels of government, the State Department consistently made it difficult for most refugees to enter this country. Let me illustrate what I mean by a brief account of the problems encountered by the nation's oldest Jewish seminary, Hebrew Union College (HUC) in Cincinnati, in its Refugee Scholars Project. These incidents, which are all too typical, are taken from Michael A. Meyer's history of the college.[5] The project, which was operational between 1935 and 1942, eventually brought eleven refugee scholars to Cincinnati. Most of those came under the provisions—already noted—of the 1924 act, which exempted from quota restriction

> an immigrant who continuously for at least two years immediately preceding the time of his application for his admission to the United States has been, and seeks to enter the United States solely for the purpose of, carrying on the vocation of minister of any religious denomination, or professor of a college, academy, seminary, or university, and his wife, and his unmarried children under 18 years of age, if accompanying or following to join him.

Although of little use to the mass of refugees or would-be refugees, these provisions seemed a godsend for the kinds of people HUC wanted to bring out. The State Department, however, and especially Avra M. Warren, head of its Visa Division, continually raised—one is tempted to say "invented"—difficulties. In most instances the college, sometimes by enlisting the support of influential individuals, managed to overcome them. In the cases of Arthur Spanier and Albert Lewkowitz, however, the difficulties proved insurmountable. Spanier, once Hebraica librarian at the Prussian State Library and later a teacher at the former Hochschule für die Wissenschaft des Judentums in Berlin (a *Hochschule* in Germany was and is an institution of higher education, not a high school), had been sent to a concentration camp after *Kristallnacht*—the night of broken glass—the anti-Jewish pogrom of November 1938. The guaranteed offer of an appointment from HUC was enough to get him released from the concentration camp but not enough to get him a visa from the American consulate. A visit to the State Department by HUC president Julian Morgenstern was necessary before the reasons for refusal could be discovered. According to Warren, Spanier was primarily a librarian. His teaching at the Hochschule, where he had taught for more than the two years stipulated by the law, was not acceptable to the State Department because in 1934 the Nazis had demoted the Hochschule from its former status as an institution of higher education, and an administrative regulation of the State Department, not found in the statute, held that the grant of a nonquota visa to a scholar coming from an institution of lesser status abroad to one of higher status in the United States was not permitted.

Spanier and the other scholar, Albert Lewkowitz, managed to get out of Germany, but to a country, the Netherlands, which soon fell under the Nazi heel. It had seemed that Lewkowitz, at least, was likely to get out since he had been a teacher of Jewish philosophy at the Breslau Jewish Theological Seminary, an institution whose status the State Department did not question. (In fact, both the Hochschule and the Seminary had been the leading institutions of their kind in the world before the Nazis came to power.) But the German bombing of Rotterdam in May 1940 destroyed the copies of Lewkowitz's records, and officials at the American consulate there insisted on the impossible: that Lewkowitz get new documents from Germany before they would issue him a visa. Five years earlier, in peacetime, Roosevelt had assured Lehman that

> consular officials have been instructed that in cases where it is found that an immigrant visa applicant cannot obtain a supporting document normally required by the Immigration Act of 1924 without the peculiar delay and embarrassment that might attend the request of a political or religious refugee, the requirement of such documentation may be waived on the basis of its not being "available."

No such waiver was made for Lewkowitz. Neither he nor Spanier ever obtained visas and both were eventually sent to the Bergen-Belsen concentration camp. Lewkowitz was one of the few concentration camp inmates exchanged in 1944, and he was permitted to enter Palestine. Spanier died in Bergen-Belsen. If relatively eminent individuals with a prestigious American sponsor such as these had difficulties, imagine what it must have been for less prominent persons.

Other horror stories could be told. In Congress, for example, the bipartisan Rogers-Wagner bill, which would have brought twenty thousand German refugee children—most of whom would have been Jewish—to the United States as nonquota immigrants, died in committee without ever having been voted up or down and without ever receiving a word of public support from the president. Its opponents argued publicly that it was just a ploy to undermine the quota system; privately they sneered that the "cute Jewish kids" would grow up to be "ugly Jewish adults." No such problems arose in 1940 when Congress expedited plans—never fully carried out—to admit fifteen thousand children from war-torn Britain. Or, for a final example, there was the 1939 case of the German vessel *Saint Louis,* loaded with nearly a thousand refugees whose Cuban visas were canceled: Attempts were made to get them admitted to the United States and the ship sailed close enough to Miami Beach for its passengers to hear the music being played at the beachfront hotels. No admission was granted and its refugee passengers were returned to Europe. Many of them died in the Holocaust.

A number of German refugees, perhaps a total of 150,000, the vast majority of them Jews, did manage to get to the United States. This was well under the German quota. For the years 1933–40 there were 211,895 German quota spaces. Only 100,987 were actually used. Prominent refugees—such as Nobel Prize winners Albert Einstein (1879–1955) and Thomas Mann (1875–1955)—had few problems entering the country. Less prominent persons had great difficulty. Finally, in 1938, in the wake of *Kristallnacht,* Franklin Roosevelt used a little of his executive authority in the interests of refugees. He ordered that any political refugees already in the United States on visitor's visas could have them extended and reextended every six months. This helped perhaps 15,000 persons here not to have to go back. At the same time, however, his spokesman, Myron C. Taylor assured the public in a radio address that:

> Our plans do not involve the "flooding" of this or any other country with aliens of any race or creed. On the contrary, our entire program is based on the existing immigration laws of all the countries concerned, and I am confident that within that framework our problem can be solved.

Taylor's—and presumably Roosevelt's—confidence was misplaced, if it in fact existed. Nothing short of a drastic revision or emergency suspension of American immigration laws could have saved a substantial number of refugees after 1938. That was a risk that FDR was not willing to take. The notion of the country being "flooded" was, of course, absurd. Neither gross nor net immigration in the 1930s even approached the quota limits. But the clutch of nativism, exacerbated by the very real problems of the depression, made such chimeras loom large in the minds of many Americans, in and out of Congress. And so, nothing much was done.

After the fall of France, in the summer of 1940, FDR asked his Advisory Committee on Refugees to make lists of eminent refugees and then ordered the State Department to issue temporary visitors' visas to those individuals. The State Department's own reports, which are not always reliable, indicate that it issued 3,268 such visas to

> those of superior intellectual attainment, of indomitable spirit, experienced in vigorous support of the principles of liberal government and who are in danger of persecution or death at the hands of autocracy.

In the event, however, only about a third of them were ever used.

Once the United States itself was at war, there was another reason advanced for not letting refugees in: Nazi spies and saboteurs might be hidden among them. Only in 1944, when the incredible dimensions of the Holocaust began to be realized, when responsible officials could no longer refuse to take cognizance of what Walter Laqueur has called "the terrible secret," did Washington take any further action. Then Roosevelt, by executive order, created the War Refugee Board (WRB). Although the WRB did not bring any refugees to the United States, it did—by using a mixture of public and private funds and unorthodox methods—assist in the rescue of the remnants of European Jewry. And Roosevelt himself, allowed one "token shipment" of 987 "carefully selected" refugees from displaced persons (DP) camps in Italy to a temporary haven in Oswego, New York. Theoretically the president had merely "paroled" the refugees into the United States temporarily, but in fact almost all of them became permanent residents and eventually citizens. Had such actions been taken earlier, the refugee story might have been substantially different, but there is no conceivable way that the United States could have rescued any more than a small fraction of the six million persons who perished in the Holocaust.

It is now abundantly clear that, as a group, the refugees who were successful in reaching the United States, have made a contribution to our culture highly disproportionate to their number. Many of those who contributed most to the making of the atomic bomb were refugees: in addition to Einstein, whose enormous prestige enabled him to convince Roosevelt that what became the Manhattan Project should be undertaken, refugee scientists such as Enrico Fermi (1901–54) from Italy, Leo Szilard (1898–1964) from Hungary, Lise Meitner (1878–1968) from Austria, did much of the crucial work that made nuclear fission in 1945 possible. In other fields as well, towering figures came to the United States: In music, for example, the composers Arnold Schönberg (1874–1951) from Austria and Béla Bartók (1881–1945) from Hungary, and dozens of topflight musicians and conductors came. Hundreds of academics in various disciplines helped transform and improve the level of intellectual life in America. In addition, many persons who came here as youngsters have made their presence felt, such as the fifteen-year-old Heinz Kissinger who arrived with his family from Germany in 1938 and grew up to become secretary of state.

Asian Americans and World War II

World War II had diametrically opposed impacts on the two major Asian American ethnic groups. Japanese Americans, in a betrayal of almost everything that America is supposed to stand for, were rounded up and shipped off to ten godforsaken concentration camps in places like "Topaz," Utah. Chinese Americans, on the other hand, saw their legal position transformed as they, but not other Asians and Asian Americans, were placed almost on a par with other immigrant and ethnic groups insofar as immigration and naturalization were concerned. Each group was only a tiny segment of the American population: In 1940 there were about 125,000 Japanese Americans and 75,000 Chinese Americans in the contiguous United States, less than two-tenths of 1 percent of the population. Nothing better symbolizes the ambiguous relationship between war and democracy than the contrasting fates of these two populations.

Although the United States fought against totalitarianism and racism in World War II, one of its most significant acts on the home front was to adapt one of the institutions of its most feared opponent—the concentration camp—for use against the Japanese Americans. In addition, of course, the largest American racial minority, blacks, were still legally second-class citizens, segregated in those states where the overwhelming majority of them lived and in the military forces raised by the federal government.

During World War I, as we have seen, the huge German American ethnic group had been badly mistreated, although none of that mistreatment came from federal statutes. (In Canada, by contrast, many naturalized German Canadians had their citizenship revoked during the 1914–18 war.) During World War II, happily, the American government and people made clear distinctions between what the propaganda called "good" and "bad" Germans (and Italians). The vast majority of unnaturalized Germans and Italians in the United States were left at liberty, even though they—and all other aliens—had been required by a law passed in 1940 to register annually as long as they remained in the United States. American entry into the war, in December 1941, transformed resident-alien Germans, Italians, and Japanese into "enemy aliens." It must be remembered that, since they were "aliens ineligible to citizenship," the Japanese, unlike the others, were noncitizens by necessity, not choice or oversight. Under the normal wartime enemy alien procedures, about two thousand Japanese aliens, almost all of them adult males, were rounded up and placed in confinement, as were proportionally fewer alien Germans and a handful of Italian aliens. (Enemy alien status for almost all the millions of unnaturalized Italians was lifted in a Columbus Day proclamation in 1942 in a move that had more to do with the forthcoming congressional elections than with honoring the Genoese navigator.) While it is clear that alien Japanese were being treated more harshly than were Caucasian enemy aliens, these incarcerations left most Japanese Americans at liberty and did no violence to the Constitution.

By February 1942, however, under the impact of a string of humiliating defeats by Japan and a rising chorus of anti-Japanese sentiment from the media, politicians, and many ordinary Americans, Franklin Roosevelt began the process whereby 120,000 Japanese Americans—more than two-thirds of them native-born American citizens—were herded into ten hurriedly built concentration camps, where some of them remained for almost four years.[6]

Over the years most Americans have come to see, as President Gerald R. Ford put it in a 1976 proclamation:

> We know now what we should have known then—not only was that evacuation wrong, but Japanese Americans were and are loyal Americans.

More specifically, the presidential Commission on the Wartime Relocation and Internment of Civilians concluded in 1983 that the wartime mistreatment of Japanese Americans

> was not justified by military necessity, and the decisions which followed from it—detention, ending detention and ending exclusion—were not driven by analysis of military conditions. The broad historical causes which shaped these decisions were race prejudice, war hysteria and a failure of political leadership.[7]

Six years later, in 1988, Congress passed, and President Ronald W. Reagan signed, a bill containing an apology to the Japanese American people for the injustice done them and authorizing the payment of twenty thousand dollars in "redress" to each of the perhaps sixty thousand surviving persons who had been in one of America's concentration camps. Almost fifty years earlier, however, the incarceration of the Japanese had been a popular wartime act. Most Americans not only approved what was done but would have approved harsher treatment. Many indicated that all Japanese should be deported after the war was over.

Thus, Japanese Americans suffered for the actions of Japan; Chinese Americans, conversely, benefited from the wartime popularity, relatively speaking, of China. American support for China, which had been one of the causes of the great Pacific war between Japan and the United States, had manifested itself off and on during the twentieth century but had not been of any real assistance to Chinese Americans. In 1943, however, as a gesture toward a wartime ally, Franklin Roosevelt recommended and Congress agreed to repeal the Chinese Exclusion Act of 1882 and all or part of fourteen other statutes that had effected Chinese exclusion. In addition, the new law gave a quota of 105 persons annually to "persons of the Chinese race" and amended the nationality act to make "Chinese persons or persons of Chinese descent" eligible for naturalization on the same terms as other immigrants. It is clear that both Congress and the president regarded this action as a kind of good behavior prize. As Roosevelt put it:

> It is with particular pride and pleasure that I have today signed the bill repealing the Chinese exclusion laws. The Chinese people, I am sure, will take pleasure in knowing that this represents a manifestation on the part of the American people of their affection and regard.
>
> An unfortunate barrier between allies has been removed. The war effort in the Far East can now be carried on with a greater vigor and a larger understanding of our common purpose.

Although American immigration law was still discriminatory toward Chinese—a Canadian-born Chinese entering the United States as an immigrant would have to secure one of the 105 Chinese quota slots—we can now see that, just as the 1882 Chinese Exclusion Act had been the hinge on which the golden door of American immigration began to swing closed, its repeal six decades later was, similarly, the hinge on which it began its renewed outward swing. In addition, a little-noticed 1946 act, which made the alien Chinese wives of American citizens admissible on a nonquota basis, had important demographic consequences for the Chinese American community. During the eight years 1945–52, when there were a total of 840 Chinese quota slots, just over 11,000 Chinese actually emigrated to the United States. Nearly 10,000 of these were women, almost all of them nonquota wives. When one considers that, as late as 1950, the Chinese American community of 117,000 contained only 28,000 women 14 years of age and older, the demographic impact of these women was obvious. In addition, this act presaged what would be a major theme of American immigration policy in the post–World War II years: family reunification.

World War II also saw the beginnings of what is usually called cultural pluralism in the ways in which most Americans defined their cultural identity. As Philip Gleason has written, the war gave unprecedented salience to ideology:

> For a whole generation, the question "What does it mean to be an American?" was answered primarily by reference to "the values America stands for": democracy, freedom, equality, respect for individual dignity, and so on. Since these values were abstract and universal, American identity could not be linked exclusively with any single ethnic deprivation. Persons of any race, color, religion or background could be, or become, Americans.[8]

Despite this sea change in American ideology, which would be intensified during the Cold War, the nation's basic immigration law remained the national origins system set up during 1924–29. It would prevail, at least on paper, until 1965. But, as we shall see, the period between 1943, when Chinese Exclusion was ended, and 1965 when the whole structure was scrapped, was one of gradual relaxation of the laws and a progressive increase in the number of immigrants who entered the country.

One other major change occurred during the war years. A labor shortage, after the glut and longterm mass unemployment of the 1930s, caused the United States deliberately to stimulate the migration of Mexican laborers to work in the agriculture of the American Southwest and West and on America's railroads. This so-called bracero program—from the Spanish *bracer,* "arm," therefore a manual worker—is an important landmark in the history of Latin American migration to the United States. In the first two decades of this century immigrants from Latin America accounted for just 4 percent of all immigrants: Since the 1960s they have accounted for more than a third of all legal immigrants, to which must be added several millions of illegal immigrants, those who simply crossed the border without the permission of the government. In the war years, and after, the notion was that Mexicans would be temporary workers, what the Germans have come to call

Gastarbeiter, guest workers. While many imported temporary workers were just that—persons who went home after the growing season or the labor shortage was over—many others stayed on to become permanent residents, with or without the proper papers. But the Mexican American experience was not something that began with the war: As we have seen, it was Spanish Mexicans who were the pioneers of the American Southwest and much of the West Coast. In the next chapter we shall pick up their story in the mid-nineteenth century and bring it up to the very recent past.

12

From the New World: Mexicans and Puerto Ricans

Immigration from the nations and colonies to the south of the United States—Latin America and the Caribbean—has long been dominated by Mexicans, although there have always been substantial numbers of migrants from other places. Until the period of World War II and after, these immigrants amounted to only a very small percentage of all newcomers. In the century of immigration between 1820 and 1930 about 1.3 million persons were recorded as entering from the whole region. More than half—some 750,000 were from Mexico and most of the rest—425,000 came from the West Indies. Central America contributed about 43,000 and the whole continent of South America only 113,000, a little over a thousand a year. Clearly most of these immigrants came from our very close neighbors and a significant but indeterminable number of those were sojourners who returned home. The bulk of Mexican immigration, 720,000 persons, had come in the years after the Mexican Revolution that began in 1909, pushed by the danger and disorder of the conflict and attracted by the booming economies of California and the Southwest. During the 1930s neither of these forces was operating: Mexico had achieved a stable government and the mass unemployment of the Great Depression repelled rather than attracted most immigrants. In addition, the United States government sponsored a repatriation program, supposedly voluntary, under which as many as five hundred thousand Mexican Americans—some of them citizens—were sent south across the border, many of them on special trains chartered by federal and local governments.

But many Mexican Americans were not recent immigrants but descendants of the founding fathers and mothers of European-style civilization in California and much of the Southwest. At the time of the Mexican War (1846–47) there were perhaps eighty thousand Mexicans living in the territory that the United States would annex—sixty thousand of them in New Mexico and twenty thousand in California. Another five thousand or so lived in Texas, which had been added in 1845. The Treaty of Guadalupe-Hidalgo (1848), which ended the war, gave those living in the territory taken from Mexico the right to stay or go to Mexico: some three thousand chose the latter course. The others could become American citizens or remain Mexican citizens and were to enjoy property rights "as if the [property] belonged to citizens of the United States." The former were promised "the rights of citizens of the United States according to the principles of the Constitution." Alas, the principles of the treaty were soon ignored in both New Mexico and California: Many Mexicans and former Mexicans were soon separated from their land and became strangers in what had been their own country, although some Mexican members of the elite in New Mexico retained power—or the trappings of power. New Mexico remained a territory until 1912 even though it satisfied the normal requirements of statehood at the time of annexation. As would be the case in Hawaii, statehood was delayed because the population was not considered "American" enough.[1]

The Mexican American community is thus both old and new. Since Mexican Americans have usually been counted by the Census Bureau as Caucasians even though most have large amounts of Amerindian ancestry, it is all but impossible to get accurate estimates of their total number. In recent years the census has resorted to a "Spanish surname" estimate, but this includes Cubans, Puerto Ricans, and other groups under a blanket heading that confuses more than it clarifies. The Cubans of Miami, the Puerto Ricans of New York, and the Chicanos of Southern California have little in common beyond the Spanish language that many but not all use in daily life. In 1980 the Census Bureau estimated that there were some 14.6 million persons of "Spanish origin," about 6.5 percent of the total population. Of these, some 8.6 million persons were presumed to be of Mexican origin. Earlier population estimates are no more precise. The *HEAEG* essay on Mexicans estimates the number of Mexican Americans in 1900 at somewhere between 381,000 and 562,000, a margin for error of almost 50 percent.[2] Comparing the midpoint of this "guesstimate"—471,500—with the census guess for 1980—8.6 million—we get a more than eighteenfold increase over eight decades in which the U.S. population increased a little less than threefold. However one manipulates and adjusts the figures, it is clear that the incidence of the Mexican American ethnic group has increased significantly and that it and other "Hispanics" are the sleeping giants of American society. In the rest of this chapter we shall consider the two most prominent Hispanic minorities, Mexican Americans and Puerto Ricans. Other Hispanic groups will be treated in the two final chapters.

Twentieth-Century Migration from Mexico

Like the Chinese, Japanese, and Filipinos who came before them, large numbers of the Mexican immigrants who began to come north after 1909 found agricultural jobs partly in the newly irrigated Imperial Valley of California, in the sugar beet fields of Colorado, and elsewhere in the Southwest and Far West. The pioneer historian of California, Hubert Howe Bancroft, argued that the Mexicans would be ideal laborers for California's ranchers and growers, because they would go back home after they were no longer needed. Others worked as copper miners in Arizona, coal miners in Colorado, and as track layers for western railroads. In the agricultural sector, in particular, they tended to fill the role that first Chinese and then Japanese agricultural workers had pioneered. In addition, there were some refugees from the Mexican revolution, ranging from those on the right, like General Victoriano Huerta (1854–1916), the man responsible for the murder of the reform president, Francisco Madero (1873–1913), to left wingers such as the syndicalist followers of the Flores Magon brothers, who bulked large among the Far Western victims of the post–World War I "Red Scare."

Between these extremes were significant numbers of businessmen, mostly small and middling, who established themselves along the once largely unsettled border between Mexico and the United States. Stretching from the cities of Matamoros/Brownsville on the Gulf of Mexico to Tijuana/San Diego on the Pacific, it runs for two thousand miles. From the Gulf to El Paso it is marked by the Rio Bravo/Rio Grande: From El Paso to the Pacific it is largely desert. While the border divides, it also unites: Its paired cities are economic units, each dependent on the other, and tens of thousands of people cross the border every day to work, to shop, and to play. The same can be said of our even longer border to the north. But there is one crucial difference. The standard of living on each side of the Canadian American border is roughly the same: Along the Mexican American border there is great disparity, with lower prices for most goods and services on the Mexican side and higher wages and more jobs on the American side. As long as those conditions prevail, the United States will continue to attract Mexicans as either temporary or permanent residents. For a long time, except for the odd customs facility, both governments largely ignored the border: Only in 1924 did the United States establish the Border Patrol.

Before that happened, World War I, which cut off the flow of European workers, eventually created unfillable job opportunities in northern American cities. Many of these opportunities were filled by southern blacks who began the "great migration" to places like Chicago at that time; others were filled by Mexican immigrants, who came not only to the Southwest and West in greater numbers but reached such places as Chicago, Detroit, and Pittsburgh. This flow continued during the 1920s, when some five hundred thousand Mexicans were counted as immigrating to the United States, there being no quota for Western Hemisphere immigrants. Between 1910 and 1930 the number of foreign-born Mexican Americans recorded by the census tripled from more than two hundred thousand to more than six hundred thousand, which suggests that there were a million or more Mexican Americans by the latter date. About 85 percent of the recorded 1930 Mexican-born population lived in the Southwest, most of them in just two states, Texas (40 percent) and California (30 percent). These immigration flows, as we have seen, were temporarily reversed during the Great Depression, both by government action and by market forces. A strict interpretation of the LPC clause, the literacy test, and the general lack of jobs, plus competition with American migrants, like Steinbeck's Joads, for the few existing jobs cut the number of incoming Mexicans on immigrant visas to just over thirty thousand for the entire decade of the 1930s, while the previously noted repatriation program sent up to half a million persons the other way.

Acute labor shortages during World War II, shortages that, in California, were exacerbated by the incarceration of the Japanese Americans, caused the United States government to negotiate the bracero program with Mexico. Under this program the United States guaranteed that Mexican workers would receive specified minimum wages and certain living and working conditions, although many complaints were filed against employers who did not meet those relatively modest standards. Conditions were particularly bad in Texas—so bad that for a time the Mexican government refused to allow any braceros to be sent there. The World War II program had as many as two hundred thousand braceros in the United States, about half in California, the rest in twenty other states. There was no program between 1947 and 1951 when a new program was created to meet Korean War labor shortages. It lasted until 1964. Its peak year was 1959, when four hundred fifty thousand braceros entered. The 1960 census reported that braceros accounted for just over a quarter of the nation's seasonal agricultural workers.

In the meantime, legal immigration, that is, those Mexicans who entered as resident aliens with permanent status, began to rise from its depression lows. Nearly 60,000 came in the 1940s, almost 275,000 in the 1950s, more than 440,000 in the 1960s, and almost 640,000 in the 1970s. The 1980 census reported that there were 2.2 million persons born in Mexico living in the United States. They were more than 15 percent of all foreign-born. Nearly three-fifths of these Mexican Americans (57.8 percent) told the census taker that they had come to the United States since 1970. The Census Bureau believes that about half of the foreign-born Mexicans it counted were illegal immigrants. And, for a final statistic, in its 1987 current population survey estimate the Census Bureau estimated that nearly twelve million of the nineteen million Hispanics in the United States were of Mexican birth or ancestry.

The whole question of illegal immigrants or, as some euphemistically prefer to describe them, undocumented persons, is a vexed one. Just how many there are in the country at any one time is a matter of great dispute. It is, as noted, a problem of counting the uncountable. Prior to the Chinese Exclusion Act of 1882 there was no such thing as an illegal immigrant, since no permission was necessary to enter the country. As more and more restrictions have been added, the reasons for entering the United States illegally have risen. Perhaps 50 to 60 percent of contemporary illegal immigrants are Mexicans, although in 1986 alone the INS apprehended alleged illegal immigrants from 93 different countries. In contrast to the popular image of the illegal immigrant as the border crosser, the wetback, the *mojado,* perhaps more than half of all illegals are not border crossers but visa abusers. That is, persons who have entered the United States legally on some kind of temporary visa and simply stayed on. As long as they have little or no contact with officialdom, such persons are not likely to get caught. Some have been found out, ironically, by being the victims of serious crime.

The question of how many illegal immigrants there are in the country at any one time is repeatedly raised by restrictionists. Nobody really knows. In two reports just two years apart (1972 and 1974) the INS told Congress that (a) there were about one million such persons, and (b) the range was between four to twelve million persons. The then INS commissioner said his personal belief—as opposed to that of his staff—was that there were between six and eight million. A private consulting firm using the "Delphi method"—you ask a bunch of experts what they think and average the replies—came up with a figure of 8 million in 1975. The Census Bureau study, which reviewed all existing studies—really another example of the Delphi method though it was not called that—concluded that as of 1978 there were between 3.5 and 6 million illegal aliens in the country. Local politicians also came up with inflated estimates: Los Angeles County officials in the early 1980s claimed that there were at least two million illegals in their county alone (and wanted federal dollars to care for them), and New York's former mayor Ed Koch argued that there were at least a million in his city. And finally, two variant, sober, educated guesses for 1980: A Census Bureau researcher has argued that at least 2.057 million illegal immigrants were actually enumerated in the 1980 census, while the National Research Council has argued that the range is 1.5 to 3.5 million. The council's report added, at a time when there was national hysteria about immigration reform (to be discussed in the final chapters):

> There is no empirical basis at present for the widespread belief that the illegal alien population has increased sharply in the late 1970s and early 1980s.

What is a reasonable person to make of all these numbers, claims, and counterclaims? It is clear that at least since the 1950s there have been substantial numbers of border crossers and other illegal entrants from Mexico and elsewhere and that they have ventured far beyond the traditional border areas where such behavior has gone on since the Treaty of Guadalupe-Hidalgo. The INS staged what it called "Operation Wetback" in the early 1950s and reported that it had deported or expelled 3.8 million Mexicans. From the 1970s to the mid-1980s the INS reported a rising number of apprehensions and expulsions of immigrants, from a little over 300,000 in 1970 to 1.8 million in 1986. These "body counts" should not be taken literally. Many individuals are caught more than once in the same day and duly added to the count: Others are casual border crossers who would not have stayed anyway, and many others are migrant workers, who, at the end of the growing season and far from home, arrange to be caught by *la migra*, as they call the INS. This is a good arrangement for each. The INS adds to its body count, and the migrants, who are willing to sign a voluntary deportation order, get a free plane ride to the border. (No one wants to put such persons in jail: It can cost up to thirty thousand dollars annually to house a single federal prisoner—more than it costs to send a student to Yale or Harvard.) I once spoke to a Mexican migrant worker in Michigan who told me that he had arranged to be deported every fall for some thirty years. He and others cross the border in the Rio Grande Valley of Texas in late winter or early spring and work their way north harvesting crops—often working for the same employers year after year—ending with cherry picking in Michigan. Other workers, following other crops, have different migration routes. These seasoned agricultural workers work in the United States from seven to ten months a year, but they live in Mexico where their families enjoy a higher-than-average standard of living. Like the Italian *golondrinas* to the Argentine, they have filled an economic niche that is satisfactory to them and their employers. Others, of course, come and try to stay, and clearly many succeed. The arguments about the effects of this pattern of migration will be considered in the final chapter. Here we are primarily concerned with noting and evaluating its existence and its effect on the Mexican American community.

THE EMERGENCE OF THE MEXICAN AMERICAN COMMUNITY

Like most other American ethnic groups, Mexican Americans have had a varied history. Their experiences in Texas, in New Mexico, in California, and in the industrial heartland of the United States have been distinct. They have even referred to themselves in various ways. Although some intellectuals and activists now insist on using the term *Chicano* to describe the entire group over its long history, the term is, in fact, of relatively recent coinage and not accepted by all. In Texas, the original term was *Tejano,* while in the twentieth century the most important Mexican American civil rights organization was called the League of United Latin American Citizens (LULAC), and an important post–World War II veterans' organization there eschewed an ethnic title and called itself the American G.I. Forum. In California, the pre-1848 settlers called themselves Californios, and by the twentieth century most would call themselves Mexican Americans or Mexicans. In New Mexico, where Mexican American influence was the strongest, the term *Spanish American* prevailed.

Whatever they were called and wherever they lived in the Southwest and Far West, Mexican Americans were dispossessed of much of their land and subject to discriminatory treatment in every aspect of their lives: in employment, in housing, and in education. Overt discrimination is probably worst in Texas, where the state law enforcement agency known as the Texas Rangers has long had a well-deserved reputation for brutality toward Mexican Americans; it is least pervasive in New

Mexico, our only bilingual state, where Spanish has a presumed equality with English. The examples of discrimination that follow all come from California, where a majority of the nation's Mexican Americans live.

After 1848 most of the few Californios who were sizable landowners—about two hundred families controlled fourteen million acres—were soon separated from their land by what the American reformer Henry George called a "history of greed, of perjury, of spoilage and high-handed robbery." The California gold rush of 1849 attracted miners from Mexico, Chile, and other parts of Latin America, as well as tens of thousands of migrants from the rest of the United States, and immigrants from Europe, China, and Australia. The new Euro-American California majority soon decided that it did not want Chinese or "greasers"—a term used for native Californians of Mexican heritage and recent immigrants from Latin America alike—to share in the bonanza and by both discriminatory legislation and mob violence sought to drive them out. The philosopher Josiah Royce (1855–1916) described ironically the treatment meted out to the Californios who were also attracted to the mining regions:

> We did not massacre them wholesale, as Turks might have massacred them: that treatment we reserved for the Digger Indians. . . . Nay, the foreign miners, being civilized men, generally received "fair trials" . . . whenever they were accused. It was, however, considered safe by an average lynching jury in those days to convict a "greaser" on very moderate evidence if none better could be had. . . . It served him right, of course. He had no business, as an alien, to come to the land that God had given us. And if he was a native Californian, or "greaser," then so much the worse for him. He was so much the more our born foe; we hated his whole degenerate, thieving, landowning lazy and discontented race.

But unlike so many of California's Indian tribelets, the Californios were not exterminated. They were, instead, swamped by the massive nineteenth-century migration—much of it immigration—to California. The slow but steady growth of Mexican American population after 1909, and even the massive repatriations of the early 1930s, created little overt friction between the Mexican Americans and the so-called Anglo population. Only after the removal and incarceration in 1942 of the minority that most white Californians found most threatening, the

Japanese Americans, was more than casual attention paid to what was California's largest minority group. Two events in Los Angeles during World War II, the Sleepy Lagoon murder and the Zoot Suit riots, can symbolize both the increase of overt discrimination and the beginnings of more aggressive resistance by the local Mexican American community.

The Sleepy Lagoon case, a romantic title fabricated by the local press to describe an abandoned gravel pit, involved the murder of a young Mexican American, José Diaz, apparently slain in a struggle between gangs on the night of August 1–2, 1942. After the killing had been sensationalized by the press as part of an alleged Mexican crime wave, the police followed with a mass roundup of suspects, arrested twenty-four young men for murder, and actually indicted seventeen of them. Beaten by police, forced to appear disheveled in court, all were eventually convicted, nine for second-degree murder and eight on lesser charges, despite what the appellate court described as the lack of any tangible evidence against them. They were, in Royce's word, *greasers,* and that was enough for the original judge and jury. However, all the convictions were reversed on appeal, and the young men were freed. This was not only a victory of justice; it was also a nodal point in the development of group consciousness and the first victory of the community over the Anglo establishment.

The blatant hostility of Los Angeles law enforcement officers to Mexican Americans, while not as notorious as that of the Texas Rangers, was pervasive. The following excerpts are from evidence presented to the Los Angeles County Grand Jury in 1943 by the sheriff's department expert on Mexican Americans, Captain E. Durand Ayres. He reported accurately enough on the discrimination that the group faced.

> Mexicans are restricted in the main to only the lowest kinds of labor . . . the lowest paid . . . [T]hey are discriminated against and have been heretofore practically barred from learning trades . . . [including in our defense plants] in spite of President Roosevelt's instructions to the contrary. . . . Discrimination and segregation . . . in certain restaurants, public swimming plunges, public parks, theaters, and even schools, cause resentment among the Mexican people. . . . There are certain plunges where they are not allowed to swim, or else only on one day of the week [and that invariably just prior to cleaning] . . . signs [read] 'Tuesdays reserved for Negroes or Mexicans.'

Ayres followed this accurate account with an all-too-typical view of the causes of Mexican American crime and delinquency.

> The Caucasian [and] especially the Anglo-Saxon, when engaged in fighting . . . resort[s] to fisticuffs . . . but this Mexican element considers [good sportsmanship] to be a sign of weakness, and all he knows is a desire to use a knife or some other lethal weapon. In other words, his desire is to kill, or at least let blood. . . . [It is] difficult for the Indian or the Latin to understand the psychology of the Anglo-Saxon or those from northern Europe. When there is added to this inborn characteristic that has come down through the ages, the use of liquor, then we certainly have crimes of violence.

This attitude of a so-called law enforcement expert helps explain the Zoot Suit riots in Los Angeles during the spring of 1943. The Zoot Suit was a fairly expensive costume worn by some Mexican American and black young men consisting of a jacket that tapered to the knees from exaggerated padded shoulders; pegged, high-waisted, narrow-cuffed trousers; watch chains almost long enough to touch the ground and porkpie hats with a feather. Although the press and some of the few popular historians who have noted the riots speak of attacks by Mexican Americans on sailors or of mutual clashes, it is clear that almost all the aggression was on the side of the servicemen and that they were aided by both civilian and military police.

As Carey McWilliams has described it, after certain clashes (almost certainly over young women and "turf") between sailors on pass or leave (not generally the most decorous group in the population) and Mexican American teenagers and youths, the sailors, with the tacit approval of the naval authorities and the police, made organized assaults not only on the bizarrely attired zoot-suiters but on any Mexican American they could catch. After receiving accolades from the press ("Sailor Task Force Hits L.A. Zooters"), for several nights informal "posses" of servicemen proceeded to beat, strip, and otherwise humiliate every Mexican American young man (and some blacks) they could find. When they could not find them on the streets, bars and movies were invaded in search of victims, all with the same kind of impunity once given to San Francisco's vigilantes. The reign of terror ended only when the local naval authorities, acting on orders from Washington, declared Los Angeles off limits to sailors. Most of the latter were neither veterans of Pacific combat nor Californians, but rather newly inducted trainees drawn from all over the nation.[3]

Although the more egalitarian climate of opinion in the post–World War II years has erased or mitigated much of the most blatant forms of discrimination, a socioeconomic profile of the Mexican American community clearly demonstrates its generally disadvantaged status. Political dreamers from Gunnar Myrdal to Jesse Jackson have envisaged a united front of what the Swedish reformer called "the American underclasses" and what Jackson calls the "rainbow coalition" in which blacks, "Hispanics," Asian Americans, and poor whites would gain political power or at least more influence in how the country is run. Those dreams have proved largely chimerical. Although black American voters have shown great cohesion and solidarity, other groups in the putative coalition, including Mexican Americans, have not. While Mexican Americans in California *have* elected local officials—mayors and city councilpersons—and a handful of congressmen, their political participation remains relatively low. In addition, many members of the community, including some of the most successful acculturators, have sought to distance themselves from blacks, a common phenomenon among immigrant groups.[4]

Part of the pattern of Mexican American life can be seen in the relatively low naturalization rates. Less than a quarter of the foreign-born Mexicans counted by the 1980 census—520,000 of 2.2 million—were naturalized. If one subtracts the number of those who told the census taker that they had arrived in the period 1975–80 and thus were not yet eligible for naturalization, 65 percent of the pre-1975 arrivals from Mexico were still not citizens (950,000 of 1.47 million), while only 34 percent of such Asians were unnaturalized (460,000 of 1.34 million).

The reasons for the relatively low naturalization rates of foreign-born Mexicans are many. Large numbers expect to return home, and many do. Relatively low rates of political participation, especially among women, are another symptom of the same phenomenon, arising from the economic class structure of Mexican American society and a long-standing reluctance on the part of many to participate in what they perceive as Anglo politics—a politics that, many of them feel, stole their country and has abused their people for a century and a half.

Mexican Americans, like many immigrant groups before them, have been slow to participate in education to the degree that many other ethnic groups have. This is in part due to the discrimination in the form of education deprivation that they have met in most American school systems in the Southwest and West, discrimination that has ranged from consistent underfunding of schools in which Mexican Americans predominate to punishing students for speaking Spanish, even at recess. The fact that many Mexican American students have been made to feel that they, their language, and their culture are not welcome in schools has only served to reinforce the cultural patterns that have led so many Mexican Americans not to complete high school, much less go on to college. Census Bureau data show that in 1987 only 45 percent of Mexican Americans twenty-five years of age and older had a high school diploma.

The strong relationship between education and income that exists in American society clearly applies to the Mexican Americans as well. A Census Bureau study for 1987 claimed that the median annual Hispanic family income was $20,310 and that the median Mexican American family income was slightly below that, at $19,970, while the median income of all white families was $32,270, almost 62 percent higher. (Had the bureau calculated the incomes of whites other than Hispanics the disparity would have been even greater.)

Although they were once largely rural and involved chiefly in agricultural labor, Mexican Americans today are overwhelmingly urban. By the 1970s some 85 percent of Mexican Americans lived in cities. Nevertheless Mexican American men are more than three times more likely to be farm laborers than are American male workers generally, and Mexican American women are even more heavily overrepresented in the agricultural labor force. At the other end of the spectrum, Mexican American men and women are underrepresented in the more prestigious and well-paying professional and technical occupations.

The persistence and continued use of the Spanish language among even native-born Mexican Americans has been pronounced and, as we shall see in the concluding chapter, has helped to trigger a nativist "English only" movement in the 1980s. There are a number of reasons for this, including proximity to the border, a continuing migration from Mexico, and a population highly concentrated into urban enclaves called barrios. The Spanish-language media in the United States—newspapers, magazines, and radio and television stations—are a multimillion-dollar industry and growing rapidly. Perhaps the most prosperous are stations KMEX and KMEX-TV in Los Angeles.

However, the perception of many outsiders that large numbers of native-born Mexican Americans cannot speak English is false. Many of them prefer to speak Spanish in some situations, and large numbers of native-born children have been and are raised in homes in which Spanish is spoken. But unlike earlier groups such as nineteenth-century German Americans, Mexican Americans have not created their own schools and few of them attend other than public schools.

Although the overwhelming proponderance of Mexican Americans is Roman Catholic, their relationship with the American church has been difficult and conflict ridden. There have been very few Mexican Americans among the religious of the American church and even fewer in the hierarchy—which, until recently, was often hostile. Well into the 1970s, for example, in the Los Angeles archdiocese with the largest number of Mexican American communicants, priests were discouraged if not forbidden to use Spanish in sermons, something most of them could not have done anyway. In addition, as was true for many devout Euro-American Catholics earlier, the formalism of the largely Irish American church turned off many Mexican Americans used to a religion that was imbued with both Spanish and Amerindian elements. Thus church attendance among Mexican Americans has been relatively low, and they are still served by proportionally few priests, churches, and other religious institutions. And, at times in the past if not in the present, church-run social institutions have made invidious distinctions between Mexican American Catholics and other Catholics. In Los Angeles in the 1930s, for example, Catholic charities routinely allocated fewer dollars per capita to needy Mexican American families than to others because it felt that Mexicans did not need as high a standard of living to get along.

Yet the piety of the bulk of the Mexican American population cannot be doubted. The most spectacular demonstration of this in recent years has been the crusadelike fervor that has accompanied the activities of the United Farm Workers in what many Mexican Americans call simply *la causa,* the cause. Led by César Chavez (1927–1993), the Farm Workers movement has combined orthodox trade union methods—organizing, strikes, and boycotts—with symbolic fasting, outdoor masses, and processions featuring the Virgin of Guadalupe, the most prominent religious symbol of Mexico. The effect of this movement in raising the consciousness of Mexican Americans, or Chicanos as many in the movement prefer to call themselves, is hard to overestimate. An anecdote will illustrate this combination of religion and politics in Southern California.

In the spring of 1968, during the heated Democratic presidential primary campaign in California, I was taken to a meeting of Chicano activists (all male, as were many such meetings). In discussion afterwards I discovered overwhelming support for Robert F. Kennedy and absolutely none for his rival, Eugene McCarthy. (Those who didn't support Kennedy advocated nonparticipation in an Anglo election.)

When I inquired what they had against McCarthy, I was told that, after all, he was an Irish Roman Catholic "like the cardinal" (James Francis Cardinal McIntyre, then archbishop of Los Angeles). When I remonstrated that Robert Kennedy was also an Irish Roman Catholic—it seemed to me a self-evident proposition—I was informed that Robert Kennedy had "taken mass with César Chavez" and that "no Irish Roman Catholic would do that." To them "Irish Roman Catholic" described not an ethnoreligious combination but a cultural style, and they perceived, quite accurately, that Robert Kennedy did not participate in that style. It is also significant that although few if any of the young men I met that afternoon attended church regularly, they saw participation by an outsider in a community religious event as something of great significance.

Puerto Rico and the Federal Government

Since Puerto Rico was annexed by the United States in the aftermath of the Spanish-American War of 1898, Puerto Ricans are not, strictly speaking, immigrants. They were American nationals and, after 1917, American citizens by birth, and their comings and goings were not affected by immigration legislation. Between annexation and the end of World War II, few Puerto Ricans migrated to the United States despite extreme poverty on the island, and no legal restraints existed to keep them out. There were no cheap means of transportation between the territory and the mainland. In 1940 there were fewer than seventy thousand Puerto Ricans or persons of Puerto Rican ancestry on the American mainland; almost 90 percent of these lived in New York City. The Census Bureau estimated in 1987 that 12 percent of the Hispanic population was Puerto Rican, about 2.25 million persons, and that more than half of them lived in New York City. The postwar development of relatively cheap air fares between San Juan and New York City—in the 1940s the fare was about fifty dollars, two weeks' wages—had enabled tens of thousands of Puerto Ricans to come annually.[5]

In Puerto Rico itself population growth since annexation has also been rapid. A census taken in 1899 found nearly a million people on the island, which has 3,435 square miles, a little over one-third the size of Vermont. By 1980, despite the massive migration to the mainland, the population had grown to more than three million. Thus, there are over five million Puerto Ricans and, given the continuing migration, there may soon be more persons of Puerto Rican birth on the mainland than in the Caribbean commonwealth. There are already more Puerto

Ricans in New York City than in San Juan, the island's capital and largest city.

A few Puerto Ricans had immigrated to the United States in the decades before the Spanish-American War. Most of them were political exiles who hoped to create an independent nation from a base in New York City. These included Francisco Gonzalo Marín (1863–97), who published his newspaper, *El Postillón,* there after it had been suppressed on the island, and Santiago Iglesias (1872–1939), founder of Puerto Rico's Socialist Party and one of the leaders who never stopped struggling for independence. American annexation aborted the main thrust of the independence movement, although it has never died. The legal relationship between the United States and Puerto Rico, dictated by the American Congress—in which Puerto Ricans are not represented—has evolved over the years.

For a brief period after the almost bloodless conquest of 1898 there was a military government. In 1900 the Foraker Act gave the island a modicum of local government: Under it Puerto Ricans were nationals, not citizens, like contemporary Filipinos. The Jones Act of 1917 declared all Puerto Ricans citizens unless they formally rejected that status. Until Rexford Guy Tugwell (1891–1979) was appointed governor late in Franklin Roosevelt's presidency, most governors had been nonentities. Tugwell, the New Deal reformer whose book *The Stricken Land* is still a good description of the socioeconomic realities of the island, helped pave the way for some political reform. His successor, in 1946, was the first Puerto Rican appointed governor, and in the following year Congress passed the Elective Governors Act. In 1948 Luis Muñoz Marín, (1898–1980), whose father had been one of the nineteenth-century political exiles in New York, was the first governor elected by the people of Puerto Rico. Two years later the Puerto Rican Federal Relations law was enacted by Congress. It enabled Puerto Ricans to draft a constitution as they saw fit, *except* that the options of either independence and statehood could not be exercised. Such a constitution was drawn up, approved by Congress, accepted by Puerto Ricans in a referendum in 1951 and put into effect in 1952. Under its provisions Puerto Rico is an *estado libre asociado,* literally, "associated free state" but translated as commonwealth. Under its provisions Puerto Rico remains a U.S. possession subject to most federal laws, including the draft. Puerto Ricans may not vote in presidential elections and have no senators or representatives, although they do elect a resident commissioner who sits with Congress but has no vote. Puerto

Ricans tax themselves and they, and businesses in Puerto Rico, pay no federal income taxes.

The status of Puerto Rico has been discussed by the United Nations more than once and will undoubtedly be discussed again. In 1953 the UN General Assembly refused to categorize the island as "non-self-governing," because of the 1952 constitution. In 1977 without significant effect, all three major Puerto Rican political groups came before the UN Trusteeship Council to protest the island's status. The ruling Popular Democratic Party complained because it did not have statehood; the Popular party, because autonomy had not been extended; and the several parties that supported independence, because that had not been granted. One wing of the relatively small independence movement has resorted to violence, most spectacularly in an attempt to assassinate President Harry S. Truman in 1950.

The most compelling reason that independence does not enjoy majority support is an economic one. Without the infusion of federal dollars, Puerto Rico would be even poorer than it now is. In 1984, for example, the U.S. Treasury recorded transfer payments to individuals and governments in Puerto Rico of $3.4 billion dollars, amounting to almost a quarter of the island's gross domestic product.

All of this raises the question of just how poor Puerto Rico is. On a Latin American/Caribbean scale, not so poor. Puerto Rico has one of the highest, if not the highest, per capita incomes of any place in the region. But by American standards it is quite low, much lower than that in the poorest American state. And it must not be imagined that all the effects of the relations with the United States have been beneficial: Some of Puerto Rico's problems are caused by, not eased by, that relationship. Not surprisingly migration and remigration between the island and the mainland is constant. A segment of that remigration consists of school-age children raised in the United States whose Spanish is so poor that they cannot function effectively in the island's schools and need remedial attention. Many of these children, in fact, cannot perform well in either language.

PUERTO RICANS IN THE UNITED STATES

Most Puerto Ricans who have come to or been born in the United States face two related problems: poverty and race prejudice. Puerto Ricans are a racially mixed group. The indigenous Amerindian population of the island was largely killed by the Spanish conquerors and the diseases they brought with them. The Spanish introduced African slavery to

Puerto Rico in 1511—it lasted until 1873—and there is a large admixture of white and black ancestry in the island's population. Although Puerto Rico is not without color prejudice, it is less pervasive and total than that existing on the mainland, and for many newcomers the experience of American-style race prejudice in which one is either black or white is a shock and a major social problem. In the view of social theorists such as Myrdal, blacks and Puerto Ricans in New York City and elsewhere ought to be political allies but in fact are more often rivals.

They are also rivals in another sense: They "compete" for places at the lower end of the poverty spectrum. This is complicated by the fact that the Census Bureau puts many Puerto Ricans in two categories: They are both black and Puerto Rican. Among Hispanic groups, Puerto Ricans are at the bottom or near the bottom according to most of the criteria by which the disadvantaged are measured. In median family income, for example, 1987 census data, which divided the American population into two groups, "Hispanic" and "not Hispanic," found the latter with twice the income of the former: $31,610 to $20,310. A different dichotomy, white and black, gave slightly more disparate figures; $32,270 and $18,100. Specifically Puerto Rican data gave the group the lowest figure, $15,190 in median family income for 1987. Mexicans had a figure of $19,970, while other Hispanic groups tended to be higher, topped by Cuban Americans at $27,290. A corollary figure, that of percentage of households headed by women, showed the Puerto Rican profile much more similar to that of blacks than to other Hispanics. Blacks and Puerto Ricans in the United States have rates that are statistically all but identical: 42 percent for blacks, 43 percent for Puerto Ricans in 1987. Mexicans and Cubans, on the other hand, have only 19 percent and 18 percent of all families headed by women. In education, 54 percent of Puerto Ricans over twenty-five were high school graduates in 1987, as opposed to 45 percent of Mexicans and 62 percent of Cubans. Many data demonstrate that second- and third-generation Puerto Ricans, like the second and third generations of most immigrant groups, have tended to improve their economic and educational standing and, as was also true for most immigrant groups, that those who moved away from the established center of immigration—for Puerto Ricans, New York City—tended to have higher incomes and so on. This was true of modest-size Puerto Rican communities in San Francisco, Los Angeles, and Lorain, Ohio. The latter community, attracted there by industry including steel mills and a Ford plant, has shown high social mobility: In the 1970s more than half the families owned their own homes, and women headed only 7 percent of Puerto Rican families there. But it is New York City—and to a lesser degree its environs—which is crucial to the Puerto Rican experience (see Table 12.1).

Although the data indicate the steady dispersal from a center or centers that is one of the characteristic patterns of ethnic distribution in the United States, that distribution is even more restricted than these tables suggest. In 1980, for example, of those first two generations of Puerto Ricans recorded as being outside New York City, the majority were in the adjoining states of Connecticut, New Jersey, and Pennsylvania and in upstate New York.

Some data suggest that the Puerto Rican community is evolving in ways that are more similar to those of earlier European migrants than to those of contemporary Mexican Americans. These include patterns of education and of exogamous marriage. Joseph P. Fitzgerald, noting that more than a third of second-generation married Puerto Ricans in the United States were married to non-Hispanic whites, and that this marriage pattern correlated strongly with increased education—more

Table 12.1

Puerto Ricans in New York City and the United States, 1950–1980

FIRST GENERATION (BORN IN PUERTO RICO)

YEAR	NEW YORK CITY	REST OF UNITED STATES	TOTAL	PERCENTAGE IN NYC
1950	187,420	38,690	226,110	82.9
1960	429,710	185,674	615,384	69.8
1970	473,300	336,787	810,087	58.4
1980	—	—	—	—

SECOND GENERATION (BORN IN THE UNITED STATES)

YEAR	NEW YORK CITY	REST OF UNITED STATES	TOTAL	PERCENTAGE IN NYC
1950	58,460	16,805	75,265	77.7
1960	182,964	89,314	272,278	67.2
1970	344,412	236,964	581,376	59.2
1980	—	—	—	—

TOTAL: FIRST AND SECOND GENERATIONS

YEAR	NEW YORK CITY	REST OF UNITED STATES	TOTAL	PERCENTAGE IN NYC
1950	245,880	55,495	301,375	81.6
1960	612,574	275,088	887,662	69.0
1970	817,712	573,751	1,391,463	58.8
1980	852,833	1,161,112	2,013,945	42.3

education means more outmarriage—predicted that "intermarriage will increase as the educational level rises, presumably reducing ethnic identification as a result."

Like so many ethnic Catholics who have come to the United States, Puerto Ricans have felt alienated from the American church. Unlike so many other groups, Puerto Ricans do not have significant numbers of priests migrating with them, and even on the island only about one-third of the Catholic clergy are ethnic Puerto Ricans. The American hierarchy has been more than reluctant, in recent decades, to establish ethnic parishes as it once did. The United States Catholic Conference, as well as a number of dioceses in the East and Southwest, has established bureaucratic organizations to try to meet the needs of its Spanish-speaking communicants, but for the Puerto Ricans these have not been particularly effective. Beginning in 1976 there have been national meetings of Spanish-speaking laity, but these, like so many national organizations of the Spanish speaking, have been dominated by Mexican Americans. Of the first six bishops of Hispanic origin, only one was in New York, the center of Puerto Rican population: He was a native of Spain. And although as many as half of the Catholics in Manhattan and Brooklyn are Spanish speaking—and Puerto Ricans are the largest segment of those—the hierarchy is, from top to bottom, dominated by European ethnics, especially Irish Americans.

As is true for many Catholics of Latin American ancestry, large amounts of what is usually called folk religion have become intertwined with more traditional Christian practices. A walk through a Puerto Rican neighborhood in New York reveals a large number of *botánicas,* places where the artifacts of folk religious practices are sold. These include incense, herbs, potions, and charms. Puerto Rican folk religious practices, often characterized by outsiders as spiritualism, center on the belief that one can communicate with the dead and that spirits from the beyond can be invoked to solve problems of daily life, console the bereaved, and so on. The followers of these practices often attribute bad things that happen to Puerto Ricans, for example, mental illness, to the influence of evil spirits.

As the data for economic status suggest, relatively few Puerto Ricans have thus far enjoyed the upward social mobility that has been the unifying theme of the American immigrant experience. Puerto Ricans who have become prominent on the mainland seem concentrated in two fields, the arts and politics, although for a while the most prominent Puerto Rican on the mainland was the baseball superstar, Roberto Clemente (1934–73). Those with entrepreneurial skills, training, and capital are more likely to remain in Puerto Rico where there are, of course, upper and middle classes of considerable size. In the arts there is an obvious and natural tendency for artists and would-be artists to come to a metropolis such as New York or Los Angeles for training and performance, just as those from, say, South Carolina would do. Among the performing artists who have been particularly successful are actors Jose Ferrer and Raul Julia and opera singers Martina Arroyo and Justino Diaz. In mainland politics, Puerto Ricans have achieved prominence in a number of state and local positions in and around New York City and in the House of Representatives; none has yet been elected a state governor or a United States senator.

This brief account of two of the earliest Spanish-speaking minorities in the United States shows how artificial the census aggregate "Hispanic" is. Mexican Americans and Puerto Ricans share, in addition to language, a common religion and a common poverty. But they share little else. They live in different regions, have different traditions, and, in the color-conscious United States, are regarded as being of different races: Most Puerto Ricans are regarded as being black and most Mexican Americans as being white, although each group is more accurately described as being of mixed racial origin.

The regions in which most of them live are vastly different, with different traditions of dealing with minority groups. And these regional traditions as well as their premigratory history help explain some of the differences in their adaptations to life in America. One way of explaining the vast difference in the school dropout rate, for example, is to note that in Puerto Rico, American-style school systems and attendance patterns have been established for almost a century and that the school system in New York City has long been conditioned to the notion of integrating a kaleidoscopic procession of ethnic groups. The California pattern was, for decades, to segregate or exclude Mexican Americans. The economic differences are at least in part explainable by the fact that most Mexican Americans live in what is now styled the Sun Belt, which has been, in recent years, the most economically dynamic part of the United States. Puerto Ricans, to the contrary, are centered in an area of slower growth and where there is little work for the relatively unskilled. We thus have the anomaly, by American standards, of better-educated Puerto Ricans—one might put that, "less poorly educated"—earning less money than less-well-educated Mexican Americans.

Both groups suffer from a common phenomenon, a phenomenon quite different from the situation that had been faced by earlier immigrant groups. Most of the latter also arrived relatively poor and uneducated, but they came to an America in which a person—particularly young adult males—could earn a modest competence without markedly improving his skills. In addition, most of his competitors in the labor force were similarly situated and he could expect more "greenhorns" to come who would be even less acculturated than he. Those conditions no longer apply. Outside of the shrinking and poorly paid agricultural sector such jobs no longer exist. And the second and third generations of Puerto Ricans and Mexican Americans (along with the native black poor) find that most of the growing number of immigrants in post–World War II America arrive with significantly more skills, education, and acculturation to postindustrial capitalism than they themselves possess.

13

Changing the Rules: Immigration Law, 1948–1980

The Immigration Act of 1924 and the national origins system it established remained the basic immigration law of the land until 1965. During the rest of the 1920s, the 1930s, and the early 1940s, the law seemed unassailable. Yet we can now see, as contemporaries could not, that the repeal of Chinese exclusion in 1943 and the granting of a token quota to Chinese was the beginning of the end for the racist and discriminatory system that was so much the product of the tribal twenties. In 1946 Congress, in separate actions, used what we can call the Chinese formula for two other Asian groups: Filipinos and "natives of India." Both groups received tiny quotas and admissibility to citizenship. For the Filipinos, as it had been for the Chinese, the change was a good behavior prize and a celebration of Filipino independence as well. The application of this formula to natives of India, who were still colonial subjects of the British Empire, was largely due to the successful lobbying of Asian Indians resident in the United States, particularly a New York merchant who called himself J. J. Singh. A prosperous Sikh businessman with great powers of persuasion and endurance (Robert Shaplen, writing in *The New Yorker,* called him a "one-man lobby"), Singh managed to get bipartisan support for his cause. Its passage left the status of Asians in a curious state. Chinese, Filipinos, and "natives of India"—who were in 1947 partitioned into the two independent states of India and Pakistan—had limited quotas and were eligible for naturalization, which meant, as had been true for Chinese, that the spouses and minor children of naturalized citizens could enter as non-quota immigrants. But all other Asians—Japanese, Koreans, and Southeast and Southwest Asians—were still "aliens ineligible to citizenship" and thus denied admission as immigrants.[1]

This anomalous situation persisted for six more years, until 1952. In that year Congress passed, over President Harry S. Truman's veto, the McCarran-Walter Act. While its other provisions will be discussed later in this chapter, here it must be noted that the act ended the total exclusion of racial and ethnic groups from naturalization and immigration and did what Charles Sumner had wanted to do in 1870: made the naturalization laws color blind. Just as the granting of the right of naturalization to Chinese in 1943 should be seen as a war measure, the further liberalization in 1952 should be seen as a fruit of the Cold War. Engaged in a struggle for the hearts and minds of what it liked to call the Free World, the United States could no longer afford a policy that so blatantly excluded so many. Despite this liberalization, however, the McCarran-Walter Act continued most of the discriminatory policies of the 1924 act and essentially preserved the national origins system. And, it should be noted, no one then even dreamed that immigration from Asia would, in just three decades, amount to approximately half of all legal immigration, or that immigration would be running at some five million a decade.

Displaced Persons and the New Pattern of Immigration

Although there had been grumblings, from both inside and outside of Congress, that making eligibility for naturalization global was "lowering the barriers," there was, by 1952, a very broad consensus in favor of it. Even Nevada's crusty Senator Patrick A. McCarran, who regarded himself as the guardian of America's gates and a staunch defender of the national origins concept, was in favor of it. But proposals to admit hundreds of thousands of European refugees in the immediate post–World War II period were a different matter and provoked bitter and prolonged controversy. Many Americans feared that the country might be swamped by refugees from a devastated Europe that was economically insecure and politically unstable, with Communist parties growing in almost every nation. Others, reflecting the fears of the twenties rather than the realities of the forties, worried that it was all a plot to let in more Jews. There were, in fact, few Jews among the millions of displaced persons in the Western European camps, and most of those were headed for Palestine/Israel, but the memory of what the United States had done—and failed to do—about refugees in the prewar period was an important part of the debate that raged in the late 1940s and early 1950s.[2]

The political debate began at the end of 1945 when President Truman, who had ignored the growing scandal about the neglect of displaced persons for months, proposed that half of the continental European quota slots, some forty thousand per year, be reserved for DPs, who would have to meet all of the criteria of existing immigration law. One often-overlooked aspect of that Truman directive is that it, for the first time, involved voluntary agencies, most of them religious, in the official sponsorship of refugees. This relationship between the VOLAGS, as they came to be called, and the immigration process has developed enormously since that time, and, as we shall see in the final chapter, they played a crucial role in amnesty after 1986. During the first nine months of 1946 only about five thousand DPs had been admitted, and Truman began to see that any effective program would have to go outside of the quota system. In his 1947 State of the Union message the president urged Congress to "find ways whereby we can fulfill our responsibilities to these suffering and homeless refugees of all faiths." The notion that the United States had responsibilities for what happened to foreigners was consonant with American pretensions to world leadership, but it had never gained majority support in the prewar period and was still controversial.

A bipartisan coalition, in and out of Congress, supported the president's position, and in April 1947, Representative William G. Stratton, a relatively conservative Republican from Illinois, dropped a bill into the hopper calling for the admission of one hundred thousand displaced persons annually *over and above* the quota numbers for four successive years. This would breach the quota system in a major way and was resisted bitterly by supporters of the old system. After more than a year of acrimonious debate and maneuvering, a compromise Displaced Persons Act of 1948 was passed at the end of June. It is the first piece of legislation in American history that set *refugee policy* as opposed to immigration policy.

The bill accepted the basic definition of displaced persons that had been adopted by the UN's International Refugee Organization (IRO) as people who were

> victims . . . of the nazi or fascist . . . or . . . quisling regimes . . . or Spanish Republicans and other victims of the Falangist regime in Spain . . . *or* persons who were considered refugees before the outbreak of the second world war, for reasons of race, religion, nationality, or political opinion . . . who have been deported from, or obliged to leave their country of nationality or of former habitual residence.

To this IRO definition, which reflected an antifascist World War II orientation, Congress added a number of categories reflecting an anti-communist Cold War orientation. These included *Volksdeutsche,* persons of German ethnicity who fled or were expelled from Eastern Europe, Italian refugees from the Venezia-Giulia, and post-1948 refugees from Czechoslovakia. In one of those absurd but face-saving compromises so typical of congressional politics, it was pretended that the quota system had been left intact. All of the newcomers were charged to some specific quota, and if the quota was not big enough it could be mortgaged. Within four years, in what was an extreme but not unique case, the Latvian quota of 286 annually had been mortgaged for over three centuries to the year 2274. The total to be admitted was also a compromise. The 1948 act cut the original proposal almost in half, allowing the admission of almost 250,000 DPs over two fiscal years. Not surprisingly, as the act was running out in 1950, it was amended to run for two more years and the authorized total was upped to 415,000, slightly more than in the original bill.

Due to the increasing Cold War orientation, only a minority of those admitted under the two acts were Hitler's victims: a larger number, perhaps a majority of all who came, were members of groups that had benefited from the policies of the Third Reich. These latter were chiefly *Ostdeutsche,* ethnic Germans from Eastern Europe and some Baltic peoples. In addition we now know that, during this period and later, certain agencies of the executive branch, notably the Department of Defense and the Central Intelligence Agency, brought hundreds of former Nazis into the country for various reasons of state, including the so-called Operation Paperclip, which concentrated on rocket scientists. To legitimize this process, a little-noticed clause was added to the 1949 act governing the administration of the CIA, giving its director a quota as large as that of many small nations to admit aliens "in the interest of national security or essential to the furtherance of the national intelligence mission." In addition, an unknown number of former Nazis and even war criminals slipped through the State Department and INS screenings, which were more and more concerned with keeping out communists than fascists.

All in all something fewer than 450,000 DPs were admitted from the end of the war in Europe until the 1948 and 1950 acts expired on June 30, 1952, which works out to something over 60,000 persons per year. Despite much talk, overt and covert, about the country being flooded by Jews, Leonard Dinnerstein has calculated that nearly 140,000 Jews were admitted to the United States in the same period, fewer than

20,000 persons per year. During that time total legal immigration into the country was 1.3 million, or just below 200,000 annually and the baby boom was adding 3.4 to 3.9 million persons to the population each year.[3]

The struggle over the DP bills helped focus attention on the immigration laws generally, with most traditional liberals arguing for scrapping or greatly modifying the national origins system and most traditional conservatives insisting on little or no change. Essentially a continuation of the national origins system, the 1952 McCarran-Walter Act's major innovations were: the previously noted liberalization of the naturalization laws; a broadening of the family reunification provisions so that female citizens could bring alien husbands into the country as nonquota immigrants; and to give preference within the quotas to alien husbands of resident alien wives. Truman vetoed the bill, which was reenacted over his veto. The president specifically approved of the major innovations and objected to what he called the "greatest vice of the quota system":

> it discriminates, deliberately and intentionally, against many of the peoples of the world. The purpose behind it was to cut down and virtually eliminate immigration to this country from Southern and Eastern Europe.

After the McCarran-Walter Act had become law, Truman appointed a Commission on Immigration and Naturalization, merely another example of the way in which his administration responded to events rather than helping to shape them. Shortly before he left the White House the commission issued a report, "Whom We Shall Welcome," which was highly critical of the 1952 act and generally consonant with the views expressed in Truman's veto message. It recommended scrapping the national origins system and instead would allocate visas without regard to national origin, race, creed, or color and issue them according to five principles: the right of asylum, family reunification, needs in the United States, needs in the "Free World," and general immigration. Although it spoke of a "maximum annual quota" it instead proposed an annual limitation on the number of immigrants to be allowed in using the same formula as did the 1924 and 1952 acts: one-sixth of 1 percent of the population annually. But where the existing act still used the 1920 census figures, the Commission proposed using the latest data (1950) with decennial updates. This would have raised the number eligible to enter annually from 154,657 to 251,162.

In addition it proposed to use 100,000 of the spaces for each of three years for the admission of refugees, expellees, escapees, and remaining displaced persons.[4]

The commission's report, at least in the short run, was an exercise in futility. The report came too late to be a factor in the 1952 election, so there was no pressure on the Eisenhower administration to do other than ignore it. McCarran denounced the report as communist inspired and insisted, falsely, that "the rock of truth is that the Act does not contain one iota of racial or religious discrimination." He pointed to the many supporters of the new law which included not only such nativist standbys as the American Legion, the Veterans of Foreign Wars, and the American Coalition of Patriotic Societies, but also organizations which spoke for new immigrant groups, such as the National Catholic Welfare Council and the Japanese American Citizens League. Modified nativism had found some strange allies, at least in terms of traditional coalitions that had prevailed in the 1920s.

Neither the liberals nor the conservatives understood the dynamics of the immigration forces in the post-World War II world. The McCarran-Walter Act, which was on the books for thirteen years, envisaged a total of just over two million immigrants in that time, while the Truman Commission's formula would have produced about 3.25 million. In fact, about 3.5 million came in legally in 1952–65, and only one-third of those admitted were quota immigrants. Both groups, essentially refighting the fights of the 1920s, focused on Europe, the traditional source of most immigration to America. For the 1950s, Europeans still represented just over half of all immigrants (52.7 percent); by the 1960s they were just a third; and by the 1980s Europeans represented only just over a tenth of all legal immigrants. Immigration from Canada, almost entirely of Europeans or the descendants of Europeans, was an important but steadily declining portion of American immigration, ranging from an all-time high of 22.5 percent in the 1920s to 12.4 percent in the 1960s, on down to 2 percent in the mid-1980s.

The great gainers, of course, were Latin Americans and Asians. Latin American immigration had been quite small until the decade 1911–20, when the combined impact of the Mexican Revolution and the wartime curtailment of European immigration pushed it to 7 percent, about half of it from Mexico. Restriction of European immigration in the next decade doubled the Latin American share, with nearly four-fifths of that coming from Mexico. The percentage dropped in the 1930s, rose back to just above the 1920s level in the labor-short World War II decade—and this figure does *not* include braceros—and then grew steadily to reach more than a fifth of all immigrants in the 1950s and nearly two-fifths in the 1960s. In the 1970s the Latin American share of legal immigration peaked at just over two-fifths and declined in the mid-1980s to just over a third. A decline in share, it must be noted, did not necessarily mean a decline in numbers, as the total numbers of immigrants has risen steadily since World War II. The two-fifths share in the 1970s brought about 180,000 Latin Americans in each year, while the percentage in the 1980s meant that just over 200,000 came in annually. Mexicans were a large but declining proportion of these immigrants: about half in the 1950s, just over a third in the 1960s and 1970s, and just under a third in the mid-1980s.

The growth of Asian immigration was even more startling. From a mere 3 percent of all immigrants in the 1940s, it rose to 6 percent in the 1950s and nearly 13 percent in the 1960s, and zoomed to more than a third in the 1970s and to nearly half in the 1980s. The changing share, by global regions and Canada, is shown in Chart 13.1.

Not all of these changes were due to the 1952 law, as it was superseded by a quite different measure in 1965. But, as the data suggest, there were long-term forces operating throughout the post-war years. Among these were changes in Europe: Few Eastern Europeans were able to emigrate and, after the postwar European economic boom began, fewer and fewer Western Europeans wanted to. In Latin America political upheaval, steep population growth, and deteriorating economic conditions produced massive pressures to emigrate, with the United States, "El Norte," the obvious target, although there were and are significant population movements within Latin America. Continued American involvement in Asia meant more, and more varied, Asian immigration, while the process of modernization in many Asian countries produced a growing educated elite who wished to, and under the new immigration dispensations were able to, emigrate to places where their hard-won skills and capital could be advantageously and safely employed. And, over and above all these factors, were imperatives dictated to American policymakers by the Cold War and by the changing attitudes toward race and ethnicity on the part of the American people.

Nor was immigration to North America the only current. V. S. Naipaul, who emigrated from Trinidad to Britain, has written:[5]

> In 1950 in London I was at the beginning of that great movement of peoples that was to take place in the second half of the twentieth century. . . . Cities like London were to change. They were to cease

being more or less national cities; they were to become cities of the world, modern-day Romes, establishing the pattern of what great cities should be, in the eyes of islanders like myself and people even more remote in language and culture. They were to be cities visited for learning and elegant goods and manners and freedom by all the barbarian peoples of the globe, people of forest and desert, Arabs, Africans, Malays.[5]

Cold War Refugees

Although the word *refugee* does not appear in the McCarran-Walter Act, under an obscure section of it the attorney general was given discretionary parole power to grant temporary admission to unlimited

Chart 13.1
Legal Immigration to the United States by Region, 1931–1984

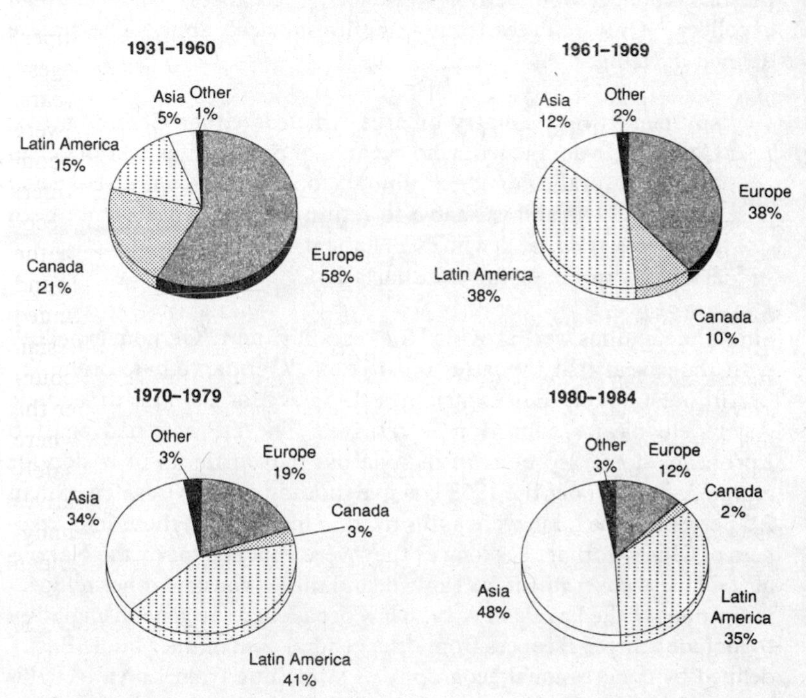

Source: Robert W. Gardner, Bryant Robie, and Peter C. Smith, "Asian Americans: Growth, Change, and Diversity," *Population Bulletin* Vol. 40, No.4 (Washington, D.C.: Population Reference Bureau, Inc., 1985), p. 2.

numbers of aliens "for emergency reasons or for reasons deemed strictly in the public interest." This was the method that Franklin Roosevelt had used, without prior congressional authority, to bring in nearly a thousand refugees in 1944. In practice, this provision meant, that the executive branch could act—for Hungarians, for Cubans, for Tibetans, for Vietnamese—and that Congress could later pass legislation regularizing that action. For the next twenty-eight years Congress passed piecemeal legislation designed to meet one refugee crisis after another, one of the reasons that actual immigration was so much higher than the 1952 act envisaged. (The other was that nonquota immigration was so much greater than the experts had assumed.)

The initial such law came in the first year of the Eisenhower administration, 1953. The Refugee Relief Act of that year authorized the admission of 205,000 nonquota persons over a two-and-a-half year period. Its Cold War orientation can be seen in its description of the persons whom it would cover: "refugees," "escapees," and "German expellees." Only refugees from communism need apply. The statute defined a "refugee" as

> any person in a country or area which is either Communist or Communist-dominated, who because of persecution, fear of persecution, natural calamity or military operation is out of his usual place of abode and is unable to return thereto, who has not been firmly resettled, and who is in urgent need of assistance for the essentials of life or for transportation.[6]

Similar definitions were provided for "escapee" and "German expellee" with the caveat that the latter must be of "German ethnic origin."

Although largely Eurocentric, the 1953 act for the first time made Asians eligible for admission as refugees. The DP acts had covered Europeans domiciled or born in Asia but had made no provision for ethnic Asians. Under the 1953 law five thousand such visas, fewer than 2.5 percent of the total, were authorized to include two thousand "refugees of Chinese origin," as long as they were vouched for by the Nationalist Chinese government on Taiwan, and three thousand other refugees indigenous to the Far East. A 1956 law broadened the general eligibility to include refugee-escapees from "the general area of the Middle East," defined by congressional geography as stretching from Libya to Pakistan and specifically including Ethiopia. Castro's revolution in Cuba brought Latin American refugees into the congressional definition in 1962, and the 1965 Immigration Act brought refugees into regular immigration legislation by providing a special "seventh preference" for refugees supposed to amount to 6 percent of annual immigration. In the thirty-five years after World War II—from 1945 to 1980, when a general refugee act was passed for the first time—more than 2.25 million persons were admitted to the United States as refugees. (Others, who were actually refugees, were admitted as quota and nonquota immigrants.) Table 13.1 lists the major categories and gives the official numbers and estimates of how many persons were involved.

Table 13.1

Persons Admitted as Refugees, 1945–1980

YEARS		NUMBER
	ADMITTED UNDER STATUTE OR PAROLE AUTHORITY	
1948–52	Displaced Persons	about 450,000
1953–56	Refugee Relief Act	about 205,000
1956	East European orphans	925
1956–57	Hungarians	38,045
1958–62	Portuguese (victims of earthquakes in the Azores)	4,811
1960–65	East European escapees (Fair Share Refugee Law)	19,745
1962	Chinese from Hong Kong and Macao	14,741
1962–79	Cubans	692,219
1963	Russian Old Believers from Turkey	200
1972–73	Ugandan Asians	1,500
1973–79	Soviet Union	35,758
1975–79	Indochinese (10 separate programs)	about 400,000
1975–77	Chileans	1,400
1976–77	Chileans, Bolivians, Uruguayans	343
1978–79	Lebanese	1,000
1979	Cuban prisoners	15,000
1980	Refugees, all sources	110,000
	ADMITTED UNDER 1965 IMMIGRATION ACT	
1968–80	Seventh preference	about 130,000
	ADMITTED WITHOUT AUTHORITY	
1980	Cuban-Haitian "special entrants" (boat people from Mariel, etc.)	140,000
Total		about 2,261,564

NOTE: Since total legal immigration 1945–80 is recorded as 10,943,489, refugees are just over 20 percent of the total.

The Immigration Act of 1965

As Nathan Glazer has astutely observed, three pieces of legislation in 1964 and 1965—the Civil Rights Act of 1964 and the Voting Rights Act and the Immigration Act of 1965—represent a kind of high-water mark in a national consensus of egalitarianism, one from which much of the country has receded in subsequent years.[7] We are here concerned only with the immigration act, but its passage, and particularly its noncontroversial passage, cannot be understood without reference to the climate of opinion of which it was a part. Little noticed at the time and generally ignored by historians of the Johnson era, who concentrate on the more spectacular and controversial programs of the Great Society and on the escalation of the misbegotten war in Southeast Asia, it is nevertheless a vitally important piece of legislation. It finally ended the national origins system and has helped to change the face of American society. Much of what it has accomplished was unforseen by its authors, and had the Congress fully understood its consequences, it almost certainly would not have passed.

Each of Truman's immediate successors, Dwight Eisenhower and John Kennedy, gave lip service to changes in the McCarran-Walter Act, but neither provided effective leadership to make change occur. Although Eisenhower pushed for and got greater refugee admissions, he made no proposals to amend the act until his last year in office when he proposed that immigration should be increased and that unused quota spaces, like those from Great Britain and Ireland, be allocated to countries, such as Italy and Greece, which had many more applicants than quota spaces. This would have blunted much of the nativist thrust of the act, but no congressional action was taken. During the 1960 campaign Richard Nixon pledged support for Eisenhower's proposals.

The winner of the 1960 election was an advocate of immigration reform. In 1958 John F. Kennedy had signed his name to a book, *A Nation of Immigrants,* strongly supportive of immigration as a source of national strength, and of immigration reform. Yet more than two years of his presidency passed before he sent a proposal for immigration reform to Congress. It had three major provisions:

First, that the quota system be phased out over a five-year period, with the released numbers being put into a pool to be distributed on a new basis.

Second, that the natives of no one country receive more than 10 percent of the newly authorized quota numbers.

Third, that a seven-person immigration board be set up to advise the president, composed of two members appointed by the Speaker of the House, two members appointed by the president pro tem of the Senate, and three by the president. After receiving their recommendations the president would be authorized to reserve up to 50 percent of the unallocated numbers to "persons disadvantaged by the change in the quota system" and up to 20 percent to refugees.

Otherwise, immigrants would be selected under a first-come, first-admitted procedure, based on registration at American consulates, subject to the 10 percent limitation noted above. Family reunification was to remain important, and the president proposed expanding the nonquota group to include the parents of American citizens, and to give parents of resident aliens a preferred status. Other proposed changes included elimination of the racist provisions by which persons of Asian ethnicity, wherever born, should be charged to the quota of Asian nations, that nonquota status be given to natives of Jamaica and Trinidad and Tobago as it was to other Western Hemisphere residents, that the preference structure for skilled immigrants be liberalized, and that certain persons with mental health problems should be admitted if their families could demonstrate the ability to support them. (This provision, one of the most detailed in the message, perhaps reflects the sensitivity of a man with a mentally retarded sister.)

The Kennedy proposal would have left much unchanged, including all of the ideological admission criteria, the lack of any quota restraint for Western Hemisphere immigrants, and even the total number of quota immigrants to be admitted annually, 156,700. No significant congressional action had been taken at the time of his assassination. Those who loved him assure us that, had he lived, immigration reform would have surely come. The historian can only wonder.

Kennedy's successor, Lyndon Johnson, had no proimmigrant track record. In fact he had joined the overwhelming majority of senators in voting to override Truman's 1952 veto of the McCarran-Walter Act. But President Johnson was not Senator Johnson, and 1964 was not 1952. In his first State of the Union message in 1964 Johnson endorsed immigration reform, called for an end to the national origins system, and generally endorsed the proposals that President Kennedy had made. Immigration reform was not the most pressing item on the Great Society agenda, and no legislative action took place that year. Nor did it play a major role in the 1964 election campaign, although the GOP vice-presidential candidate, William Miller, warned that the Democrat's immigration program would open the "floodgates," he quickly backtracked when his position was attacked.

When Congress took the matter up after the election, some kind of major revision of immigration law was clearly in the cards. There was no major effort to defend the national origins concept, although a few southern senators raised the old slogans. Chief of them was North Carolina Democrat Sam Ervin, who insisted, as was the wont of those who supported the status quo, that rather than being discriminatory, the McCarran-Walter Act was

> like a mirror reflecting the United States, allowing the admission of immigrants according to a national and uniform mathematical formula recognizing the obvious and natural fact that those immigrants can best be assimilated into our society who have relatives, friends, or others of similar background already here.

What Ervin and others never explained was that the "mirror" was badly distorted, like those at amusement parks, and tended to reflect not the population of the 1960s but that of a much earlier time, before the turn of the century.

The meaningful debates in Congress, however, were not over whether to change the national origins system but over how to change it. The two most disputed provisions were over whether the chief thrust of the new law would be skills presumably needed by the United States or family reunification, and over the question whether to continue the practice of having no set limit on immigration from the Western Hemisphere. Contrary to the original proposal submitted by the Kennedy administration, family unification took precedence and, in fact, dominated the bill, and a ceiling was placed for the first time on Western Hemisphere immigrants. The bill retained most of the other barriers that Congress had been erecting to immigration since the late nineteenth century, including the LPC clause, the requirements of physical and mental health, and the various ideological tests.

President Johnson signed the new law on October 3, 1965, in a ceremony held on Liberty Island in New York Harbor with the decaying and unused buildings on Ellis Island in the background. The Texan, for once, underplayed the importance of a piece of legislation for which he was responsible.

> This bill that we sign today is not a revolutionary bill. It does not affect the lives of millions. It will not reshape the structure of our daily lives, or really add importantly to our wealth or our power.

The bill, in fact, changed the whole course of American immigration history, although it did so along lines that were already apparent for the few who had eyes to see. In addition, it facilitated a great increase in the volume of immigration. In no year since the bill passed have fewer than a third of a million legal immigrants entered the United States, and since 1978 half a million or more have entered in most years. The most striking effect of the new law has been further to increase the share of immigration slots going to Asia and Latin America.

It is doubtful if any drafter or supporter of the 1965 act envisaged this result. In his Liberty Island speech Lyndon Johnson had looked to the past, not the future, stressed the fact that he was redressing the wrong done to those "from southern or eastern Europe," and although he did mention "developing continents," there was no other reference to Asian or Third World immigration. Members of his administration, almost certainly in good faith, had testified before Congress that few Asians would come in under the new law, and this view was shared by the nation's major Asian American organization, the Japanese American Citizens League. Although it supported the bill as a step forward, as it had supported McCarran-Walter in 1952, it complained that

> although the immigration law eliminated race as a matter of principle, in actual operation immigration will still be controlled by the now discredited national origins system and the general pattern of immigration which exists today will continue for many years yet to come.

The basic thrust of the new law, which was, technically, a series of amendments to the 1952 act, was to scrap national origins and to substitute in its place overall hemispheric caps on visas issued: 170,000 for persons from the Old World, 120,000 for persons from the New. No country in the Old World was to have more than 20,000 visas. (In 1978 this was changed to a global ceiling of 290,000, with the 20,000-per-country caps applying everywhere). But a parallel system, essentially that of the old nonquota immigration system, kept a large number of persons exempt from preference requirements and numerical limitation. Table 13.2 shows the preference systems established under the 1952 and 1965 acts. It clearly shows that there was more continuity than change from one system to the other, once the discriminatory concepts of national origins and the Asia Pacific Triangle had been removed, and remembering that the basic numbers had been upped.

In practice, the law has worked quite differently from the way its

Table 13.2

Preference Systems: 1952 and 1965 Immigration Acts

IMMIGRATION AND NATIONALITY ACT, 1952

Exempt from preference requirements and numerical quotas: spouses and unmarried minor children of U.S. citizens.

1. Highly skilled immigrants whose services are urgently needed in the United States and the spouses and children of such immigrants. 50 percent.
2. Parents of U.S. citizens over age twenty-one and unmarried adult children of U.S. citizens. 30 percent.
3. Spouses and unmarried adult children of permanent resident aliens. 20 percent.
4. Brothers, sisters, and married children of U.S. citizens and accompanying spouses and children. 50 percent of numbers not required for 1–3.
5. Nonpreference: applicants not entitled to any of the above. 50 percent of the numbers not required for 1–3 above, plus any not required for 4.

IMMIGRATION ACT OF 1965

Exempt from preference requirements and numerical quotas: spouses, unmarried minor children and parents of U.S. citizens.

1. Unmarried adult children of U.S. citizens. 20 percent.
2. Spouses and unmarried adult children of permanent resident aliens. 20 percent (26 percent after 1980).
3. Members of the professions and scientists and artists of exceptional ability. 10 percent: requires certification from U.S. Department of Labor.
4. Married children of U.S. citizens. 10 percent.
5. Brothers and sisters of U.S. citizens over 21. 24 percent.
6. Skilled and unskilled workers in occupations for which labor is in short supply in the U.S. 10 percent: requires certification from U.S. Department of Labor.
7. Refugees from communist or communist-dominated countries, or the Middle East. 6 percent.
8. Nonpreference: applicants not entitled to any of the above. (Since there are more preference applicants than can be accommodated, this has not been used.)

sponsors expected. The great fallacy was the belief that there were large numbers of European immigrants ready, qualified, and able to come to America. Most seemed to think that twenty thousand annual arrivals from many European countries would absorb most of the Old World immigration visa slots. Had the 1965 act, or something like it, been passed in 1952, when Truman recommended revision, such a result would have probably been obtained. By 1965 most of those Western Europeans who wished to come were not likely to meet LPC and other restrictions. Growing numbers of Asians had been coming to the United States since 1943 and many other well-trained Asians were able to qualify under the various preference provisions of the 1965 act. Once an Asian (or anyone else) came here as a permanent resident alien, a whole cohort of other persons became eligible to enter the country as third-preference immigrants. And as soon as the original immigrant became a United States citizen, as large numbers of them did after the minimum five-year waiting period, more persons became eligible as second-preference immigrants, while others could come in exempt from numerical requirements. And, of course, those brought in as preference immigrants could start the same procedure. Since the 1965 act went into effect, the preponderance of all nonrefugee migration has been the chain migration of relatives, a process likely to continue as long as the law stays as it is and conditions in Asia and Latin America remain as they are.[8]

Although the law speaks of a global ceiling of 290,000 annually, that applies to those subject to numerical limitation. The actual numbers have been much greater. In the ten years between 1976 and 1985, for example, recorded legal immigration ranged from a low of 398,089 in 1977 to a high of 796,356 in 1980 and averaged some 546,000, 88 percent higher than the presumed ceiling. Table 13.2 shows the categories of immigrants who were officially admitted in the low and high years, 1977 and 1980. The two great variables are those exempt from numerical limitation and refugees. The former has grown steadily: in the period 1976–85 from 113,083 to 210,761 and, if the trend continues, may soon exceed the number admitted under numerical limitation. The number of refugees fluctuates. The special Cuban-Haitian entrants figure for 1980 refers to the so-called Mariel exodus and other boat people from Cuba and Haiti who were apprehended (see the discussion in chapter 15).

The question of needed professions and skills, which bulked so large in the thinking of immigration reformers in the 1960s, has proved, partially because of the slower than anticipated growth in the American economy, of minor significance. It has amounted, in recent years, to less

Table 13.3

Numbers and Categories of Immigrants Admitted, 1977 and 1980

CATEGORY	1977	1980
Subject to numerical limitation	275,531	289,479
Exempt from numerical limitation	117,857	165,325
Refugees	4,701	201,552
Cuban-Haitian entrants	—	140,000
Total	389,089	796,356

NOTE: Most, but not all, immigrants are included in the year they actually arrive in the country, but there are exceptions. Those who come as temporary immigrants but have their status changed to permanent residents are included in the year that the change of status takes place. Refugees are included in the year of their arrival, but asylees, a category added in 1980, when (and if) their status is approved.

than 4 percent of immigration. Because so many immigrants come in without regard to numerical limitation, the 20,000 cap on persons from any one country have proved chimerical. In 1985, for example, there were 61,000 legal Mexican immigrants, 48,000 Filipinos and 35,000 Koreans.

Clearly, the 1965 law has not worked out as its proponents expected. The experts simply did not know what they were talking about and had not, obviously, paid close attention to what had been happening in immigration patterns to the United States. Two major adjustments to the law, the Refugee Act of 1980 (to be discussed later in this chapter) and the Immigration Reform Act of 1986 (to be discussed in the final chapter) have not materially changed the functioning of the law. The 1965 law was enacted at a time when immigration was only peripheral to the major political and social concerns of most Americans. Under its auspices immigration grew rapidly, but it was a growth that had been going on since the 1930s, when immigration all but disappeared and amounted to just half a million persons. In the 1940s a million came, 2.5 million in the 1950s, 3.3 million in the 1960s, 4.5 million in the 1970s, and about 6 million in the 1980s. By the beginning of the 1980s, the country had witnessed the Mariel episode and was experiencing a general feeling of helplessness that came to a head during the last months of Jimmy Carter's presidency but had its roots in the war in Southeast Asia and the economic dislocations it triggered. For these and a variety of other reasons, immigration matters moved toward the center of public consciousness and began, again, to take on largely negative connotations. Before that happened or, at least, fully crystallized, the Carter administration put the first comprehensive refugee law on the statute books.

The Refugee Act of 1980

The Refugee Act of 1980 was an attempt to solve, once and for all, the problem of refugee admissions. As we have seen, no refugee policy existed in statute law before World War II, and the first refugee law came only in 1948. The 1965 Immigration Act attempted to incorporate refugee policy into a general immigration act by setting aside a constant portion of regular admissions (the seventh preference: 6 percent of a presumed 290,000, or 17,400 a year) for refugees. The number was probably unrealistic in any event. On the very day that Lyndon Johnson signed the 1965 act he also exercised his parole authority—an authority that every president since Franklin Roosevelt had exercised—to authorize a resumed flow of airlifted Cubans to the United States, a flow that would approach four hundred thousand fifteen years later when Congress was considering what became the 1980 Refugee Act. But it was the end of the Vietnam War that demonstrated clearly the inadequacy of the seventh-preference allocation. Beginning in 1975, first the Ford and then the Carter administration stitched together ten makeshift parole programs that brought four hundred thousand Vietnamese and other Southeast Asians to the United States between 1975 and 1979. It was the experience with Southeast Asian refugees that largely shaped the 1980 law, and, as events were quickly to demonstrate, that experience, in which most refugees were initially in countries of "first asylum," was not a good guide to the immediate future.

By 1979 there seemed to be general agreement that the gap between what American law authorized in terms of refugee admissions and what was actually being done was so great that a general refugee law would have to be created right away even though Congress had created a Select Commission on Immigration and Refugee Policy to survey immigration and refugee policy and make broad policy recommendations in 1981. A Carter administration bill proposed to increase the annual total of refugees to 50,000 and to reduce the annual quota numbers, by steps, to 270,000, making the total annual admission numbers 320,000 plus however many came in "without numerical limitation" from 1981 on. But the bill provided that, although 50,000 would be the normal number, the president might raise that number after annual consultation with Congress. This was, to all intents and purposes, a continuation of the president's parole power in the tradition set by Franklin Roosevelt. In the event, as table 13.4 indicates, the 50,000 number quickly proved illusory.

Table 13.4

Number of Refugees and Asylees Admitted, 1981–1985

YEAR	REFUGEES	ASYLEES	TOTAL
1981	155,291	1,175	156,466
1982	93,252	4,731	97,983
1983	57,064	8,333	65,397
1984	67,750	11,627	79,377
1985	62,477	6,514	68,991
Total	435,834	32,380	468,214

The definition of "refugee" was broadened to conform with the United Nations 1968 definition:

> any person who is outside any country of his nationality or in the case of any person having no nationality, is outside of any country in which he last habitually resided, and who is unable or unwilling to return to, and is unable or unwilling to avail himself of the protection of that country because of persecution, or a well-founded fear of persecution, on account of race, religion, nationality, membership of a particular social group, or political opinion.

This was broader than the previous statutory definition, although the Senate had "advised and consented" to the UN protocol in 1968, so this part of the new law was not a significant departure from previous policy.

But one aspect of the law did break new ground: For the first time the laws of the United States recognized the right of asylum and created a new legal category of refugee, an "asylee." Asylees are refugees and must meet the criteria for refugees. But unlike other refugees, an asylee is a person who applies for entry into the United States while already here, either legally, such as a person who came in on a student or visitor's visa or, as has often been the case since 1980, arrived illegally. The 1980 act put a cap of five thousand asylees annually, although almost immediately, partially because of federal court orders, that cap was exceeded and in 1984 more than doubled. The law also provided different ways of counting refugees and asylees. The former appear in immigration statistics in the year in which they arrive, the latter in the year in which they are approved. Both may adjust their status a year after arrival/approval and, five years after that, are eligible for naturalization with all the rights of citizenship status including, of course, the right to bring in relatives. Most aspects of this law were relatively uncontroversial at the time of passage. The original bill sailed through the Senate 85–0 and had the cosponsorship of that political odd couple, Edward M. Kennedy on the left and Strom Thurmond on the right. At the end of World War II, making provision for any refugees had been highly controversial. Thirty-five years later the notion that the United States was obligated to take in refugees and have a generous policy toward them—or some of them—was totally uncontroversial. Having a refugee policy was part of the new American consensus.

The Mariel Crisis and the Ambiguity of Refugee Policy

The ink was hardly dry on the 1980 act—it was enacted on March 17—when the Mariel crisis began on April 21. It was an outgrowth of a complicated three-way dispute between Cuba, Peru, and the United States over the right of asylum for some 3,500 dissident Cubans in the Peruvian Embassy in Havana. The United States agreed to take in these people if they passed a screening and were allowed to go first to Costa Rica. Fidel Castro then declared that any persons who wished could leave, but they had to go directly to the United States. The dramatic boatlift from the Cuban port of Mariel to Key West and other places in South Florida began and got massive coverage on television. Cuban exiles in Miami chartered boats and violated United States law by going to Cuba to pick up relatives, friends, and anyone else there was room for. In a matter of weeks some 125,000 Cubans had arrived, shattering all notions about an "orderly" refugee policy.

The Carter administration found itself on the horns of a dilemma. On the one hand Cold War imperatives made it wish to accept those seeking freedom from communism and to embarrass its enemy, Fidel Castro; on the other hand, it wanted the bureaucratic order suggested by the 1980 Refugee Act. The result was strenuous vacillation. First there were stern warnings about breaking the law and a few fines were levied on some of the earliest boats. Then, after the public reaction to such moves seemed hostile, President Carter himself proclaimed, two weeks after the exodus had begun, that the United States would welcome with "an open heart and open arms" the tens of thousands of freedom seekers coming to Florida. Toward the end of May, amid reports that Castro was opening his jails and mental institutions and allowing the former inmates to leave, and with a noticeable cooling in public support for the boatlift, the Carter administration again "got tough," reimposing fines and even seizing some boatlift vessels, and the exodus came to a halt. In a wonderful piece of bureaucratese the government decided that these people (and the Haitians discussed below) were neither refugees nor asylees (the new law had gone into effect in April) but were rather "Cuban-Haitian Entrants (Status Pending)." The problem of what to do with them was complicated by fears of the then newly discovered plague of AIDS among both groups (for the Cubans, it was one of the fruits of their misbegotten war in Angola). The Reagan administration dithered about as badly as had the Carter people, but less publicly. In July 1985—more than five years after the crisis—all but about 2,500 of the Marielitos were allowed to adjust to permanent resident alien status, and in 1990 most of them would begin to be eligible to bring in family members.

Haitians, on the other hand, who began arriving in their own or chartered boats well before the Mariel crisis but in much smaller numbers, were treated in a very different way. Fleeing from a succession of corrupt authoritarian regimes from the nation with the lowest per capita income in the New World, Haitians were declared to be fleeing from poor economic conditions. Their boatlifts, which had through 1980 brought perhaps fifteen thousand persons to the United States that the government knew about, was largely halted by the Coast Guard towing their boats back to Haiti.

The government—in this case *both* the Carter and Reagan administrations—argued that a distinction had to be made between economic and political entrants lest the United States be "flooded" with refugees. There were millions of refugees in the world: One 1987 estimate spoke of 13.3 million worldwide, up from 9.9 million in 1984. But most of those were in countries of first asylum—Laos, Pakistan, and several African nations—from which entry to the United States for purposes of asylum was all but impossible. The United States government can pick and choose, through screening, the few thousands it admits annually from among these millions. The asylum seekers, the Mariel Cubans, the Haitians, the Central Americans (who will be discussed in the final chapter), the occasional elite defector, all present themselves at times and places of their choosing. In the case of refugees in faraway countries of first asylum, the executive branch has a relatively free hand: In the case of asylum seekers, the federal courts and sometimes public opinion can play a part, and Congress is more likely to intervene. The debate is an ongoing one, not likely to be settled soon. The treatment of Haitians has been ascribed to race prejudice. Representative Bruce Morrison (D., CT), chair of the House subcommittee on immigration, told a reporter in mid-1989: "There's been a lot of discrimination with Haitians. They are black, they are from a nation close to ours, and their country isn't Communist." Government officials naturally deny this since, after the 1960s, official discrimination has no longer been permissible, another example of how the American consensus has shifted. The Miami district director of the INS, Perry Rivkind, even resorted to fantasy in his denial of what seems obvious to many: "I've always said I wish a boatload of blue-eyed Anglo-Saxon Protestants tried to enter the U.S. illegally. They too would be subject to exclusion."

Rivkind's fantasy masks a different reality. Immigration officials can no longer confidently say who is and who is not illegal. American law now legalized a certain kind of unauthorized entry: the entry of an asylum seeker with "a well-founded fear of persecution." Just what amounts to such a fear, as Congressman Morrison suggests, depends both on what kind of regime is doing the persecuting and on who is making the judgment. Such complexities seem now to be a permanent part of the once-simple matter of immigrating to America, and it is clear why some officials resort to fantasy and a nostalgia for what must seem to them, if not to others, to have been the good old days, when categories were simple and the fiats of INS officials were only rarely subject to review.

14

The New Asian Immigrants

In the years since the end of World War II, and especially since the passage of the 1965 Immigration Act, no aggregate group has benefited more from the changes in American immigration law than have Asian Americans. Many, if not most, of the Latin Americans who came, it can be argued, would have come anyway, and Cold War mandates dictated the acceptance of Cuban refugees. Asian Americans also were beneficiaries of the general trend toward a more egalitarian society, at least in terms of race and ethnicity, that peaked in the mid-1960s. As we saw in chapter 11, shifting attitudes toward our Chinese ally during the war triggered the first positive change in the law in 1943. Two other factors helped bring about the changes. The belated admission of Hawaii to statehood in 1959 has meant the presence of Asian American senators and representatives providing political clout in Washington. In addition, the juxtaposition of these so-called model minorities with the perceived nonachievement of black internal migrants and most of the "Hispanic" population, hastened the increased acceptance of Asians, especially vis-à-vis other contemporary newcomers.

In demographic terms the growth of Asian American population has been startling. In 1940 Asian Americans were less than four-tenths of 1 percent of the American people: This incidence increased to half of 1 percent in the next twenty years. By 1980 Asians were 1.5 percent, and one responsible projection for the year 2000 has hypothesized Asian Americans as 4 percent of the population. This would represent a tenfold increase in incidence in sixty years. Table 14.1 ignores projections and shows Asian American population by major ethnic group at each census beginning in 1940.

If one looks only at the aggregate data, Asian Americans are younger than the average American. They also have fewer children, are less

Table 14.1

Asian American Population, by Major Ethnic Group, 1940–1980

YEAR	CHINESE	JAPANESE	FILIPINO	KOREAN	ASIAN INDIAN	VIETNAMESE	TOTAL	PERCENTAGE OF U.S. POPULATION
1940	106,334	285,115	98,535	—	—	—	489,984	.4
1950	150,005	326,379	122,707	—	—	—	599,091	.4
1960	237,292	464,332	176,310	—	—	—	877,934	.5
1970	436,062	591,290	343,060	69,150	—	—	1,439,562	.7
1980	812,178	716,331	781,894	357,393	387,223	245,025	3,466,421	1.5

NOTE: The data here include Hawaii, so they are not directly comparable to the data in tables 9.1 and 9.2.

There were a few thousand Koreans and Asian Indians in the United States prior to their appearance in the national census data in 1970 and 1980, respectively, going all the way back to the first decade of this century. There may have been 75,000 Asian Indians in 1970, most of them recent immigrants. There were only handfuls of Vietnamese before 1960. Other groups enumerated in 1980 included Laotians, 47,683; Thai, 45,279; Kampuchean, 16,044; Pakistani, 15,792; Indonesian, 9,618; Hmong, 5,204, and 26,757 others.

likely to be unemployed or in jail, and are more likely to get higher education than the average American. These and other characteristics have led to Asian Americans being called the model minority, a conception that takes certain middle-class norms as the ideal. But if one looks at the various ethnic groups, one sees all kinds of differences, both between and within groups. Taking the simplest and least value-laden category, we find the aging Japanese American population, which is overwhelmingly native born, with a median age of 33.5 years as opposed to a U.S. median of 30.0. Although recent immigrants tend to be much younger, Asian Indians are slightly above the national norm at 30.1. Other Asian groups range downward from Chinese at 29.6 to Vietnamese at 21.5. By comparison, black Americans have a median age of 24.9, while Hispanics were at 23.2. To understand these and other differences, it will be necessary to encapsulate briefly the history of each Asian ethnic group, something Harry H. L. Kitano and I have done at greater length in *Asian Americans: Emerging Minorities* (2001).[1]

Japanese and Chinese

Sparked by a heavily female immigration in the years after 1952, Japanese American population growth, which had slowed perceptibly in the 1920s and 1930s—it actually declined in the latter decade in the contiguous United States—has been significantly more rapid than that of the United States, but much less steep than that of other Asian ethnic groups. Despite an annual quota of just 185 annually until 1965, some 45,000 Japanese immigrated to the United States between 1952 and 1960. About 40,000 of them (85.9 percent) were female and a majority of them were married to non-Japanese soldiers and former soldiers. After the early 1960s, immigration slowed. When the 1965 Immigration Act, which opened the door for so many Asian ethnic groups, was passed, few Japanese wished to emigrate. Had its provisions been on the books immediately after the war, undoubtedly emigration from war-devastated Japan would have been heavy. But, by 1965, as was also true for most Western Europeans, the economic motive to emigrate was no longer urgent for most Japanese. Japanese immigration slowed, absolutely and relatively, after 1960, and so did Japanese American population growth, most of which, in recent years, has been due to natural increase, the excess of births over deaths. The group's population growth was 14 percent in the 1940s, 42 percent in the 1950s, 27 percent in the 1960s, and 21 percent in the 1970s. (Comparable figures for the whole population were, 14 percent, 18 percent, 13 percent, and 11 percent.) There is every reason to believe that the slowing of Japanese American growth will continue. Japanese were the largest Asian American ethnic group from 1910 to 1970, by 1980 they were third, and one projection for the year 2000 puts them sixth, after Filipinos, Chinese, Vietnamese, Koreans, and Asian Indians.

As we have seen, Japanese were, by 1980, an aging population, slightly older than the American population as a whole or than any of its major aggregates, and since they had had fewer children and a slightly higher life expectancy than the general population, that aging process will almost certainly continue. Japanese American population predictions are complicated by the increasing amount of out marriage, mostly with Caucasians, for the third and subsequent generations. Data for Los Angeles County show such marriages there at or above 50 percent of all marriages involving Japanese since the 1970s. Japanese Americans, in 1980, were more likely to finish high school and to go on to college than were white Americans, something that was true for most Asian American groups. Japanese American income was not as high as the group's educational profile might suggest, but it was higher than that of most Americans. The median income of full-time Japanese American workers in 1979 was $16,829 as opposed to $15,572 for whites. But since so many Japanese Americans lived in the West (80 percent) where living costs are higher than average, the difference is even smaller than it seems. At the other end of the scale only 4.2 percent of Japanese American families lived in officially defined poverty, while 7 percent of white families did. The profile is very much of a largely middle- and lower-middle-class group, which is what Japanese Americans have become.

The Chinese American experience has been quite different. Population growth and immigration have increased in every decade. Population grew by 41 percent in the 1940s, 58 percent in the 1950s, 85 percent in the 1960s, and 86 percent in the 1970s; it is no longer possible to say exactly how many Chinese have immigrated in any one period, as Chinese come from many places: Taiwan, Hong Kong, Southeast Asia, and China itself are the major sources. Many who came in under the Vietnamese refugee programs were, in fact, ethnic Chinese, and now identify themselves as such. A substantial proportion of the more-than-half-million increase in the Chinese American population between 1960 and 1980 represents immigrants, and they and their children represented an absolute majority of that population.

When compared to Japanese Americans, Chinese Americans are younger (median age 29.6 years), less concentrated in the West (52.7 percent), have a slightly lower median income per full-time worker ($15,753), and more than twice as high a percentage of families in poverty (10.5 percent). It is clear that the presence of large numbers of recent immigrants depresses the income figures and adds to the poverty percentage. The 1979 income data for full-time workers who migrated from Taiwan or Hong Kong show much lower incomes, with the amount being directly proportional to the time spent in the country. Those who came after 1975 had median incomes of $9,676 and a family poverty rate of 22.8 percent; those who came between 1970 and 1974 had incomes of $12,392 and a poverty rate of 6.2 percent, while those who came before 1970 had incomes of $13,692, and a poverty rate of only 2.8 percent. The seeming paradox that persons, such as pre-1970 immigrants, earn less money but have a lower poverty rate than do all Chinese Americans is easy to explain. First of all, immigrant households tend to be larger and contain more unrelated individuals and fewer old persons and thus more wage earners. Second, the median for all Chinese is raised by having large numbers of relatively affluent persons. Third, there are obviously large numbers of native-born American Chinese with low incomes who are in poverty in spite of their model minority status.

A further difference between Japanese Americans and Chinese Americans can be seen in the education data. Using the 1970 census, Chinese and Japanese Americans seemed to present a similar educational profile. Looking at those persons who were twenty-five and older, 68.1 percent of Chinese Americans were high school graduates as opposed to 68.8 percent of Japanese Americans; median years of schooling completed were all but identical, 12.4 for the former and 12.5 for the latter. This similarity in overall data masked great communal differences. More than a quarter of the Chinese adults had not gone beyond the seventh grade, and more than a quarter were college graduates. For Japanese Americans the comparable figures were about a tenth not going beyond the seventh grade and almost a sixth graduating from college.

This and other data clearly show the bifurcated nature of the Chinese American community, a community that, despite great achievement, still contains considerable poverty and deprivation. Dr. Ling-chi Wang, a San Francisco community activist and chair of Asian American Studies at the University of California, Berkeley, testified before a U.S. Senate committee about the "silent" Chinese of San Francisco and presented data that showed conditions in San Francisco's inner-city Chinatown that were anything but model. Unemployment was almost double the citywide average, two-thirds of the living quarters were substandard, and tuberculosis rates were six times the national average.

Although many Chinese Americans speak of two kinds of Chinese—the ABCs (American-born Chinese) and FOBs (fresh-off-the-boat recent immigrants, although almost all arrive by plane)—the immigrant/native born dichotomy does not explain all of the differences. There is, however, a greater tendency for recent immigrants to be poorly educated, deficient in English, and to work in the low-paid service trades, such as laundries, restaurants, and the sweatshop enterprises typical of the inner city. Conversely more and more of the ABCs tend to be college educated, have middle-class occupations, and live in integrated or relatively integrated housing outside the inner-city Chinatowns that, until recently, were home for the vast majority of Chinese. Manhattan's Chinatown has become so crowded that satellite Chinatowns have developed in Queens, once the domain of white ethnics for whom Archie Bunker became the stereotype. To be sure, the dichotomy is not polar. Many immigrants are well-to-do, more than a few wealthy. The computer magnate An Wang was more than a billionaire, according to *Forbes* magazine's list in 1983, and was reported to be the largest single contributor to philanthropic causes in Boston. The world-reknowned architect I. M. Pei was probably the most famous Chinese American immigrant.

American-born Chinese are also making important contributions to American culture, most notably in literature. Particularly important has been the work of two women, Maxine Hong Kingston (1940–) and Jade Snow Wong (1922–). The former's magnificent first two novels—*The Woman Warrior* (1976) and *China Men* (1980)—although difficult, are must reading for anyone interested in the nature of Chinese life in America. The latter's two memoirs—*Fifth Chinese Daughter* (1950) and *No Chinese Stranger* (1975)—are more accessible, and, when read together, can show the increase in confidence of Chinese Americans in the quarter century that separates them.

The projection, noted above, that Chinese Americans would drop behind Filipinos in number by the year 2000 was called into question by the traumatic events in Beijing's Tiananmen Square in the spring of 1989. In the short run, at least, larger numbers of well-educated and well-to-do Chinese from Hong Kong are likely to emigrate than one would have expected initially, and some in Congress quickly moved to increase the number from Hong Kong eligible to come. Whether or not that happens, the Chinese American population can be expected to continue to grow significantly faster than the general population for the foreseeable future.

Filipinos

Three distinct increments of Filipinos have come to America. Shortly after the American annexation of the Philippines in 1898 came groups of students, with and without government support, some of whom stayed on and founded the Filipino American community. They settled chiefly in the Midwest and East. Then, in the 1920s and early 1930s, came farm workers who filled the same kinds of jobs in Far Western agriculture that Chinese and Japanese had pioneered. As American nationals, Filipinos could not be prevented from migrating to the United States although they, too, were aliens ineligible to citizenship until 1946. Most of these Filipinos were in Hawaii and California. In 1930 there were more than sixty thousand in the former, where they were 17 percent of the population, another thirty thousand in California, and perhaps fifteen thousand more in the rest of the United States. In California male Filipinos outnumbered females by fifteen to one. The current migration of Filipinos, coming largely after 1965, has been educated and consists of upwardly mobile professionals and would-be entrepreneurs, similar in occupational profile to many other recent Asian immigrant groups. Unlike most other "new" Asian immigrant groups, the majority of newcomers have been female. In 1960 just 37 percent of Filipino Americans were female; by 1980 nearly 52 percent were. No Filipinos have been in refugee programs, but a few elite Filipinos have been in political exile in the United States—most notably Corazon and Benigno Aquino and, after the Aquinos returned to the Philippines, Imelda and Ferdinand Marcos. The majority of each increment, like most Filipinos, have been Roman Catholics (hardly any Muslims from the Southern Philippines have come to America), but religion does not seem to play a major institutional role in the lives of most Filipino Americans.

The students were so few in number that they created hardly a ripple in the American consciousness. One scholar, Benicio T. Catapusan, a Filipino who earned a Ph.D. at the University of Southern California in 1940, estimated that between 1910 and 1938 some fourteen thousand Filipinos were enrolled in American schools. Most did not graduate, but many of those who did became leaders in government and business after returning to the Philippines. Many of those who did not finish stayed in the United States, usually plying the low-paying dead-end jobs that employed most Filipino Americans in those years. A number of Filipino intellectuals came to and lived in America. The most notable of them, the writer Carlos Bulosan (1911–56), wrote of the tension between the democratic ideals they learned about in the American-style school system in the Philippines and the harsh realities of American life. As Bulosan wrote to a friend:

> Do you know what a Filipino feels in America? . . . He is the loneliest thing on earth. There is much to be appreciated . . . beauty, wealth, power, grandeur. But is he part of these luxuries? He looks, poor man, through the fingers of his eyes. He is enchained, damnably to his race, his heritage. He is betrayed, my friend.[2]

Like the Japanese before them, Filipino laborers first migrated to Hawaii, where the Gentlemen's Agreement of 1907–8 had cut off new supplies of Japanese labor. According to Mary Dorita Clifford, between 1909 and 1934 the Hawaiian Sugar Planter's Association arranged for nearly 120,000 Filipinos to come to Hawaii: 86.6 percent were men, 7.5 percent women, and 5.9 percent children. On the expiration of their contracts, some went back to the Philippines, others stayed in Hawaii, and still others moved on to the West Coast of the United States. As Filipino population grew in California in the 1920s, an anti-Filipino movement flourished, fed by the same forces that had attacked Chinese and Japanese. By 1928 the national convention of the American Federation of Labor resolved that:

> Whereas, there are a sufficient number of Filipinos ready and willing to come to the United States to create a race problem . . . we urge exclusion of the Filipino race.

Persistent discrimination, including a change in the antimiscegenation laws of California and three other Western states to include "members of the Malay race," and a good deal of violence directed against Filipinos, made the anti-Filipino movement similar to the anti-Chinese and anti-Japanese movements which preceded it. One added theme was a persistent tendency to depict Filipinos as savages, a tendency that probably stemmed from the exhibitions of Igorot and other tribal peoples from the islands at American world's fairs in the years following annexation. In fact, most immigrants were not "primitive" but modernized Filipinos who spoke two European languages, Spanish and English, as well as a Philippine language such as Tagalog, Visayan, or Ilocano. But one California judge, in 1930, declared from the bench that Filipinos were but ten years removed from savagery while another, after the depression had set in, gave his opinion:

> It is a dreadful thing when these Filipinos, scarcely more than savages, come to San Francisco, work for practically nothing and obtain the society of white girls. Because the Filipinos work for nothing, decent white boys cannot get jobs.

The special status of Filipinos as American nationals caused the political side of the anti-Filipino movement to become, at least superficially, anti-imperialist, since Filipinos could be excluded only if the Philippines were independent. In 1934 the Tydings-McDuffie Act promised the Philippines independence in 1945 and gave a quota of fifty persons per year, half of the previous minimum for any quota nation. Devout exclusionists, such as California's senior senator, Hiram W. Johnson, had wanted total exclusion. Filipinos still remained "aliens ineligible to citizenship." Between 1934 and the end of World War II, Filipino population in the United States aged and dwindled, but the war years changed the image of the Filipinos to that of loyal allies against the Japanese while wartime prosperity in the United States somewhat improved their economic status. In 1946 Filipinos were made eligible for naturalization and the islands' quota was doubled to one hundred annually. Although Filipino immigration was not large in the immediate postwar years, it was much larger than the quota would suggest. During the thirteen years of the McCarran-Walter Act, when there were a total of 1,300 quota spaces, the INS recorded 32,201 Filipino immigrants, most of them, obviously, nonquota immigrants. In 1963 for example, 3,618 Filipino immigrants were recorded. They were outnumbered by 13,860 "nonimmigrants"—tourists, businesspeople, students, and the like—many of whom were eventually able to change their status to immigrant. In addition, one category of Filipinos was given special status: Those who had served in the American armed forces—including the militialike Philippine Scouts—during World War II could become American citizens while still in the Philippines and thus enter the United States with their immediate family members. The United States Navy, which had traditionally recruited Filipinos as messmen and to perform other menial tasks, continued to do so even after the Philippines became independent. In 1970, for example, there were some 14,000 Filipinos serving in the American navy, more than in the Filipino navy. Thus, between 1950 and 1960, Filipino population in the United States increased 50 percent, most of it on the mainland. After the passage of the 1965 act, Filipino immigration increased sharply—in many years since 1965 Filipinos have been the largest or second largest nationality immigrating—and the population zoomed, as table 14.2 indicates.

The post-1965 migration, although it continued trends already manifested in Filipino immigration, was in many ways different from that which had gone before. The geographical distribution of Filipino shifted east and south. Whereas in 1950 about half of all Filipinos lived in Hawaii, in 1980 only about a sixth did. California had become the home of nearly 46 percent of Filipinos. Southern states, where previously few Filipinos had lived although a handful had been reported in Spanish New Orleans in the late eighteenth century, now contained about a tenth of Filipino Americans, as did the Midwest and Northeast.

Most, perhaps two-thirds, of recent Filipino immigrants have been professionals, most notably nurses and other medical personnel. During the 1970s, for example, the fifty nursing schools in the Philippines graduated about two thousand nurses annually. At least 20 percent of each year's crop of graduates soon migrated to the United States, where shortages of trained nurses, especially those willing to work the long and uncomfortable hours demanded by public hospitals, provided instant employment. Many hospitals recruit nurses in the Philippines, and hospital administrators have testified that many of our urban medical treatment facilities could not continue to operate if all the foreign medical personnel were removed. Almost seven out of ten foreign-born Filipino women over sixteen were in the labor force in 1980, a higher proportion than that of any other Asian ethnic group. Nonmedical professionals, however, often are employed at jobs well below their skill levels, a phenomenon common among immigrant professionals generally, except at the very highest levels. (Albert Einstein and other Nobel

Table 14.2

Filipinos in the United States, 1960–1990

YEAR	TOTAL	PERCENTAGE OF INCREASE	MALE	FEMALE	PERCENTAGE FEMALE
1960	181,614	48.0	114,179	67,435	37.1
1970	336,731	85.4	183,175	153,556	45.6
1980	774,652	130.0	374,191	400,461	51.7
1990	1,419,711	83.3	656,765	762,946	53.7

Prize winners got appropriate academic employment even in the 1930s, but many well-qualified refugee academics of that era were never able to secure positions appropriate to their training and expertise.) Filipino lawyers may find work as clerks, teachers as office workers, dentists as dental technicians, and engineers as mechanics. However, they *do* find work, and the lower-status jobs in the United States often pay better than do higher-status jobs in the Philippines.

Assuming that the immigration laws do not change, there is no reason to expect that Filipino immigration will slacken, as economic instability in the Philippines seems endemic. And should political oppression again occur there, a number of refugees might seek to claim asylum privileges provided by the 1980 Refugee Act.

Asian Indians

Although some Asian Indians came to the United States in the late nineteenth and early twentieth centuries, all but a few thousand of the more than six hundred thousand Asian Indians in the United States at the end of the 1980s are post–World War II immigrants and their children, and the vast majority of these are persons who have come since 1965. The awkward term *Asian Indian* was adopted by the Census Bureau for the 1980 census at the urging of the immigrant community: This was to avoid confusion with Amerindians on the one hand and Pakistanis and Bangladeshis on the other. Previously the government—and the *HEAEG*—used the term *East Indian.*[3]

The early migrants, ten thousand or so, may be divided into two groups: poor laborers in the American West who comprised the bulk of the migrants, and several small elite groups distributed across the country. The former were almost all Sikhs; the latter were a cross section of India's elite groups, including many Sikhs and a few Muslims.

The Sikhs, from the fertile Punjab, first came to the United States around the turn of the century from Western Canada and worked in lumber mills and on railroad gangs in the Pacific Northwest and California. In addition to the previously mentioned mob attack on them in Bellingham, Washington, they were persistently discriminated against: The polite, if inaccurate, term for them was *Hindu* or *Hindoo,* although they were commonly called "ragheads," for the turbans their religion demanded they wear. Most who stayed, perhaps five thousand at most, found niches in two California localities: the Imperial Valley, near the Mexican border, and the northern Sacramento Valley, where the more successful of them became farmers.

There were three distinct groups of elite migrants: swamis, students, and merchants. The swamis, Hindu missionaries to America, began to come in the late nineteenth century. Historians of religion in America say that the most important of these was the Swami Vivekananda, who spoke at the World's Parliment of Religions, part of the Chicago World's Fair of 1893. Vivekananda stayed for two years, founded the Vedanta Society, and returned in 1899, bringing two monks with him to preside over societies in New York and San Francisco. The second most important missionary was the Swami Yogananada who first came in 1920 to attend the Pilgrim Tercentenary Anniversary International Conference of Religious Liberals. He, too, stayed to found a religious organization, the Togoda Satsanaga Society in Los Angeles. Both groups emphasized the philosophy and practice of yoga, including posture and breath control. Both groups appealed mostly to middle- and upper-middle-class Protestants and had attracted some 35,000 adherents by the 1930s. The latter, which evolved into the Self Realization Fellowship, was the most extensive nonethnic "Hindu" organization in the United States until the appearance of the Hare Krishna movement in the 1960s.[4]

The early Indian students in the United States were largely rebels with at least emotional ties to the Indian Freedom movement. The natural place for Indians and other British colonials to study was at the seat of empire in England. Har Dayal, for example, resigned a fellowship at Oxford to come to the United States to do revolutionary work and, although he was primarily interested in freedom for India, had time to become an official of the Industrial Workers of the World (IWW). Some students, mostly at Berkeley, where there may have been as many as thirty-seven in the World War I era, became part of the quixotic and fatal Gadar Movement.

Gadar, which may be translated as "revolution" or "mutiny," was an attempt by Indian exiles in the United States to overthrow the Raj by sending revolutionaries and arms back to India. They were financed by the German government. The movement, betrayed by informers within its ranks, was nipped in the bud. No arms reached India: The revolutionaries who did were arrested, and some were killed. Since the unmasking of the conspiracy took place before the United States had entered the war, those who were caught (35 of 123 indicted were tried and convicted in a San Francisco court) were given only minor prison terms ranging from 30 days to 22 months for violating American neutrality laws. About half were Asian Indians: The others were Germans and Irish Americans.

Before this happened, the migration of poor Indians had been largely stopped by stringent application of the LPC clause. In 1917 the so-called "Barred Zone" act made the immigration of Indians impossible, although students, scholars, ministers of religion such as Yogananada, and merchants could come in and sometimes stay.

Among the small merchant elite, clustered on both coasts, the outstanding figure was Jagjit Singh who led the successful fight for Asian Indian citizenship discussed in chapter 13. The 1946 statute and other relaxations of immigration law and regulations before the passage of the 1965 act led to a regeneration of the existing Asian Indian communities. Fewer than seven thousand East Indians entered the United States as immigrants in the period 1948–65, almost all of them nonquota immigrants. But the individual who seems to have benefited the most from the change in the law was a man who had been here since 1920.

Dalip Singh Saund was born into a family headed by an illiterate but prosperous contractor just outside the Sikh holy city of Amritsar. Shocked by the massacre there in 1919 when—as shown in the film *Gandhi*—soldiers of the Indian army machine-gunned a nonresisting crowd that failed to disperse when ordered, Saund decided to leave India. Already possessed of a degree in mathematics, he continued his studies at Berkeley, earning three more degrees—an M.A. and a Ph.D. in mathematics, and, more practically, an M.S. in food processing.

Despite his qualifications—which he could have used had he returned to India—he concluded that "the only way that Indians in California could make a living" was to join with compatriots who were successful in farming. He settled in the Imperial Valley Sikh enclave, worked first as a foreman on an Indian-run cotton ranch, and then became a rancher and businessman. He acculturated with a vengeance—he had begun shaving and stopped wearing a turban shortly after emigrating—and in 1928 he married a woman from an upper-middle-class Czech American family. He and his wife, besides being successful and prosperous ranchers, became civic activists for a whole range of causes, including freedom for India and citizenship for Asian Indians in the United States.

Soon after he became a citizen Saund was elected to a judgeship, and in 1956 he was elected to Congress as a Democrat. He was the first Asian American elected to Congress and to this day the only one born in Asia.[5] (The only other foreign-born Asian American congressperson, the one-term Republican senator from California, S. I. Hayakawa, was born in Canada.) Saund's election was a mild if quickly forgotten sensation, and he was sent to India by the United States Information Agency to advertise ethnic democracy. Twice reelected, he was defeated in 1962 after a stroke confined him to a hospital bed.

The 1965 Immigration Act spurred a renewal of Indian immigration and has created a thriving and diverse community that has few obvious links, other than nationality, with the earlier settlers. Most are not Sikhs, few are in California or the West, and almost none of the newcomers are in agriculture. The recent population data are shown in table 14.3.

The vast majority have been well-educated and trained professionals who have quickly found comfortable employment niches in American society albeit, like most immigrants, they tend to be overqualified for the positions they hold. Many others have become entrepreneurs—sari shops and tandoori restaurants are the most visible enterprises—and they have found other idiosyncratic occupations. These include motel operation, with most of the operators sharing a common surname, Patel, and hailing from the state of Gujarat. Another set of entrepreneurs gained the contract to run and staff all the newspaper kiosks in the New York subways. What these operations have in common is a need for large numbers of low-paid employees, which is often filled by newly arrived relatives who enter as chain migrants. In the United States, Indian immigrants have not specialized in convenience food stores, as have Indians in London and Pakistanis in Copenhagen, but large numbers of newspaper kiosks in London have been owned and staffed by Indians.

The statistical profile presented by the available data from the 1980 census shows a reasonably well-off middle-class community. The median income of a full-time worker was $18,707—almost $2,000 higher than that of the next highest Asian American group, Japanese. But 7.4 percent of all Indian families were below the poverty level, as opposed to 4.2 percent of Japanese families. Characteristically, the more recent immigrants were less well off. Those who immigrated after 1975 earned

Table 14.3

Asian Indians in the United States, 1970–1990

YEAR	TOTAL	PERCENTAGE OF INCREASE
1970 (est.)	75,000	—
1980	387,223	416
1990	815,447	111

much less—about $11,000 per full-time worker—and more than one post-1975 family in ten was in poverty.

Asian Indian households were quite small, averaging 2.9 persons, as opposed to 3.4 for Koreans, 3.6 for Filipinos, and 4.4 for Vietnamese. The families that made those households were remarkably cohesive: 92.7 percent of all Asian Indian children under eighteen lived in a two-parent home, the highest for any Asian ethnic group. All of the other Asian groups, save Vietnamese, have figures in the eighteenth percentile, higher than the white American rate of 82.9 percent, and well above the Hispanic rate of 70.9 percent and the black rate of 45.4 percent.

Although many Indian women are well educated and have professional positions, they are less likely to be in the labor force than other Asian American women and have markedly less education than Asian Indian men. Whereas white, black, and Hispanic women were slightly more likely to be high school graduates than were men of the same groups, in all six of the most numerous Asian American groups the men were more likely to have diplomas. In some ethnic cohorts the difference was marginal (among Japanese Americans aged twenty-five–twenty-nine the gender gap was 0.1 percent); in many instances the differences were sizable. The widest gap revealed in the data published by the Population Reference Bureau was between Asian Indian men and women aged forty-five to fifty-four: 87.9 percent of the men but only 62 percent of the women had diplomas.

The data we have suggest a conservative middle-class community, but large numbers of that community have been here such a short time that it is dangerous to make other than tentative conclusions about it. Indian temples are probably proliferating more rapidly than Islamic mosques, though with less publicity and without the foreign subsidies that the latter enjoy, as Japanese Buddhist temples once did. In Cincinnati, for example, a community of some four hundred largely well-to-do middle-class families has pledged to raise seven hundred thousand dollars in three years to build a temple that will serve as a community center as well. Like most other Asian immigrants from the Third World to the United States, Indians represent a talent and capital drain from India, a talent and capital infusion for the United States.

Koreans

Almost all of the eight hundred thousand contemporary Korean Americans are either post–Korean War immigrants or their descendants.

Early in this century a few thousand Koreans migrated to Hawaii and the American mainland: As late as 1930 there were fewer than nine thousand, with about three-quarters in Hawaii. After the Korean War, a significant number of Korean women came to the United States as war brides, almost all of them married to non-Asian-American servicemen. During the 1960s, for example, some 70 percent of all Korean immigrants were female, and women outnumber men significantly in almost every age cohort. The 1965 immigration law set off the same kinds of Korean immigration chains as have been commented upon for other groups. The figure for overall Korean American population are shown in table 14.4.[6]

One unique aspect of the early Korean immigrants is that most of them, traditionally Buddhist, were recent converts to Christianity, usually Protestant Christianity. The role of American—and to a lesser degree Canadian—missionaries in stimulating and facilitating this emigration is very important. After the occupation of Korea by Japan following the Russo-Japanese War (1904–5) and total Japanese annexation, announced in 1910, the government in Tokyo made emigration very difficult. Some political refugees did manage to get to the United States where an exile movement was kept alive. Like the Asian Indians of Gadar, or various Irish American organizations harking back to the Fenian movement of the post–Civil War era, the Koreans were sometimes violent.

In 1908, Durham W. Stevens, a Caucasian American adviser to the Japanese administration in Korea, made a number of statements belittling Korean nationalism and supporting the Japanese takeover. He was assassinated in the lobby of San Francisco's Palace Hotel by a Korean patriot, Chang In-hwan. The latter was convicted and sentenced to twenty-five years in prison. Released in 1919, he died in California in 1930. In 1975 his remains were flown to Seoul, and he was reinterred as a hero in the national cemetery.

The most significant of the Korean exiles was Syngman Rhee (1875–1965). A converted Christian who had been in prison from 1897 to 1904

Table 14.4

Koreans in the United States, 1970–1990

YEAR	TOTAL	PERCENTAGE OF INCREASE
1970	69,150	—
1980	354,593	417
1990	798,849	125

for political activity in Korea, Rhee came to the United States to try to persuade President Theodore Roosevelt to protect Korea from the Japanese, as had been vaguely promised in an 1882 treaty between Korea and the United States. Roosevelt was not amenable to this; later he noted privately to his secretary of state that "We cannot possibly interfere for the Koreans against Japan. They couldn't strike one blow in their own defense."

Rhee stayed in the United States, except for one brief visit to Korea in 1912 under the protective aegis of the YMCA, for the next four decades. He took three degrees from American universities, culminating in a 1910 Ph.D. in international relations from Princeton, and in 1919 was chosen by some Hawaiian exiles as the first (and only) president of the provisional government of Korea. After the end of World War II, Rhee returned to Korea with American occupation forces and became the first president of the Republic of Korea in 1948. He was forced into a second exile in 1960—again to the United States—where he died in 1965. Some other, less exalted exiles also returned after World War II, but most of the few thousand Koreans in Hawaii and California, who came as sojourners, had by then put down permanent roots.

Between the end of the war and the passage of the 1965 immigration act, four separate categories of Koreans came to the United States. Most numerous were the war brides married to servicemen, Peace Corps volunteers, and other American citizens. One count found more than twenty-eight thousand such persons before 1975. These, in Harry Kitano's phrase, "assimilated before they acculturated" and often have little contact with the larger Korean American community, particularly those whose husbands are still in military service or who settle away from the major centers of Korean American population. A smaller group of Koreans have been orphans or other children adopted by American citizens, most of whom are not Koreans. One study found over six thousand such adoptions between 1955 and 1966: Almost 60 percent had non-Korean fathers, (46 percent white, 13 percent black). These Korean Americans tend to have even less contact with the Korean American community than do the brides. Of the thousands of Koreans who have come to the United States on student visas—there are perhaps five thousand here at present—some have not returned to Korea but have, in one way or another, been able to change their visas from student to immigrant or long-term visitor. And, finally, there were a few, after 1952, who came as regular quota immigrants—Korea had, as we have seen, an annual quota of 100—or as the relatives of American citizens. All these groups combined (including the old settlers and their children) certainly numbered fewer than twenty-five thousand when the 1965 law changed the rules of the game.

As the data in table 14.4 show, the Koreans have played the new game very well. The communal profile provided by the 1980 census figures shows the same kind of "model minority" performance as many other Asian groups. The Koreans were young (median age 26 years), well-educated (over 90 percent of males twenty-five–twenty-nine and forty-five–fifty-four were high school graduates), with full-time workers having a median income of $14,224, lower than that of Japanese or Chinese, higher than that of Filipinos or Vietnamese. Korean families seem quite stable: Almost 90 percent of all children under eighteen live in two-parent households. As is true for all Asian groups except the Indians, Koreans are concentrated in the American West (42.9 percent, 28.7 percent in California), but nearly a fifth lived in the Northeast, the South, and the Midwest. One conspicuous economic niche that Koreans have filled, particularly in New York City and Philadelphia, is that of operating small fruit-and-vegetable stores that stress high quality and personal service. Others have become professionals: Koreans are probably overrepresented on the faculties of American colleges and universities: More than a quarter of the employed native born and more than a fifth of the employed foreign born in 1980 were "managers, professionals, executives." Among Asian Indians in the same census, 23 percent of the native born and a startling 47 percent of the foreign born, were in that category.

The most visible group of Koreans in America are the residents of the bustling Koreatown in Los Angeles, a vast and growing enclave just north of the central city, with Olympic Boulevard as its axis. Unlike the contemporary enclaves of Central Americans in Los Angeles, not to speak of the long-established Mexican American barrios or black neighborhoods such as Watts, Koreatown exudes an air of prosperity and growth, although it must be noted that the 1980 poverty rate for Korean American families—13.1 percent—was almost twice that of whites. In addition to innumerable business enterprises, the Los Angeles Korean community has, in a very few years, erected an impressive number of community institutions: newspapers, schools, churches—215 Korean Christian churches in Southern California by 1979 according to one count—and a number of cultural organizations including a Korean American Symphony orchestra.

Korean parents are much concerned with the education of their children. A celebrated example is the parents of the brilliant Korean

American musician, Myung-Whun Chung, recently appointed conductor of the Opéra Bastille in Paris. Convinced that their children could get a better musical education in the United States, the family migrated for that reason. The results justify the move: The eldest child, violinist Kyung-Wha Chung was until recently better known than her brother; and, with a second sister, cellist Mung-Wha, the three have performed as a well-regarded piano trio.

Vietnamese

The Vietnamese are different from all the other major groups of recent immigrants from Asia. Rather than self-selecting immigrants reasonably well-qualified for success in America, Vietnamese, or many of them, have been poorly equipped for life in an urban society. To use the vocabulary of Ravenstein, they are mostly "push" rather than "pull" immigrants. Had they not been refugees—and refugees about whom the United States, with good reason, had a guilty conscience—most could not have qualified for admission.

Vietnamese have no long history of immigration to the United States: The few who left Asia usually went to France, including, for a time, the man known to history as Ho Chi Minh. Only after the United States took over from the French responsibility for resisting communism in Southeast Asia in the mid-1950s did a trickle of Vietnamese come here, chiefly as students. As American participation in Vietnam grew, so did Vietnamese migration to the United States, although as late as 1970 there were probably not even 10,000 Vietnamese in America and few of those had immigrant status. As the war went badly, more refugees began to arrive, and after the final debacle in 1975 the numbers became quite large. The 1980 census counted 245,000 Vietnamese, almost all of whom were post-1974 arrivals. In addition there were increasing numbers of Laotians and Kampucheans (Cambodians) who are also refugees from Southeast Asia and, essentially, from the Vietnam War and its aftermath. An estimate for 1985 put their numbers at 218,000 and 160,000 respectively. When smaller groups such as the Hmong, are added the total number of Vietnamese War refugees and their children in the United States by 1990 will exceed 1.25 million.[7]

The 1980 census data give an entirely different statistical profile for Vietnamese and other refugees from the Vietnam War than for other large Asian groups. They resemble not so much a model minority as the more traditional disadvantaged minority groups. They are very young (median age 21.5 years), not well-educated, and very poor. The median wage for a full time worker was $11,641 in 1979: more than a third of all Vietnamese families were below the poverty line, and more than a quarter of all Vietnamese families received some form of public assistance. Despite a determined attempt by the United States government to distribute Vietnamese evenly across the country, the 1980 census showed a clustering in the West (46.2 percent) and especially in California (34.8 percent). Within California the greatest area of concentration is Orange County, just south of Los Angeles, and the 1990 census will probably show an even greater concentration in those places. A federal refugee official in 1988 estimated that the percentage in California had risen to 39. Another reason some California officials give for the increased internal migration to California is the state's more liberal welfare system. In California, for example, a family may receive welfare payments under the federal Aid to Families With Dependent Children program even though both parents live at home, but this is true in twenty-five other states and not true in Texas, where many Vietnamese live. Most of the rest of the Vietnamese were in the South, particularly along the Gulf Coast, where some fisherfolk were resettled and equipped with modern gear by the federal government. The conflict that this created between the newcomers and established fishermen has been sensitively treated in Louis Malle's film, *Alamo Bay* (1985).

Those who have read the success stories that the press loves to run—the Vietnamese girl who wins the spelling bee—will wonder about the bleak statistics. Those statistics mask the fact that there is a very successful segment of the Vietnamese population here. One employed foreign-born Vietnamese in eight was a "manager, professional or executive." Most of these, to be sure, were proprietors of small businesses, and most were from the more Europeanized sector of Vietnamese society. Former Air Vice Marshal Ky is representative of this group. Large numbers of high-ranking Vietnamese officials came here and to Canada, where a number are established in Montreal and other parts of Quebec, with significant amounts of capital. The same thing happened when the Nationalist Chinese government was driven from the mainland in 1949 and occurred in the late 1980s when many of the Nicaraguan contra leaders took up permanent residence in and around Miami.

But most Vietnamese refugees come with no significant amount of money, and many of the so-called boat people who continue to flee are stripped of what little they have by pirates and/or venal officials in the countries of first asylum. These are the persons who are now in poverty. Even poorer, as groups, are the Laotians, the Cambodians, and such premodern peoples as the Hmong. Few Laotians and Cambodians and no Hmong were really equipped to cope with modern urban society before they left Southeast Asia, and the transition has been quite painful and difficult. If the isolated success stories become more representative is something that only time can tell, but many of those most directly involved with these refugees fear that they, or most of them, will become a permanent part of that other America where poverty and deprivation are the rule rather than the exception.

Table 14.5

Southeast Asians in the United States, 1990

Vietnamese	614,547
Laotians	149,014
Cambodians	147,411
Thai	91,275
Hmong	90,082

15

Caribbeans, Central Americans, and Soviet Jews

Migration from Latin America was long almost exclusively a Mexican affair. The 1960 census showed fewer than 400,000 persons "of foreign stock," that is first and second generation, from all the rest of the Spanish-speaking countries of this hemisphere, as opposed to almost 1.75 million from Mexico. Since there were 34 million persons of foreign stock in the United States, these other Latin Americans made up just 1 percent of that group and two-tenths of 1 percent of the whole population. In each of the next two decades more than 200,000 Cubans came, as did more than 250,000 South Americans per decade, and more than 100,000 Central Americans. From the tiny Dominican Republic, which had 17,000 persons of both generations in the United States in 1960, more than 90,000 came in the 1960s and more than 140,000 in the 1970s. In each of those decades Mexicans amounted to less than half of the recorded immigration from Spanish-speaking America. (This does not, of course, include Puerto Ricans, who are not legally immigrants.) In addition, to complete our hemispheric survey, there are growing numbers of immigrants from the non-Spanish-speaking islands of the West Indies, chiefly Jamaica, Trinidad and Tobago, and Haiti. The whole group sent just over 30,000 immigrants in the 1950s, 130,000 in the 1960s, and 271,000 in the 1970s. Except for the Cubans, there is much more sojourning among these immigrants than among Asians, few of whom return. These newcomers thus have migration patterns that resemble those of late-nineteenth-century Europeans, except that they come largely by air and over much shorter distances. There is every reason to believe that the migration of persons from the Caribbean and from Latin America south of Mexico will become increasingly important in the foreseeable future, particularly if American law continues to favor chain migration.

The following tables, based on the admittedly rough estimates of the Census Bureau's "current population survey" estimates for 1987, gives some notion of the numerical dimensions and geographical distribution of the various elements of what some have called Hispanic America, whose total was judged to be about 19 million. It should be noted that this figure does *not* represent persons living in Puerto Rico, whose population was about 3.5 million at that time.

Cubans

Cuba, only ninety miles away, long tempted expansionist Americans. Thomas Jefferson thought that Cuba would naturally become part of the United States and, in the decades before the American Civil War, schemes for the annexation of Cuba were concocted by Southern politicians. After the Civil War, the United States became the base and sanctuary for Cuban revolutionaries. In 1898, when Cuban liberation from Spain was achieved partly under American auspices during the Spanish-American War, Cuba became independent, but was, in fact, an American protectorate for six decades, a protectorate ended only by Fidel Castro's revolution in 1959.[1]

Table 15.1

Hispanic/Latino Population, United States, 2000 Census

Total	35,305,818	12.5% of U.S. population
Mexican	20,640,711	58.5% of U.S. Hispanic/Latino population
Puerto Rican	3,406,176	9.6% of U.S. Hispanic/Latino population
Cuban	1,241,685	3.5% of U.S. Hispanic/Latino population
Other Hispanic/Latino	10,017,244	28.4% of U.S. Hispanic/Latino population

Table 15.2

Hispanic/Latino Population, by State, 2000 Census

		% STATE POP.	% HISPANIC/ LATINO POP.
California	10,966,566	32.4%	31.1%
Texas	6,669,666	32.0%	18.9%
New York	2,867,583	15.1%	8.1%
Florida	2,682,715	16.8%	7.6%
4 States	[23,186,520]		65.7%
Other 46	[12,119,289]		34.3%

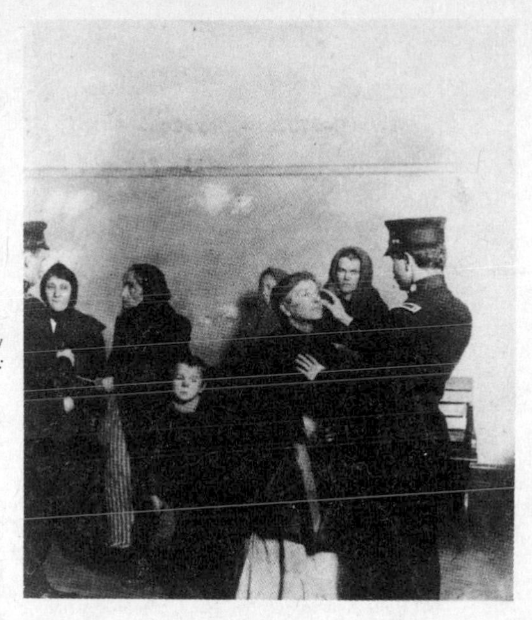

Trachoma examination, Ellis Island, early twentieth century. (*National Park Service: Statue of Liberty National Monument*)

One of the ferries that brought immigrants to the promised land from Ellis Island. (*National Park Service: Statue of Liberty National Monument*)

Four opponents of immigration: Carroll D. Wright, who called French Canadians "the Chinese of the eastern states," in a 1902 photo. (*Library of Congress*)

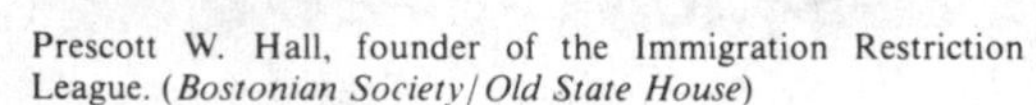

Prescott W. Hall, founder of the Immigration Restriction League. (*Bostonian Society/Old State House*)

Albert Johnson (R., Wash.), chief author of the 1924 Immigration Act. (*Library of Congress*)

Henry Cabot Lodge (R., Mass.), whose tactics led to the abrogation of the Gentlemen's Agreement. (*Chicago Historical Society*)

Two cartoon views of immigration restriction: The *Newark Daily News* saw restriction as a guard against the "refuse" of Europe, while Dutch-born Hendrik Willem van Loon fantasized about a possible American Indian immigration policy. (*National Park Service: Statue of Liberty National Monument*)

A group of middle-class Germans in Hamburg about to embark for America in the 1920s. They have been staying in a church-sponsored hospice. (*Staatsarchiv Hamburg*)

Filipino American farm workers posed in a field near Delano, California in 1936. (*Visual Communications, Asian American Studies Central, Inc.*)

The rise of totalitarian regimes in Europe drove many from their homes. Some illustrious immigrants enriched American life. Here Albert Einstein, flanked by his secretary, Helene Dukas (*left*), and his daughter Margot (*right*), are sworn in as citizens in Trenton, New Jersey, on October 1, 1940. (*National Archives*)

The Hungarian composer Béla Bartók. (*Library of Congress*)

The German-born architect Ludwig Mies van der Rohe. (*Chicago Historical Society*)

Not all who wished to come were admitted. In 1939, the German motor ship *St. Louis* got so close to shore that its hundreds of Jewish passengers could hear the music played at Miami Beach resort hotels, but had to return to Europe. (*Mariner's Museum, Newport News, Virginia*)

While fighting against tyranny abroad, the United States badly abused some of its citizens and resident aliens. This woman in a War Relocation Center "apartment" was one of more than 120,000 Japanese Americans, more than two-thirds native-born American citizens, who were incarcerated for varying periods of time between 1942 and 1946. (*National Archives*)

With the passage of the Displaced Persons Acts after World War II, various volunteer agencies, many of them church-related, assisted with refugee resettlement. Here Mrs. Claire Rassel of Traveler's Aid (*right*) makes arrangements for expected refugee arrivals at Chicago's La Salle Street Station with a representative of Lutheran Charities, Irene Tocas. Tocas, herself a refugee only four months in the United States, had previously worked for the UN's International Refugee Organization in Munich. (*United Nations*)

A family reunion of Hungarian refugees in New Jersey in 1957. The American government advertised its assistance on the luggage it provided. (© *Will Gainfort/Special Collections and Archives, Rutgers University Libraries*)

Mexican American farm workers pitting apricots near Canoga Park, Los Angeles County, 1924. (*Security Pacific National Bank Photograph Collection/Los Angeles Public Library*)

Immigrant workers have always been employed at the dirtiest and most dangerous jobs. Here a contemporary Mexican American worker is filling storage batteries. Although his eyes and hands are protected, he is exposed to lead and sulfuric acid. (*Earl Dotter*)

A Cuban refugee woman waiting to be processed at the immigration facility at Miami's airport, January 1974. (*Michal Heron/Woodfin Camp and Associates*)

A scene in Miami's "Little Havana" in 1981. Shown is part of the prosperous business district on "Calle Ocho," SW 8th Street. (*Michal Heron/ Woodfin Camp and Associates*)

Haitians sailed this twenty-foot boat to a point just off Ft. Lauderdale, Florida, before being seized by the U.S. Coast Guard and held for return to Haiti. (*United States Coast Guard*)

An American civilian official punches a would-be Vietnamese refugee in the face to force him to release his grip on an aircraft about to depart Nha Trang, Vietnam, April 1, 1975. North Vietnamese troops captured Nha Trang shortly afterwards. (*UPI/ Bettmann*)

These Vietnamese "boat people" were among the luckier ones, having safely crossed the South China Sea and arrived in Manila on January 8, 1979. Many if not most of them were eventually allowed to come to the United States. (*UPI/ Bettmann*)

As was the case with earlier immigrants, contemporary Asian American immigrants try to maintain some of their cultural traditions. Here, Ohio governor Richard Celeste poses with a group of costumed Asian Indian dancers in Columbus. The other men in business suits are officials of national and state Indian American organizations. (*Inder Singh, National Federation of Indian-American Associations*)

Although most Vietnamese refugees arrived without significant monetary assets, a few came better provided. Here is Nguyen Cao Ky, a former South Vietnamese leader, in his American liquor store–delicatessen. His wife, Dang Guyet Mai, has her back to the camera. (*UPI/Bettmann*)

One of the economic niches filled by Korean Americans is that of fruit and vegetable vendors. Shown is a small New York City market in the late 1980s. (*Hee Sun Ko/Korea News, Inc.*)

Ethnic succession is apparent in many lines of work. The International Ladies' Garment Workers Union was originally formed by immigrant Jewish and Italian garment workers early in this century. Today most of its members are either Hispanic or Asian, like these Chinese American members of a New York City local. (*Earl Dotter*)

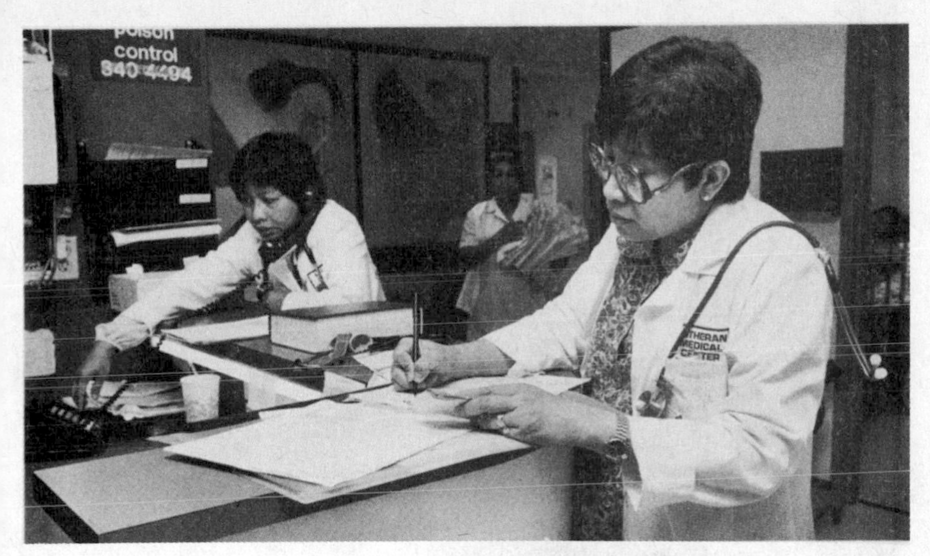

Many of today's immigrants bring badly needed professional skills to the United States. American hospitals, particularly big-city public hospitals, are heavily dependent on foreign-trained medical professionals, largely Asian. Here two registered nurses from the Philippines are shown in the Emergency Room of the Lutheran Medical Center in Brooklyn, New York. (*Earl Dotter*)

Many of the Hmong people are highly skilled at sewing. Here the immigrant artist Yer Yang Mua plies her needle at the 1987 Festival of Michigan Folk Life. (*Michigan State University Museum*)

Although these Japanese American girls are probably third-, fourth-, or even fifth-generation Americans and are unlikely to be able to speak any significant amount of Japanese, they still enjoy dressing up in fancy ethnic costumes, as in the 1985 "Nisei Week" parade in downtown Los Angeles, 1985. (*Craig Aurness/Woodfin Camp and Associates*)

"Liberty Weekend," July 1986, which celebrated both the 210th anniversary of the Declaration of Independence and the unveiling of the refurbished Statue of Liberty, was a televised extravaganza that celebrated the contributions of immigrants to American life even while Congress was again debating further immigration restriction. (*Steven M. Falk/Gamma-Liaison*)

The Statue of Liberty under reconstruction shortly before its reopening in July 1986. Most of the expenses of its reconstruction, and of the creation of a major museum of immigration on neighboring Ellis Island, were borne by a private foundation, the Statue of Liberty/Ellis Island Foundation. The work was supervised by the National Park Service. (*National Park Service: Statue of Liberty National Monument*)

Not surprisingly, during that long period, Cuban American communities were planted and sent down roots. Although individuals constantly came and went, the first Cuban American community we know anything about was in Key West, Florida, where a Cuban-owned cigar factory provided the nucleus for a settlement of fifty persons in 1831. For the rest of the nineteenth-century tobacco and politics account for almost all Cuban immigration. Cigar-making centers developed in New York City and Tampa, initially using Cuban tobacco and Cuban labor. The New York industry soon came to be dominated by European immigrants, but in Tampa the Cuban presence has remained important to the present day. George Pozzetta and Gary Mormino have written a rich account of how Italian immigrants joined Cubans in the Tampa suburb of Ybor City where cigar making was centered.[2] Cuban exiles, including the most important Cuban thinker of the nineteenth century, José Martí (1853–95), conducted much of their campaign against Spain from American cities and drew support from these communities. After Cuban liberation, there was little immigration for half a century. The 1950 census found only thirty thousand foreign-born Cubans in the whole country: thirteen thousand in New York, eight thousand in Florida, and a little over a thousand in California.

Growing political unrest in Cuba during the 1950s coupled with the economic dislocations that both helped cause and resulted from that unrest, sent more Cubans to the United States than in all previous history. By 1960 the census found eighty thousand foreign-born Cubans and forty thousand of the second generation. These immigrants were a varied group, ranging from members of the elite who were out of favor or who saw the end coming, to persons seeking work and economic opportunity. It was in the 1950s that Miami became the Cuban American population center: During the formative years of Cuban American settlement, it was largely a swamp.

Immediately after Fidel Castro came to power on January 1, 1959, a small exodus of supporters of the former regime came to the United States, but the mass movement of exiles and refugees began only in 1960 when it became increasingly clear that the Cuban revolution was determined to reshape Cuban society from top to bottom. Although the Eisenhower administration severed diplomatic relations with Cuba in January 1961, and the Kennedy administration launched the fiasco invasion of Cuba a few months later, commercial airline traffic between the two countries continued uninterrupted until the Cuban missile crisis of October 1962. More than 150,000 Cubans came to the United States between Castro's coming to power and that interruption, after which there were no direct flights for more than three years. During that time, some 30,000 Cubans managed to get here, either through third countries, most often Mexico and Spain, or by clandestine trips in small boats. The Johnson administration signed a "memorandum of understanding" with Castro's government at the end of 1965 that established a Cuban airlift to Miami, usually one flight a day. Between then and the airlift's ending in 1973 more than 250,000 Cubans entered the United States. From then until the chaotic Mariel boatlift discussed earlier, relatively few Cubans came. All told, something over 800,000 Cuban refugees have been enumerated as entering the United States since 1960.

Despite federal government efforts to distribute Cuban refugees throughout the country, consistent with its similar policy toward Vietnamese, the Cubans have clustered in southern Florida in general and Miami in particular, where they and their descendants have played an increasingly important role in the area's economic, social, and political life. Of the nearly one million Cuban Americans, perhaps half a million or more were clustered there. Dade County (1980 population 1.6 million) approached the kind of concentration of immigrants many American urban areas had in the late nineteenth and early twentieth century. In 1870, for example, New York, Chicago, San Francisco, and six other cities had immigrant concentrations of over 40 percent of the population. Little Havana was spectacular, but not unique in the nation's history. Other concentrations of Cubans, who were 99 percent urban, were in the greater New York area, particularly New York City, Newark, and Jersey City, and in Los Angeles and Chicago, all places with large numbers of other Spanish-speaking residents.[3]

Unlike most earlier immigrants, however, most Cuban Americans were at least lower middle class, as their statistical profile suggests, although skilled Cuban immigrants, like others, often entered the labor force well below their skill levels. David Reimers uses the story of Carlos Arboleya to illustrate that point. When Arboleya, who had been the chief auditor of Cuba's largest bank, arrived in Miami with his wife and two-year-old son in 1960, he had only forty dollars, plus the allowances that the federal government gave to all such refugees. The city's banks found him overqualified, and some may well have been unwilling to hire a Hispanic foreigner. He took a job as an inventory clerk in a shoe factory at forty-five dollars a week and within a year and a half had become controller and vice president of the company. He later returned to banking, and, eight years after he had come to Miami was president of the Fidelity National Bank. He and other Cubans skilled in financial management helped spark a boom in South Florida. By the

1970s, about a third of Miami's commercial bank employees were Cubans, including 16 of 62 bank presidents, 250 vice presidents, and over 500 other bank officers.[4]

Although Arboleya and the other bank officers were clearly well above the norm, the gross data for Cuban families, from the 1987 Census Bureau estimate, show a largely middle-class pattern. The bureau divided all American families into a Hispanic and non-Hispanic dichotomy, and arrived at a figure of $20,310 median family income for Hispanics and $31,610 for "not Hispanic." (A black-white dichotomy, with some Hispanics in each group but most counted as white, produced figures of $18,100 and $32,270.) When Hispanic income was broken down into ethnic groups, Cubans led the way, with median incomes of $27,290; Mexicans received $19,970, and Puerto Ricans $15,190.

These data, like the income data for some Asian groups, conceal a poor and deprived minority among Cuban Americans. A significant number of white Cuban Americans are in poverty, particularly the elderly and those in households headed by women. In addition, there are the more generally disadvantaged black Cubans. Although black Cubans make up a significant part of the island's population, they have been underrepresented in the refugees who have come to the United States. A pre-Castro Cuban census in 1953 showed some 72 percent of the population as white. But more than 90 percent of Cuban refugees to the United States have been white reflecting, first of all, their domination of the Cuban middle class. In addition, many Cuban blacks perceive themselves to be major beneficiaries of the Cuban revolution and also believe that there is significant color discrimination in the United States. (Blacks are also significantly underrepresented among Cuba's present ruling elite, but that is another matter.) Black Cuban refugees have reported housing and other discrimination in the Miami area, including discrimination in predominantly Cuban areas and from Cuban American landlords. The majority of black Cuban Americans—there may be as many as 50,000—live in the Northeast and have significantly lower incomes than do other Cubans. On the other hand, Asian Cubans, most of whom are of Chinese heritage, have been overrepresented among the refugees. That 1953 Cuban census reported 0.3 percent of the population (16,657 persons) as "unmixed persons of the yellow race," while the 1970 U.S. Census indicated that about 2 percent of Cuban Americans were of Asian ancestry. This probably reflected both their participation in the Cuban middle class, mostly as petty proprietors, and their perception of the position of Chinese Americans.

The Cuban Refugee program, like other refugee programs, has been expensive. Before the Mariel exodus in 1980, the Cuban program cost the federal government some $1.4 billion dollars, about $2,000 per refugee, in direct costs. Other costs were borne by state, county, and local governments and by the VOLAGs. This has been greatly resented by native American poor, particularly blacks, who watched Cuban families being offered a full panoply of social services as soon as they arrived in the United States. The newcomers received medical care, job and educational counseling, as well as direct cash payments at federal standards, while the blacks had difficulty getting the same kinds of services and, if they received cash public assistance, got it at the much lower state rate. This has helped to spark continuing racial unrest and disorder in the Miami black community into the 1980s.

It is clear that Cuban Americans represent a success story, but a success story whose dimensions are still developing. Many thousands of Cubans, especially among the earliest refugees, still regard themselves as exiles and are prepared to return to Cuba as soon as the hated regime falls. More and more Cuban Americans, however, regard themselves as Americans first and say they would not go back even if Castro were overthrown tomorrow. This is increasingly true of those born and educated in the United States. For a time, the exile mentality greatly retarded some aspects of Cuban American acculturation, a retardation that is not completely ended. But the earliest refugees have been here thirty years, and some of the first refugee children born in the United States now have children of their own. For a long time a concentration on exile politics and a relative reluctance, vis-à-vis contemporary groups like Asian Indians, to become naturalized citizens, inhibited their ability to function at full strength in American politics, but that day seems to be ending. Cubans are beginning to win Florida elections. Most notably, in August 1989, Ileana Ros-Lehtinen, a Dade County Republican, became the first Cuban American elected to Congress and one of the few woman immigrants ever elected. But Cuban American politics is likely to be quite different from Puerto Rican or Mexican American politics: The Cubans have essentially middle-class agendas and rarely see themselves as having much in common with other Spanish-speaking groups.

Dominicans

The situation of Dominican Americans is vastly different from that of the Cubans, although their island homes are only a few miles apart.

They have been little studied and much of the information about them is impressionistic, but their situation seems similar to that of many other Caribbean peoples and needs to be considered here. First of all, their numbers are uncertain. The figure of a quarter of a million immigrants by the beginnings of the 1980s refers only to those listed as immigrant entrants; from 1966 to 1976, for example, more than a million Dominicans were recorded entering the United States on tourist visas, an extraordinarily high number for a nation of perhaps seven million persons. Many believe that large numbers of Dominicans have remained in the United States as illegal immigrants. If that is in fact the case, very few, just over fifteen thousand, applied for amnesty under the 1986 act (to be described in the next chapter).

We do know that, except for a few highly paid professional baseball players (Mario Soto, the former fastball pitcher for the Cincinnati Reds, is a good example) most work at low-paying jobs. Most of the immigrants who live in the greater New York area seem to come from the cities and have significant amounts of education. In the New York City area, where they are not distinguishable from other Spanish speakers to most casual observers, they live in enclaves, particularly on the Upper West Side of Manhattan and in Queens, as well as in northern New Jersey and some areas of Long Island. Two of the Spanish-language newspapers in New York are slanted to appeal to Dominican readers, printing much news of the island as well as of the local community.

It is not possible to paint a statistical profile of the Dominican American population because that kind of detailed information does not exist. It is clear that they, along with most of the other post-1965 immigrants, are conducting chain migration, that large numbers return to the Dominican Republic, and that some of those return to the United States. There is among many of them the belief that they are working in the United States merely to earn enough money to provide a better life for themselves back home but, as has been true from the start of our history, many who came to sojourn almost certainly will not leave.

Dominicans have entered the low-paying service industries and the garment industry in the New York area, and many women first came as live-in maids. One would imagine from reading the fragmentary accounts that exist that Dominicans in New York City have a higher standard of living than do Puerto Ricans, but this impression may not be accurate. Some Dominicans who have been interviewed by scholars have been concerned with keeping social distance between themselves and Puerto Ricans in New York, and they seem to be more likely to send their children to parochial schools.[5]

Dominicans have been included here not so much to describe them, but to let them stand as surrogate for the large number of Spanish-speaking immigrant groups in the United States who are *not* refugees. The 1980 census counted, for example, more than 140,000 Colombians, all but about 30,000 of whom were post-1965 immigrants. And although few if any authorities believe that there are more Colombians than Dominicans here, it is a fact that significantly more Colombians—more than 25,000—applied for amnesty than did Dominicans.

Haitians

In stark contrast to the Dominicans are the Haitians. Although they share the same island, Hispaniola, they have little else in common. Haitians speak French, have a long history of emigration to the United States, and have been studied in a more systematic way. If the Dominican Republic can be called a poor country, Haiti is a desperate one. It is routinely described as the poorest nation in the Western Hemisphere, and its average life expectancy is said to be in the low thirties, although, as is true throughout the Third World, the data that exist can hardly be called statistics. The World Bank estimated a few years ago that, in the mountainous countryside where most of the people live, nine out of ten Haitians live below what it calls the "absolute poverty level" of $135 a year, or $.37 a day.

Haiti has a long and sometimes glorious history, having its origin as a nation in a series of slave revolts against French masters nearly two hundred years ago. Before that some eight hundred Haitian "men of color" fought on the American side during the American Revolution. But, since the early nineteenth century its history has been tragic. The Haitian people have suffered from almost perpetual misrule, including a period, 1915–34, when United States Marines ran the country. Misrule became brutal repression after 1957 when, in succession, François Duvalier, called "Papa Doc," and his son Jean-Claude, called "Baby Doc," ruled as "presidents for life." Since the son fled the island in 1986 to live in luxury in the South of France, taking much of the treasury with him, their brand of stable state terrorism has been replaced by unstable varieties.

Relatively large-scale migration from Haiti is a very recent phenomenon: The 1980 census found 92,000 persons, 88 percent of whom said they had come after 1965. Some persons knowledgeable about the community thought that the 1980 population was really about 300,000. Prior to 1965 small numbers of Haitians had been long established in the United States. As early as 1925, according to Professor Michel Laguerre of the University of California, Berkeley, there were about 500 Haitian immigrants in New York's Harlem. They developed import businesses, became merchants, and taught French and Spanish in the public schools. This kind of middle-class community, augmented by growing numbers of political exiles from Haiti's small elite, continued to predominate into the 1960s. At that time about half of the Haitian American community was employed in professional or white-collar occupations. Since then lower-class immigrants, less skilled and less educated, have predominated. Not New York but Miami has become the capital of Haitian America, with perhaps 50,000 or more Haitian Americans. But Little Haiti is very different from Miami's largely prosperous Little Havana. The few blocks in Northeast Miami where the Haitian enclave has developed are among the city's most dilapidated: worn-out plumbing, rotting floorboards, two or more families sharing two or three rooms.[6]

Unlike the Cubans, who are welcomed as refugees from communism, the Haitians are unwanted refugees from hunger. And those who do flee from right-wing political oppression in any country—whether black Haitians or white Chileans—have not generally been recognized or admitted to the United States as refugees. If the Duvaliers and their successors had been Marxists, the Haitians might have found a better reception. Yet, since 1980, American law has provided that persons suffering persecution or with a "well-founded fear of persecution" are eligible for asylum. U.S. government policy has been to make a distinction between those who come for economic reasons and those with political motivations, a judgment that often requires the wisdom of Solomon—a quality few, if any, bureaucrats are likely to possess. The official argument, which is not without merit, holds that if the government automatically accepted the claims of asylum seekers, even larger numbers of poor Caribbean or Latin American peoples would gain admittance, since many of them, unlike the Asians and Africans who make up the bulk of the world's thirteen million refugees, can often get to American territory without undergoing screening in countries of first asylum. Opponents of the government's interpretation of the law, particularly those in the sanctuary movement, insist that the rules are being applied selectively. In addition, many argue that the Haitians are discriminated against because they are black.

It is striking to note the difference in the way that the United States government, in the early 1980s, treated two groups of Caribbean boat people who each came uninvited and without papers: the Cuban

Marielitos and the Haitians. A small minority of the Cubans were put into long-term detention; the majority of them were not only paroled fairly quickly, but most were soon receiving reasonably adequate federal government stipends. The majority of the Haitians were placed in long-term detention, often in Miami's notorious Krome Avenue Detention Center; those who were paroled were "free" but often without income. They were generally not eligible for federal support, and Florida public assistance eligibility was both hard to obtain and at the relatively low amounts typical for a southern state. Cubans continue to be generally welcome, while the United States Coast Guard and Navy now try to intercept Haitian boats before they get into American territorial waters—and the jurisdiction of American courts—and escort or tow them back to Haiti.

The reality of Haitian emigration is more complex than either the United States government or most of its critics are willing to admit. Rather than a simple dichotomy, those who flee Haiti because of fear of persecution as opposed to those who flee because of poverty, many Haitian emigrants, like millions of emigrants throughout our history, have come for a variety of motives. As Ravenstein would say, they are pushed and pulled. American law, however, demands that officials decide what the chief motivation is and act accordingly: Clearly, it was not just in Charles Dickens's England that "the law is a ass." It is also instructive to note that, despite very different rhetoric, there has been no noticeable difference in the reception of Haitian refugees by the three administrations (those of Carter, Reagan, and Bush) that have been confronted with the Haitian asylee problem, although it was the Reagan people who hit upon the effective expedient of having the boats towed back to Haiti. The American captains are not required to make judgments about motives: If the boat has Haitians in it, they turn it back.

Despite all the difficulties, the Haitian American community is growing and will continue to do so. Tens of thousands of Haitians are now legally domiciled here, and as long as the law remains essentially the same, they will be able to institute and continue migration chains that are even now building the community.

Central Americans

Until recently, few Americans paid much attention to Central America, a region often described, sneeringly, as "banana republics." What American interest there was traditionally focused on Panama, where Theodore Roosevelt helped to foment a revolution to secure a government willing to let the United States build a canal on its own terms. Otherwise American economic interests in the region were chiefly those of one large American corporation—the United Fruit Company—and a few banks. American marines did occupy Nicaragua for a time in the 1920s and, in the 1950s, the Eisenhower administration used the CIA to overthrow a left-wing government in Guatemala.[7] But only in the 1980s did Central America, or most of it, become a major concern for most Americans. With the exception of Costa Rica, none of the nations of Central America enjoys a stable, democratic government. All are poor countries, and Guatemala, Honduras, El Salvador, and Nicaragua are desperately so. Authoritarian regimes, most often of the right, have existed there since the region gained independence from Spain in the early nineteenth century. Most of the current concern about Central America arises out of Nicaragua's Sandinista revolution, which toppled the Somoza family dictatorship in 1979 and soon aligned itself with the Soviet Union and Cuba, much to the horror of most American politicians. The unsuccessful attempt to overthrow the Sandinistas by helping the Contras, civil war in El Salvador, and persistent repression in Guatemala, brought the region to national wareness in the 1980s, as refugees, asylees, and illegal entrants from Central America became numerous enough to attract attention.

If Caribbean boat people have represented a problem to recent American administrations, what Ronald Reagan called "feet people" seemed to be a nightmare and, like most nightmares, based more on fantasy than on reality. As President Reagan put it, while arguing for more aid to the Nicaraguan Contras:

> If we allow Central America to be turned into a string of anti-American dictatorships the result could be a tidal wave of refugees—and this time they'll be "feet people" and not "boat people"—swarming into our country seeking safe haven from communist repression to the south.

To back up the president's warnings, the State Department's coordinator of refugee affairs predicted that 2.33 million refugees would be created by a communist takeover of all Central America. Critics of the Reagan and Bush administrations' policies have pointed out that very few refugees have fled Nicaragua, the one Marxist government in Central America, and that the vast majority of those Central Americans who have got here have been from El Salvador and Guatemala, where anti-Marxist governments supported by the United States hold sway.

Yet even nightmares have some basis in reality, and Central American refugees are a growing element of the American population. A stroller down Sunset Boulevard in Los Angeles—not its swank Pacific Palisades part but the section that leads to downtown—can see the new Central American enclaves. Unlike Koreatown, not far away, the area is not a booming one but another large pocket of poverty in a rich city that can be called the capital of Third World America.

How many Central Americans have come? Clearly a growing number, but one whose precise dimensions are highly questionable. The 1980 census recorded 94,000 Salvadorans and 64,000 Guatemalans, 88 percent of whom told the census taker that they had come since 1965. There were not enough Nicaraguans to merit a separate listing in the published census data on foreign born. The 1986 amnesty data give us some picture of the illegal dimensions: almost 138,000 Salvadorans, 51,000 Guatemalans, and 15,000 Nicaraguans formally applied to the program. All of those, presumably, were persons who came before 1982, and many more have arrived since then. Note that the figure for Salvadorans exceeded the 1980 census figure by almost 50 percent and we must presume that most of those were persons not included in the census count. By the later 1980s the then president of Salvador, José Napoleon Duarte, asked the United States not to deport Salvadorans because the remittances that they were sending back from "El Norte" were such an important part of the national income, more valuable than American foreign aid.

The Central Americans who have reached the United States comprise only the tip of the iceberg: There are an estimated two million Central American refugees, perhaps one person in eight of the region's population. However, most of these are refugees within their own country. Of the half to three-quarters of a million who have left their own country, most are in other Central American nations or in Mexico, which has perhaps a quarter of a million, most of them in camps and other settlements near Mexico's southern border. Although Mexico gets some aid from the United Nations High Commission for Refugees, these camps are a strain on Mexico's economy. For Central Americans, Mexico is the country of first asylum and, like most such countries in the world today—Thailand for refugees from Southeast Asia, Pakistan for Afghans—it increasingly resents the burdens it has to bear.

Although supporters of the sanctuary movement and others point out that it is almost impossible for Salvadorans and Guatemalans to get asylum privileges, it is not at all easy for Nicaraguans to do so, even though the United States is virtually at war with the Sandinista government. At the height of the Reagan administration's aid to the Contras, in 1984, 8,292 Nicaraguans applied for asylum: only 1,018 were accepted, about 1 in 8. According to Laura J. Dietrich, deputy assistant secretary of state and the person responsible for administration's refugee policy:[8]

> Most Nicaraguan asylum applicants state that they do not like the Sandinistas, do not want to serve in the Nicaraguan military, and have come to the United States because we are a free country. However sympathetic this administration may be to their plight, under our laws, these are not grounds for granting asylum.

Yet the same story, almost certainly, would get a Cuban acceptance.

Central American refugees have been classified by one specialist, Elizabeth G. Ferris, into three main types: political exiles, urban refugees, and peasants. The first and smallest group, usually well-educated and from middle- or upper-class backgrounds, conforms to previous notions about refugees. These can usually cope most successfully with life in the United States and are most likely to get asylum. The third group, the peasants, is the most numerous and has the fewest resources, either in money or experience. These rarely get to the United States.

It is the second group, the urban refugees, who are most often the unsuccessful asylum seekers and who seem to make up the bulk of the illegal immigrants. They are usually of working-class or lower-middle-class background. The first to come are usually urban young men, who are likely to be pressed into the army or urged by the left to join guerrilla forces. Such pressures, plus the ever-present fact of violence, corpses in the streets, tales of torture and death from relatives and friends, and the worsening economic situation, all encourage them to leave. Most of those who come to the United States, unlike many first-time Mexican migrants, look for and usually find jobs in urban areas because this was the kind of life that they knew in the cities and small towns of their homeland. Often when established, they will send for relatives and, as near as we can tell, they are much less likely to return home than Mexicans. They believe that it is dangerous to return and know that it would be difficult to come back again.

Many Americans have protested their government's policies and attitudes toward Central American migrants. Some have involved themselves with the religiously oriented sanctuary movement. This group deliberately provides housing, food, and jobs for those it considers legitimate refugees but who are in this country illegally. In addition, some help them to get into the country and assist them in evading immigration and other law enforcement officials. The sanctuary movement and its supporters insist that their cause is righteous and that they are in an old American tradition going back to the Underground Railroad of slavery days, which sheltered and otherwise assisted runaway slaves. They point out that Harriet Tubman, the black woman "conductor" of the underground railroad who was so successful that some called her Moses, is now honored on a U.S. postage stamp.

The government, of course, takes a different view, and launched a few well-publicized prosecutions of sanctuary activists, including one case in Arizona in which clergy were actually sentenced to jail terms, but has largely ignored most of them. Near my home in Cincinnati, for example, there is a Quaker meetinghouse that has long been used as a sanctuary for Central Americans without any attempt at concealment and with no apparent interference from law enforcement authorities. A number of places in the United States, including the state of New Mexico and the city of Los Angeles, have formally declared themselves in sympathy with the sanctuary movement and have said that they will not cooperate with federal officials seeking to prosecute its leaders.

Soviet Jews

In striking contrast to the position of Haitians and Central Americans, and in contrast to the Western European Jews of the 1930s and the Holocaust survivors of the 1940s, Soviet Jews have gone right to the head of the queue. In addition, the United States government has put enormous pressure on the Soviet Government to let its people go. Beginning in 1972 the American Congress adopted the so-called Jackson amendment (for Senator Henry M. Jackson, a Washington Democrat), which linked favorable trading privileges for the USSR with more liberal Soviet immigration policies. Those policies have fluctuated with the state of U.S.-USSR relations: When détente, or later *glasnost,* flourished, emigration flourished, at least relatively speaking. When the Cold War heated up, it withered. Perhaps nothing illustrates better the changes in the climate of American opinion about Jews than the contrast between the attitudes of the 1930s and 1940s, when the admission of Jewish refugees was resisted, and those of the 1970s and later when Congress actually pressured both the executive branch and the USSR to let more Jews out of the one country and into the other.

There are perhaps 2,000,000 Jews in the Soviet Union today, as compared to 3,600,000 in Israel and perhaps 6,000,000 in the United

States. Clearly, Soviet Jews comprise the last sizable group of Jews in the world likely to emigrate to Israel if allowed to do so but, much to the frustration of the Israeli government, an increasing percentage of Soviet Jews who succeed in leaving the USSR come to the United States.

The mechanism under which that immigration takes place is a curious and complex one. All Jews leaving the Soviet Union have emigration visas that state their destination as Israel. But you can't go to Israel directly from the Soviet Union, so the Jews fly to Vienna. Once in Austria, which is technically a country of first asylum, they are asked where they wish to go. In 1988 some 19,000 Jews were allowed out of Russia. Only 7 percent chose to go to Israel; the rest chose the United States, despite pleas and promises from Israeli officials who had access to them. The Israeli government has been trying to get the Soviet government to change the system and funnel the emigrants through Bucharest, where they would presumably not have a choice, but the USSR has not agreed. Once such persons are in Israel, which is a country of refuge rather than a country of first asylum, they would no longer be eligible to enter the United States as refugees but would have to go through regular immigration procedures. An official of the Israeli government said recently that if Soviet Jews did not want to come to Israel, Israel did not care whether they got out or not, but surely few Israelis would share that sentiment.

Those Jews who have come to the United States, perhaps one hundred thousand of the more than two hundred thousand who have left, have been assisted by the large and well-organized American Jewish community. Although some Soviet Jews have resettled in nearly every part of the United States, where local Jewish communities have sponsored them, there have been the same kinds of clustering as with other immigrant and refugee groups. The largest concentration of Soviet Jews is in Brighton Beach, at the southern tip of Brooklyn next door to Coney Island. Its residents, who call it Odessa on the Atlantic, make up a third or more of the fifty thousand residents. A recent Hollywood movie, *Streets of Fire,* in which Klaus Maria Brandauer played an improbable Soviet Jewish refugee who had been a world-class boxer, was set in this community. There the restaurants now tend to have Russian names, there is a Black Sea bookstore, and all kinds of community businesses have Russian or Russian-English names in Cyrillic letters. Russian is often heard on the streets.

A refugee journalist, Yevgeny Rubin, tried to explain to an American reporter what it was like to come to a free country.

> We are like animals from a zoo suddenly freed. . . . Imagine you were born in prison and lived in it all your life and then were set free and you don't know what to do and where to go—you are in the jungle of freedom.

There are sometimes problems in the jungle. Another Soviet immigrant, Alex Zayats, told the *New York Times* that most Soviet Jews "imagined America as a fairy tale story." Many so mistrusted the Soviet press that they "assumed that the opposite of whatever it said about America was true." His father didn't believe what he had read about violence in New York until, as Alex tells it, "exactly two weeks after we arrived here my father was robbed in downtown Manhattan."

Despite, or perhaps because of, the enormous pressure that the American community and its allies have exerted to help get the refugees released, there are often conflicts between them and the established community after they arrive. Many of the newcomers, especially the younger and better educated, are not particularly interested in Jewish religious or communal life, much to the chagrin of some of their sponsors. This conflict is ironic. As one of the secular Jews put it: "The paradox is that in Russia I was a Jew and now I am a Russian." So severe have been the conflicts and disillusionments that a very few mostly older persons have chosen to return to the USSR. This should come as no surprise. After all, some nineteenth-century Russian Jewish immigrants chose to return to the land of the czars. One suspects that if they were now able to do so, some of the returnees would reemigrate to America.

The latest of many ironies of this Russian Jewish emigration is that the relaxation of emigration rules under *glasnost,* beginning in 1988, has caused for the first time a real logjam in Western Europe. Gorbachev's USSR was releasing Jews faster than Reagan's and Bush's America could process them. In all of 1986, six hundred Jews were allowed to leave Russia. In 1988, nineteen thousand were, and the pace quickened further in the first half of 1989. The United States bureaucracy takes seventy-five days or more to clear each application. And, for the first time, a few would-be migrants to America are being turned down by the State Department representatives, presumably on the grounds that they have not demonstrated a "well-founded fear of persecution" in what Ronald Reagan used to call the "evil empire." It was once enough just to be from the Soviet Union. It seems that the easier it is to get out of the USSR, the harder it is going to be to get into the United States.

In the meantime, the logjam in expensive Vienna has forced thousands of those waiting in transit to be shifted to the rundown Italian seaside resort of Ladispoli where, in early 1989, there were seven thousand such persons with more expected every day. Some had been there for as long as six months. (If they wish to go to Israel there are no delays.) They are supported there by the American charity the Jewish Joint Distribution Committee, which is also feeling the pinch caused by *glasnost* but is not complaining. The JDC paid about a million dollars for the maintenance of Soviet Jews in Europe in 1987, twelve million in 1988, and expects to spend fifty-five million dollars in 1989, assuming that about thirty-eight thousand will pass through Vienna and opt for America. What would happen if all bars to Soviet Jewish emigration were lifted and, say, five hundred thousand Jews decided to leave, is impossible to predict. Even more fascinating is what one young Jew in Ladispoli told Roger Boyes, a reporter for the London *Times:* "You know what? There are some people in Russia pretending to be Jews so that they can get out. Pretending to be Jews! This *perestroika* is crazy."

16

The 1980s and Beyond

During the 1980s, for the first time since the mid-1920s, immigration took up a central position on the American social agenda. While both the volume and incidence of immigration continued the steady increase that had begun just after World War II, anti-immigration attitudes, which have always lurked near the surface of the American mind, again emerged. These nativistic attitudes had been largely quiescent during the fifteen years between the enactment of the Immigration Act of 1965 and the Mariel boatlift, a period in which the United States as a whole moved from the euphoric confidence of the Great Society to the resentful feelings of impotence best typified by the Iranian hostage crisis, whose denouement, at the moment of Ronald Reagan's assumption of the presidency, seemed to symbolize a sea change in American life. A society that had aspired to conquer space and abolish poverty had seemed to focus for more than a year on the safe return of a handful of people. Although, as we have seen, the last Carter year saw the passage of the first comprehensive refugee act in our history, that act may be regarded as one of the last artifacts of the liberal consensus that had dominated American life for a half a century. Two years earlier Congress had taken a step that was to prove more congruent with the mood of the 1980s: It created a body to reexamine American immigration policy.

In setting up the Select Commission on Immigration and Refugee Policy, with instructions to report in March 1981, Congress was consciously following an earlier tradition. The only previous such body, the United States Immigration Commission, had reported in 1911. Six years later, Congress enacted the literacy test and barred a large proportion of the world's population from immigrating to the United States, and, seven years after that, the 1924 act represented a high-water mark for American nativism. In some ways, history did seem to be repeating itself in the 1980s: five years after the select commission's report, a major piece of immigration legislation resulted: the Immigration Reform Act of 1986. But in other ways there were clear differences. Whereas the first fruits of the 1911 report were unambiguously restrictive—reflecting both the commission's report and the national mood—those of the 1981 report were ambiguous, reflecting the tone of both the report and the nation. The earlier set of "reforms" sought to reverse existing American immigration policy, the second set merely to modify it. Congress, too, was ambiguous: Having authorized the select commission, it codified refugee law a year before the commission reported.[1]

And the commissions themselves were different. The 1911 body consisted of members of Congress who in many instances overrode the findings of its staff; the commission that reported in 1981 was a blue-ribbon body, bipartisan, and chaired by Father Theodore M. Hesburgh, longtime president of Notre Dame and, according to the Guinness Book of Records, the world champion collector of honorary degrees, with more than a hundred and still counting. The 1911 body was conservative if not reactionary, that of 1981 decidedly liberal. Hesburgh has insisted that his commission attempted to "walk the fine line between honoring America's tradition of being a land of opportunity for the world's downtrodden and dealing with today's harsh realities." He believed, with most of his advisers, that the United States must "regain control of its borders," although most serious students of the matter have long realized that the United States has *never* controlled its borders and is not likely to do so. Yet the image of gaining control, which Hesburgh did not invent, has proved to be immensely popular with press and politicians and fitted well with the new national mood of attempting to gain or regain control, not just of immigration but of the economy, of welfare, and of drugs.

Hesburgh, a humanitarian, stressed economic arguments in a way that also mirrored the national perception of limited possibilities:

> During the next 15 years, assuming a persistently strong economy, the United States will create about 30 million new jobs. Can we afford to set aside more than 20 percent of them for foreign workers? No. It would be a disservice to our own poor and unfortunate.[2]

Other spokespersons for the poor and unfortunate echo Hesburgh's views and add to them an ugly antiforeign bias that the priest himself, whose church has a long tradition of defending immigrants, never used. Columnist Carl Rowan, for example, wrote in 1986:

> The United States is a nation without meaningful control of its borders. So many Mexicans are crossing U. S. borders illegally that Mexicans are reclaiming Texas, California, and other territories that they have long claimed the Gringos stole from them.

What Hesburgh and Rowan see as "job stealing" others see as economic growth. Economists as ideologically discrete as John Kenneth Galbraith and Ronald Reagan's Council of Economic Advisors take the latter view. Galbraith has consistently urged less restrictive immigration policies and has celebrated governmental impotence: "Fortunately the power of government is small so [illegal immigration] continues. The Council of Economic Advisors, unable to countenance directly illegality, nevertheless argued that it didn't harm American workers. In its 1986 annual report it held:

> There is evidence that immigration has increased job opportunities and wage levels for other workers. . . . Immigrants come to this country seeking a better life, and their personal investments and hard work provide economic benefits to themselves and to the country as a whole.

At about the same time, liberal economist F. Ray Marshall, who had been secretary of labor in the Johnson administration, told a congressional committee:

> There is some debate over the effects of immigration on employment, but there is little doubt in my mind that it depressed wages of U.S. residents, especially in the absence of policies to maintain economic growth.[3]

The argument was an old one: Germans, Irish, Chinese, Japanese, French Canadians, Italians, Jews, and almost every conceivable immigrant group have been, at one time or another, accused of lowering the American standard of living, and the question of the impact of immigration on the standard of living is a debate without end. What is not debatable is that the argument has much more force and appeal in an era of economic stringency and lowered expectations. Thus the Hesburgh commission's 1981 report fell on fertile ground. Its basic recommendations were ambiguous. On the one hand, it urged that the liberalized immigration policies of the 1965 law remain essentially unchanged. On the other, it recommended a tightening of border controls, and, in a suggestion that sent chills up and down the spines of traditional civil libertarians, urged the government to conduct research in order to produce "forgery proof" identify cards to be carried by all Americans. What the report's advocates saw as just another identity card, like a driver's license, opponents regarded as an internal passport. And finally, in a recommendation that seemed unprecedented, the commission urged that the government set up an amnesty program to allow illegal immigrants who were already established in the United States to regularize their status and become United States citizens. To many of its readers the report seemed to be saying that there were good and bad immigrants: Its recommendations for tightening immigration policies were chiefly aimed at Latin American migrants and would have left Asian immigration patterns largely alone.

The Immigration Reform Act of 1986

For the next five years Congress debated immigration and the national media gave it more attention than it had received since the 1920s. A key figure in the debate was Republican Senator Alan Simpson of Wyoming. Simpson had been a member of the Select Commission and obviously thought that its proposals did not go far enough. His arguments were clearly nativistic, but, as contrasted with the nativism of the 1920s and 1930s, not overtly racist. He hailed the achievements of some nonwhite immigrants and insisted that he was not prejudiced against anyone. His arguments stressed cultural homogeneity and said, in more sophisticated language, what nativists had been saying since the days of Benjamin Franklin. In a letter to the *Washington Post* shortly after the commission report was published, Simpson wrote:

> A substantial proportion of these new persons and their descendants do not assimilate satisfactorily into our society. . . . [They] may well create in America some of the same social, political, and economic problems that exist in the countries from which they have chosen to depart. Furthermore, if language and cultural separation rise above a certain level, the unity and political stability of our nation will—in time—be seriously eroded.[4]

He was the cosponsor and prime mover of the so-called Simpson-Mazzoli bills of the early 1980s, which never became law although in one form or another they passed each house. (His cosponsor was

Romano L. Mazzoli, a Democrat from Louisville, Kentucky.) Simpson's bill, like the commission report, proposed to raise the numerical limits. But, unlike the commission recommendation, Simpson's bill mandated placing all relatives within the quotas so that the actual number of immigrants would decrease. The bill also would have abolished the fifth preference, brothers and sisters of United States citizens, and thus significantly inhibited the chain migration of recent immigrants. Simpson justified this change at a congressional hearing by arguing that:

> I do feel that family preference categories should be based on the U. S. concept and definition of a nuclear family and not on the definition of such a family as expressed in other nations.

After five years of debate Congress finally passed an immigration act in 1986 following some last minute compromises. The Immigration Reform Act of 1986 left the basic structure of American immigration law untouched, and legal immigration, as shown in table 16.1, continued to grow in the 1980s. Although the bill was hailed by its sponsors, the INS, and the media as a solution to the problem of illegal immigrants, most serious students realized that, despite the ballyhoo, the law would change little. The 1986 act contained four major provisions: amnesty; requirements that employers verify the eligibility of all newly hired employees—whether resident aliens, naturalized citizens, or native-born Americans—to work in the United States; provision of seemingly "tough" sanctions, including prison sentences, for employers who hire illegal aliens; and special provisions to make it easier for growers, mainly in Texas and California, to import foreign agricultural workers. The last provision was clearly antithetical to the thrust of the act—which was to cut down on immigration from Mexico—but it was necessary to secure the votes of congresspersons representing those interests.

The highly controversial amnesty provisions were quite complex, and it is not yet possible fully to evaluate their impact. The general amnesty provisions affected only illegal aliens who could prove that they had been in the United States continuously since December 31, 1981. To be considered for amnesty, an illegal alien had to apply formally to the Immigration and Naturalization Service and be able to document continuous residence. Since illegal aliens would naturally be loath to declare themselves directly to the immigration authorities, it was sensibly provided that immigrant advocacy agencies—nonprofit VOLAGs such as religious social agencies and Travelers' Aid—could serve as intermediaries and assist would-be applicants for amnesty to assemble their documentation. The INS, after some initial foot dragging, soon predicted that some 3.5 million illegal aliens would apply; many immigrant advocates predicted that very few would do so. By the time that the statutory period for applying had expired, 1.7 million persons had their applications for regular amnesty accepted. In addition another 1.4 million had qualified under the easier provisions that applied to those working in perishable agriculture. They had only to demonstrate that they had worked in U.S. agriculture for at least ninety days between May 1, 1985, and May 1, 1986.

The 3.1 million aliens accepted into the program represent a success story, but one whose dimensions are not yet known. Each successful applicant has two more major hurdles to clear. Within two and a half years after acceptance into the program each amnesty recipient must file and successfully complete an application for permanent residence. To obtain that status an applicant must have resided in the U.S. continuously since being accepted into the program, have no criminal convictions or pending prosecutions, submit a negative test for AIDS antibodies, not have been on welfare and otherwise demonstrate financial responsibility, and demonstrate a knowledge of the English language and United States history. If the second application is not completed in two and a half years, the amnesty lapses and the individual is, again, an illegal alien subject to deportation. The final hurdle comes five years after permanent residence has been achieved. The recipient of amnesty may then—and only then—make application for United States citizenship. In theory, any applicant who receives permanent resident status under the program can lose it in the ensuing five years by going on welfare, getting convicted of a crime, or leaving the United States, but it is difficult to see how these provisions of the law will be enforced; in addition some of the special requirements are sure to be challenged in the courts if and when the INS tries to enforce them.

Chart 16.1 shows the national origin of the successful applicants for acceptance into the amnesty program. Nearly 70 percent were Mexicans, and more than 90 percent were from the New World. They were predominantly male (68 percent), young (59 percent under thirty, fewer than 10 percent over forty-four), and chiefly from California (54 percent) and Texas (14 percent).

Problems with the program have become apparent. A woman in Houston from El Salvador, who came to the United States in 1981 and thus could qualify for amnesty, now has a dilemma: Her five children,

Chart 16.1
Applicants for Amnesty, 1986–1988

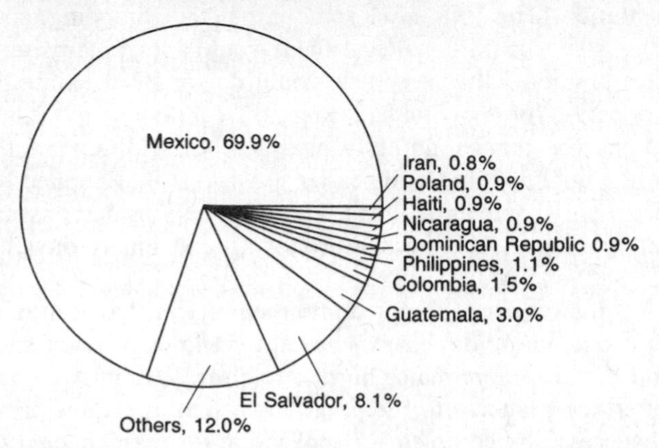

ten to eighteen years of age, joined her after the 1982 deadline and are thus ineligible for amnesty. "It is a great worry for me," she told a *New York Times* reporter, "because my two oldest have graduated from American high school. Their home is with me here, but they cannot get real jobs. What is their future?"[5]

Amnesty was the carrot of the 1986 act: The stick was supposed to be provided by the new eligibility for employment requirements and sanctions against employers who persisted in hiring illegal aliens contrary to law. Beginning June 1, 1987, every large-scale employer had the responsibility for determining the employment eligibility of all newly hired persons. The law specified three different kinds of documents: "A," "B," and "C." "A" documents, which established both employment eligibility and identity, were the best, but relatively few Americans possessed them. They were a United States passport, or a certificate of United States citizenship, or a certificate of naturalization, or an unexpired foreign passport with a current "authorization to work" endorsement, or a resident alien registration card, *if* it contained a photograph and evidence of current authorization of employment. For those who did not have such documents, the "B" and "C" document lists were provided, establishing employment eligibility and identity, respectively. Applicants for employment without a list "A" document had to produce, Chinese restaurant fashion, one document from list "B" and one document from list "C" before they could be hired legally. List "B" consisted of a social security card (unless it specified that it did not authorize employment), or a birth certificate showing

U.S. birth, or a newly created INS Form I-94 with an employment authorization stamp. List "C" consisted of a driver's license or similar document issued for identification by any state provided it contained a photograph, or an original identity document issued by any state with a photo or physical description, or a discharge certificate (DD Form 214) from the United States armed forces. Employers are required to fill out and maintain a separate form for each employee hired, listing which documents they examined for each employee. Administrators at my own university in Cincinnati estimated that the cost of creating, administering, and maintaining such files would run to fifty thousand dollars annually. For each employee hired someone must sign the following certification:

> I certify, under penalty of perjury, that I have examined the above described documents which were presented to me by the individual named [above] that the documents appear to be genuine, that they appear to relate to the individual named, and that the individual is a United States citizen, a legal Permanent Resident, or a nonimmigrant alien with authorization to work.[6]

Within a year millions of such documents were reposing, uninspected, in the files of employers all over America. The law enforced employers and prospective employees everywhere to deal with costly, time-consuming and, in the final analysis, useless precautions. While some of the congresspersons who voted for the law may have believed that these regulations would be effective, the more realistic understood that they could not possibly work, and the law also provided money for research to create a "forgery-proof identity card," illustrating the naive faith of even the sophisticated in the ability of technology to solve problems.

The essential hypocrisy of the law can most easily be seen in its employer sanctions provisions. Everyone knows that agricultural employers in two states—California and Texas—have been hiring illegal aliens since the World War II era and before. Since most of these are migrants who move from job to job, the only effective way for INS agents to apprehend those who have crossed the border has been to conduct raids on the fields at harvesttime. Due to pressure from congresspersons representing grower interests, the new law specifically forbids INS agents from "interfering" with workers in the fields, although spot raids on places of urban employment, such as garment manufacturing shops, are still permitted. The disproportionate political clout of well-organized agricultural minorities is a little-remarked phenomenon of the modern, industrialized "first world." Some of the greatest strains within the European Economic Community are caused by measures designed to protect farmers, while similar measures taken by Japan are a constant bone of contention between it and United States trade representatives.

In addition, the law favored growers in a most blatant way, by authorizing up to 250,000 additional "green cards" for temporary agricultural workers, and providing a mechanism for these new braceros to become permanent residents if they labored in the fields for as much as three years. Thus, despite the tough rhetoric about gaining control of the border, the new law provided an expanded legal method for more farm laborers to enter.

Nor are employer sanctions likely to work in agricultural areas. Juries in the agricultural regions of California and the Southwest simply will not convict their friends and neighbors for hiring illegal aliens, which is something that has had community approval since time out of mind. The fact of the matter is that it has been all but impossible for the civil rights division of the Department of Justice to obtain convictions in the few peonage prosecutions that have been launched against the tiny minority of agricultural employers who indulge in such practices.

The foolish notion that the various provisions of the law would combine to solve the problem of illegal aliens is already apparent to almost everyone except high officials of the Immigration and Naturalization Service. Representative Charles E. Schumer, a Brooklyn Democrat who was instrumental in achieving the compromise that got the 1986 law enacted, admitted ruefully in mid-1989: "The legislation has had some effect but not close to what it should have been." Typically, he blamed the Republican administration for not providing enough funds for the INS and insisted that "the law really has not been given a fair test."

Early studies by scholars conducted on both sides of the border are even more negative. Some claim that, not surprisingly, a small industry in Mexico enabled thousands of persons to enter the United States *after* the passage of the 1986 act with forged documents that got them into the amnesty program. Even more importantly, the mass of newly legalized residents is clearly attracting relatives and friends who enter illegally, as they had done before. The fact that nearly 70 percent of the amnesty recipients are male assures that women and children will follow. Studies in Mexico by Professor Wayne Cornelius of the University of California, San Diego, show "no significant return flow of illegals who suddenly found themselves jobless in the United States," and he speculates that the new law may have had the paradoxical effect of keeping more Mexican illegals in the United States than it kept out, as many who would have returned home after a season of work were now afraid that they would not be able to return to the United States. The notion that nothing much has changed is supported even by the INS's own estimates of the number of illegal aliens. It argued, in mid-1989, that there were only 3.5 to 4 million illegal aliens in the United states, as opposed to 6.5 to 7 million before the law. But if one adds the 3.1 million legalized illegals in the amnesty program, it is apparent that the law has merely divided the illegals into two groups: Those who have legalized their status and those who have not. Most authorities not connected with the INS believe that the law had an initial inhibitory effect as prospective border crossers took a wait-and-see attitude to the much-heralded new law. These experts now believe that the old patterns of border crossing are reasserting themselves. As Cornelius put it: "We have found no evidence that the 1986 immigration law has shut off the flow of new undocumented immigrants."[7]

The Newest Nativism

All this would be merely ironic, or even comic, if it only concerned bureaucrats, politicians, and border crossers. But, unfortunately, it may well have other, more far-reaching effects. The long and sometimes hysterical debate over immigration policy in the 1980s, in which even the staid and scholarly Population Reference Bureau could refer to illegal immigration as "a cancer in American society," had helped to create a climate of opinion in which, after a long period of quiescence, ugly nativism was again on the rise. This was not just an American phenomenon but could be seen in Europe, in the Soviet Union, and in the Far East. In countries as diverse as West Germany and Norway, right-wing parties gained unexpected strength in 1989 by campaigning on anti-immigrant platforms. And in crowded Hong Kong, violence erupted between residents and unwanted Vietnamese boat people. In the United States, the Immigration Reform Act heralded as a way to "regain control over our borders" clearly has not had that result. Thus, whatever its framers intended, it raised false hopes among those already concerned about immigrant questions and thus contributed to further anti-immigrant animus.

This animus was chiefly directed against Hispanic immigrants, and a host of interlocking and well-financed pressure groups were stepping up their campaigns against immigrants in general and Spanish-speaking migrants in particular. Many of the most prominent organizations were subsidiaries of a nonprofit organization called "US Inc," including FAIR (the Federation for American Immigration Reform), the Center for Immigration Studies, Californians for Population Stabilization, and, perhaps the best known, US English. While clearly nativist, these organizations do not usually indulge in the crude racism that characterized their predecessors in the 1920s. If one examines the testimony of FAIR spokespersons, such as the Washington lawyer Roger Connor or the historian Otis Graham, there is none of the overt prejudice that was characteristic of the earlier era, although they are arguing for the same end. During the debate over the Mazzoli-Simpson bills FAIR and its spokespersons were prominent in the ranks of its supporters, arguing chiefly for an overall reduction of immigration.

By the mid-1980s, US English, whose chief stated goal is a law making English the official US language, had eclipsed other members of the coalition in notoriety and effectiveness. In a number of states and cities from coast to coast, it sponsored successful referenda to make English the official state language, to forbid public employees from talking to one another in a foreign language on the job, and, in Miami, recommended the abolition of publicly funded Spanish-language emergency services. A number of American states had already adopted English as their official language, many of them, such as Nebraska, having done so as part of the wave of anti-German hysteria accompanying World War I. Such declarations are usually found in state constitutions cheek by jowl with provisions designating the state flower or bird. In many of the states which have recently adopted English-only provisions, such as Indiana, no numerically significant foreign-language speaking population exists, but in states such as California and Florida the voters approving them have specific, human targets in mind.

Founded by the distinguished semanticist and former United States senator, Samuel I. Hayakawa, and an obscure Michigan ophthalmologist, Dr. John Tanton, US English amassed a large advisory board including such respected figures as Saul Bellow, Bruno Bettelheim, Alistair Cooke, and Walter Cronkite, as well as many who were merely well-known such as Walter Annenberg and Arnold Schwarzenegger. The multiethnic backgrounds and reputations of many of its sponsors were evidence of what I have called "white collar nativism." Its popular appeal was based both on its reflexive and mindless cultural nationalism—its literature constantly pointed to the supposedly horrible example of Quebec as an object lesson in what could happen here—and on a genuine concern on the part of some of its supporters for the improvement of the communication skills of many Americans. But a recent exposure of the true views of its cofounder, Tanton, caused some disarray among its supporters. Combining a pseudoerudition with a vulgar appeal to prejudice, Tanton asked, in a recent essay:

> In this society, will the present majority peaceably hand over its political power to a group that is simply more fertile? Can *homo contraceptivus* compete with *homo progenitivo* if our borders aren't controlled? . . . Perhaps this is the first instance in which those with their pants up are going to get caught by those with their pants down. As whites see their power and control over their lives declining, will they simply go quietly into the night? Or will there be an explosion?

If Tanton's vulgarity was merely in bad taste, his use of the term *white* was an appeal now impermissible in American society. Walter Cronkite and Linda Chavez, who had been the most prominent Spanish-surnamed person in the Reagan administration, publicly resigned from the US English board while others may well have simply gone gently into the night.[8] Yet it is clear that Tanton revealed the not-too-well-hidden agenda of most of the white-collar nativists of the 1980s. In the 1920s the chief targets were Catholics and Jews from Europe; in the 1980s it was the Spanish speaking who found themselves in the nativists' sights.

But different targets were not the only differences between the 1920s and the 1980s. The general alignments within society are quite different. The 1920s were dominated by struggles between rural and urban elements in American society, between the forces of Protestantism and Catholicism best symbolized by the election of 1928, in which Herbert Hoover defeated Al Smith. On almost every issue that divided society in the 1920s, the same forces were aligned. The 1980s are culturally much more complex. While many observers persist in seeing the 1980s as solely a struggle between the forces of New Deal liberalism and its conservative opponents, that dichotomy no longer governs many issues. Many of the supporters of FAIR, including its chief spokesman, Roger Connor, are eloquent in their support of the rights of working people. While FAIR liberals support movements to reorient the American immigration system away from the family-oriented provisions that form its core, conservatives, like Republican Senator Orrin G. Hatch of Utah, insist that family preferences be maintained. Hatch, on the other hand, opposed any kind of family planning and all forms of abortion, while many of those most acutely aware of population problems also oppose most measures to increase or maintain immigration, although few are willing to go as far as demographer Kingsley Davis and bar it entirely. In 1974 Davis argued:

> So dubious are the advantages of immigration that one wonders why the governments of industrial countries favor it. . . . One will find few clarifications, but official statements hint that the goals are to fill essential jobs and to stimulate population growth. One suspects that the actual causes are government inertia and pressures by employers to obtain cheap labor.

The purely economic arguments for continued immigration are better than Davis would admit. The popular writer on economic and social questions, Ben J. Wattenberg, has argued in his 1987 book, *The Birth Dearth,* that what America needs is not fewer but more immigrants. Only additional immigration, he feels, can reverse the otherwise seemingly inevitable population decline.[9] In addition, the obvious "graying" of the major industrial nations, including the United States, will place almost unbearable pressure on social welfare systems unless either such trends are reversed—an eventuality that seems highly unlikely—or more young, able-bodied, and perhaps highly skilled immigrants are brought in. A very "iffy" population projection by the U.S. Census Bureau in 1989 projected that the part of the population aged sixty-five and over would rise from a current 12 percent to nearly 22 percent by the year 2038. In the long run, despite much recent handwringing about America in decline, West Germany and Japan, among advanced nations, seem much more vulnerable to the problems of graying than does the United States.

But the major arguments about immigration restriction are more cultural than economic. Many Americans and many members of Congress are dismayed that Europeans make up such a small percentage of contemporary immigration. Relatively few Western Europeans wish to emigrate: The prosperity of Western Europe has removed the major push factor, and the image many Europeans have of the contemporary United States as a nation awash with crime, drugs, pollution, and economic problems has diminished its pull and caused many of them to regard the United States as an interesting place to visit but not one they want to live in. There is a conspicuous exception to this generalization: Relatively large numbers of Irish not only want to come to America but in fact have done so, albeit illegally.

There may be as many as 150,000 illegal Irish immigrants in the urban corridor that runs from Boston to Philadelphia, a region in which Irish American neighborhoods and potential support systems are plentiful. The problem for most Irish is that is that the consanguinal chains have become too weak to support further migration under American law. While many or perhaps even most of the new Irish immigrants have cousins in America, they do not have the parents or brothers and sisters that the law demands. They are thus barred from the family preference system that dominates current immigration, and, since they are economically motivated immigrants, like the Haitians, they cannot claim refugee status either.

Unlike the Haitians, the Irish tend to be well educated and have been able to blend into existing Irish American neighborhoods in the northeastern United States. Anyone who wishes to find them can do so, as Francis X. Clines pointed out in the *New York Times,* in a pub like the County Cork Benevolent and Protective Association on Greenpoint Avenue in Long Island City, or, in fine weather, at picnics in Gaelic Park in the Bronx, or at soccer games, particularly when Irish or Irish American teams play. An illegal immigrant I interviewed in a pub in Boston's South End recently (I promised not to mention its name) told me his story, which is not atypical of the contemporary Irish illegals. A university graduate in accounting, he told me he could find no job at all in Ireland and had heard from a friend that jobs were to be had for the asking in Boston. He flew over and entered the United States legally with a tourist visa in 1984 and has stayed on ever since. He works as a relief bartender in several Boston bars and does the books for two of them. He is paid in cash. He intended, he told me, to marry an American citizen, leave the country, and come in legally as her husband.

Hardships do, of course, stem from illegal status. My Boston informant told me that he could make twice as much money working as an accountant for an American corporation, but the money he was making was more than he could have earned in Ireland. The insecurity of illegal status prevents people from acting in a normal way. Here, according to the *New York Times,* is part of the story, from a twenty-six-year-old woman who identified herself as Rosie Kiernan:

> "I had a death in the family but couldn't risk going back home."
> The expired tourist visa in her passport might have been noted by

> immigration officials and her name and passport number recorded by INS officials to prevent a return entry. "I don't know when I'll see [my family] again, and I know a lot of young Irish here in my situation."[10]

Since most of the Irish illegals came here after the January 1, 1982, cutoff date in the 1986 amnesty law—only a few hundred Irish were granted amnesty—an Irish Immigration Reform Movement has begun to agitate for another amnesty whose progress can be traced in the columns of New York's *Irish Echo,* now the house organ of the new Irish immigrants. The IIRM's green-and-white posters proclaim "Legalize the Irish," a demand that many American politicians would like to grant. But there is no way, in current American law, to extend an amnesty deadline for the Irish without including every other nationality. Thus, with the ink on the 1986 Immigration Reform Act hardly dry, the United States Senate passed another major immigration revision in 1988. After that measure died in the House, a similar bill passed the Senate in mid-1989. Cosponsored by liberal Democrat Edward M. Kennedy and conservative Republican Alan K. Simpson, the bill sailed through the upper house by a vote of 81–7.

The lopsided vote on final passage concealed fierce divisions about various parts of the bill. And, despite the anti-immigrant mood of many in the Senate, the final measure actually expanded the total number to be admitted. It provided a new "national ceiling" of 630,000 annually—not counting refugees and asylees. But—and this made it more palatable to many senators—that figure was to apply for only three years. Then a new national ceiling would be established by Congress after getting recommendations from an independent commission. The bill proposed that this be repeated every three years. But the numbers in the bill were deceptive. Of the 630,000, 480,000 were allocated to close relatives of U.S. citizens and permanent residents. All spouses and minor children of U.S. citizens were guaranteed admission, as were at least 216,000 other "family preference" admissions each year. That these two categories might well exceed the 480,000 slots supposedly reserved for relatives was noted by many senators, but family-oriented conservatives, led by Orrin Hatch and Arizona Democrat Dennis DeConcini, insisted on the 216,000 figure as their price for support of the bill. So the ceiling is really expandable upward.

The other 150,000 slots were reserved for immigrants without family connections who would qualify because of skills, education, and job-related assets. The 1988 bill had contained a provision giving priority within these slots for those with English-language proficiency, and this had "pointed" the discriminatory aspects of the bill. A provision that would have included it in the 1989 bill, proposed by Simpson and opposed by Kennedy, was defeated, 56–43. Within the 150,000 a special pool of 54,000 slots annually were to be awarded on a point system giving highest priority to young, educated and skilled persons in high-demand occupations. Many senators apparently believed that this would attract white immigrants. The headline in the *International Herald Tribune* put it nicely: "U.S. Senate Votes Immigration Shift / Stress Put on Skilled West Europeans." But, in fact, the bill could do no such thing. Substantial numbers of those quota spaces, had the Senate bill become law, would have gone to Hong Kong and India and the Philippines, whose educated young people are anxious to come to the United States and whose educational achievements, including doctorates, are likely to outpoint those of most potential European immigrants.

Other provisions of the bill also balanced the wish lists of various senators. Illegal aliens were barred from the census figures to be used for the 1990 reapportionment, illegal aliens were barred from receiving federal benefits, and the deportation of close family members of persons being legalized under the 1986 amnesty provisions were prohibited. This last provision, if enacted into law, would solve the dilemma of the amnesty recipient from El Salvador whose five children were unable to qualify. It would also boost, substantially, the number of those effectively legalized by the 1986 amnesty. Eventually, if the woman from El Salvador becomes an American citizen, her children would go to the head of the queue. After a period as "undeportable illegal aliens"—this would be a new category in American law—they would become children of a U.S. citizen.[11]

Whether or not the 1989 bill becomes law before the expiration of the 101st Congress in January 1991, it is clear that, for the foreseeable future, immigration issues will be fairly prominent on the American social agenda. The relatively high level of immigration in the 1980s, as shown in table 16.1, saw more legal immigrants arrive than in any decade since the peak immigration years after the turn of the century. Yet, despite the perception of many Americans that the country was full of immigrants, the incidence of foreigners in the American population was well below the levels maintained from the Civil War to the era of World War I, when each census showed between 13 and 14 percent of the population of foreign birth. That figure fell steadily for half a century, and began to rise only after the 1970 census, as shown in table 16.2. The 1990 census will surely show another increase, perhaps as

Table 16.1

Legal Immigration to the United States, 1981–1990

YEAR	NUMBER
1981	596,600
1982	594,131
1983	559,763
1984	543,903
1985	570,009
1986	601,708
1987	601,516
1988	643,025
1989	1,090,924
1990	1,536,483
Total (est.)	7,338,062

Table 16.2

Foreign Born as a Percentage of Total Population, 1910–1980

YEAR	PERCENTAGE
1910	14.7
1920	13.2
1930	11.6
1940	8.8
1950	6.9
1960	5.5
1970	4.7
1980	6.2

high as the 8 percent level. Assuming a total population, then, in the neighborhood of 250 million persons, that would be 20 million persons of foreign birth. If the foreign born were present in the same proportion as in 1910, there would be 35 million of them. To be sure, immigration is not a question of numbers alone, and the United States of 1990 is not the United States of 1910, but these comparative numbers should be kept in mind when the subject of immigration policy is discussed.

As has always been the case, immigrants today are not spread evenly around the country but are clustered in certain states and cities. As we saw in discussing the geographical distribution of amnesty applicants, the overwhelming majority of immigrants, legal and illegal, come from just a few states: California, New York, Florida, and Texas contain a majority of all the foreign born, and three cities, Los Angeles, New York, and Miami, and their hinterlands, are the immigrant cities par excellence. New York and, to a lesser degree, Los Angeles, have a long tradition of dealing with foreign populations, but Miami—in my youth a sleepy little city in which the "foreigners" were winter tourists from the Northeast—has no such tradition. The stresses and strains of a variety of ethnic enclaves there have made it not only one of the immigration capitals of the 1980s but also one of the centers of anti-immigrant feeling.

Whatever Congress does, the basic patterns that have been established since 1965, with most immigrants coming from Asia and Latin America, are likely to continue. Unless there is a disastrous depression in the United States, a depression whose severity and length rivals that of the 1930s, there will be many more persons wanting to immigrate to the United States than the country will be willing or able to accept. There will, of course, be changes within the basic pattern, and these changes are more likely to be dictated by events in the rest of the world than by purely American forces. As I write, in mid-1989, there are three particular situations that may put new pressures of the whole American immigration system. These stem from events in Central America, in China, and in Eastern Europe.

In Central America the apparent failure of the Contras to overthrow the Sandinista government in Nicaragua and the lack of active support for them from the Bush administration seemed to doom their movement. Thousands of their rank and file are potential asylum seekers—many of the leaders are already established in Miami—toward whom many in the American government will feel a special obligation. In addition, the continuing turmoil in Central America all but guarantees that immigration from that region will become increasingly important.

In China the brutal crushing of the prodemocracy movement in Tiananmen Square caused a reaction that created sympathy and provided asylum for those of its student leadership who managed to elude the security forces of the People's Republic. It also called into question, in the minds of many of Hong Kong's 3.5 million Chinese, their future as a part of China after 1997. While the relatively poor majority of Hong Kong's people will remain there no matter what, tens of thousands of its well-educated, skilled, and affluent elite will manage to emigrate somewhere. Although Britain and other nations of the Commonwealth are the most likely places of refuge for most, large numbers will surely eventually arrive in the United States. Emigration of well-to-do Hong Kong Chinese has already resulted in strong and virulent reactions to their presence in Canada. Many Canadian universities have placed quotas on the number of students from Hong Kong they will admit, and in Vancouver, British Columbia, where there are now perhaps 130,000 Chinese in a population of perhaps 750,000, thousands of the recent arrivals have driven up the price of houses and apartments beyond the means of many of the city's middle-class residents and an ugly anti-Chinese campaign has resulted. In the United States the immediate Congressional reaction to the horror in Tiananmen Square and the nationwide repression that followed was to put provisions into the 1989 immigration bill earmarking five thousand slots a year for persons from Hong Kong and allowing Chinese students in the United States to extend their visas. Although President Bush vetoed a separate bill extending Chinese students' visas, he promised to extend them administratively.

Of even greater possible consequences were the stunning events in Eastern Europe triggered by Mikhail Gorbachev's *glasnost* and *perestroika.* The initial effect, as we have seen, was a steady increase in the number of Soviet Jews allowed to emigrate. But the spread of the "freedom" to some Eastern bloc countries—particularly Poland and Hungary—has created new strains in countries like East Germany, Czechoslovakia, and Romania. Hungary's opening of its border with Austria and the ensuing flow of East Germans to the West through that border initially affected West Germany, which, despite a constitutional provision that guarantees any "German" automatic entry and full citizenship in the Federal Republic, felt compelled to close temporarily its offices in East Germany and Hungary and stop issuing the passports, which had been automatic. And, within the Soviet Union itself, ethnic tensions, particularly in the former Baltic republics of Lithuania, Latvia, and Estonia, have created potential crisis points for American immigration policies. On the one hand, if the liberalization within the Soviet Union continues, Balts, many of whom have relatives in the United States, may become a numerically significant immigrant group; on the other hand, should there again be a massive wave of repression within the Soviet Union and Eastern Europe, the possible consequences for American immigration policy are also considerable. It is clear that, unlike the situation in the 1930s, the United States could not and would not ignore the new refugees such repression would surely create.

To be sure, not every world crisis has a significant impact on American immigration. The long-term Chinese repression in Tibet has had little demographic impact because very few Tibetans have had the means to emigrate to the United States. The same is true for all of black

Africa, but a complete or partial triumph of the African National Congress in South Africa would create a white emigration movement whose effects would surely include emigration to the United States, although, as in the case of Hong Kong, the immediate impact would be more on Britain and the Commonwealth.

There is one further reason to believe that immigration will retain its central position on the American agenda: the rediscovery of the immigrant tradition by the American people. In the 1920s, and after, the United States denied or forgot where it came from. It was our colonial past that was celebrated, a past that featured "settlers" rather than "immigrants," a past whose essentially elitist nature can be symbolized by the too-meticulous, too-prettified recreation of colonial Williamsburg, Virginia, largely through the financial support of the Rockefeller family. The so-called ethnic revival of the 1950s and 1960s, in part a mimetic reaction to the black revolution, helped pave the way for a deeper appreciation of our immigrant roots that came in the 1970s and 1980s. While this was in part a manifestation of Hansen's law—the third generation wants to remember what the second generation tried to forget—it seemed to affect most of the American people, not just the descendants of those who came in the peak immigration years before and after the turn of the last century.

As early manifestation of the new appreciation of immigration could be seen, curiously enough, in the bicentennial celebrations of the American Revolution in 1976. The sesquicentennial celebrations, in the 1920s, had all but ignored immigration and immigrants, but the celebrations in the 1970s tried to embrace Americans of every race and ethnicity. Clearly, the image that American society had of itself and wished to perpetuate had changed significantly. This trend was even more apparent in the celebrations of the centenary of the Statue of Liberty, culminating in Liberty Weekend in New York Harbor in July 1986. Although the statue had originally been called "Liberty Enlightening the World" and had little to do with immigration, it had become the symbol, par excellence, of the nurturing aspects of American society. Its refurbishment, by a quasi-public commission using private funds and headed by a flamboyant second-generation Italian American, Lee Iacocca, was the centerpiece of the celebration, which was presided over by President Ronald Reagan and included honoring a hundred distinguished immigrants of almost every conceivable ethnicity. The Statue of Liberty–Ellis Island Centennial Commission and its satellite Statue of Liberty–Ellis Island Foundation, Inc., also undertook the rehabilitation of the old immigrant reception center on Ellis Island and the creation there of a huge museum devoted to American immigration, past, present, and future. The opening of that museum, in the fall of 1990, two years before the centennial of Ellis Island as an immigrant reception center, will surely serve as a capstone to the revival of interest in the immigrant past, which has characterized the past decade and a half.[12]

Yet, as we have seen, the euphoria of Liberty Weekend could coexist with renewed nativist fears about the threat from "feet people" and the danger of the United States becoming Hispanicized. The nativist tradition is almost as old as the immigrant tradition in America, and notions about its demise are, as Mark Twain observed about a false report of his death, greatly exaggerated. Some who celebrate their own immigrant roots the loudest are among those who express fears about present or future immigrant invasions. This should not be surprising. One of the glories of the United States is that it is, as Nathan Glazer recently put it, a "permanently unfinished country," and one of the hallmarks of its unfinished state is the constantly changing mix of peoples who come here.[13] Appropriately, an unfinished country has not only an unfinished past but one whose perceived shape changes as well. The historical vision that created Williamsburg is quite different from that which is creating Ellis Island. It would be difficult to deny that the vision that conceived Ellis Island in the 1980s was more catholic than that that conceived Williamsburg in the 1930s. But there is no reason to assume that the Ellis Island view is a final one. Four or five decades hence surely another view of the past will have arisen, but that view will no more be able to exclude Ellis Island than the Ellis Island view was able to exclude Williamsburg. The colonial past and the immigrant past cannot be erased. How they are seen in the twenty-first century will depend on the needs of that century. In our time a heightened awareness of our immigrant past has accompanied the revival of immigration. Whether or not that revival continues, the twelve million who have come in the quarter century since 1965 will, with their descendants, be an important factor in American life for the foreseeable future and will continue to contribute to its growing diversity.

17

Immigration in an Age of Globalization

American immigration in the years since 1986 has mostly demonstrated continuity, but there have been some startling changes. The major continuities have been in the increasing volume of immigration, its sources, and in the apparently incorrigible incompetence of the Immigration and Naturalization Service. The changes have been in attitudes toward immigration and the eventual reflection of those changes in immigration law and regulations. The incompetence of the INS, long assumed by critics on both the right and the left who spoke of the regulation of immigration as a "broken system," received some pointed underlining in 1999 when the respected Maxwell School of Government at Syracuse issued its first government performance report card grading twenty government agencies that the Clinton administration had identified as having a "high impact" on the American public. The INS received the lowest overall grade (*C*−) and was the only one of the twenty agencies that did not get a grade of better than *C* in any of the five judged categories. The school's crisp summary of the agency's problems was: "Infusion of resources doesn't solve mission conflicts, system deficiencies."[1]

America's Immigrants: A Twentieth-Century Snapshot

Whereas in the decade ending in 1990, 7.3 million legal immigrants were counted, in the first seven years of the next decade the tally grew to 7.6 million and the decade's total would approach 10 million. In addition, the twenty-year period, 1981–2000, has been the heaviest such period in our immigration history, although the heaviest ten-year period of all time remains 1905–1914 when 10.1 million immigrant entries were recorded. But those immigrants came to a nation whose total population was 92 million (1910); the end of twentieth-century immigrants came to a nation of some 280 million. The adjacent table and chart (17.1) show the volume and some of the complexity of twentieth-century immigration. One of the ways to assess the impact of immigration is to check the incidence of foreign-born persons in the population. (For the nineteenth century see tables 6.4 and 6.5., pg. 125 and pg. 129) During the twentieth century foreign-born persons amounted to as much as 15

Table 17.1

Immigration and Its Major Sources, by Decade, 1901–1998

DECADE	NUMBER	MAJOR SOURCES
1901–10	8,795,386	Europe—91.6%; North America—3.9%; Asia—3.7%
1911–20	5,735,811	Europe—75.3%; North America—19.2%; Asia—4.3%
1921–30	4,107,209	Europe—60.0%; North America—35.9%; Asia—2.7%
1931–40	528,431	Europe—65.8%; North America—28.8%; Asia—3.1%
1941–50	1,035,039	Europe—60.0%; North America—32.2%; Asia—3.6%
1951–60	2,515,479	Europe—52.7%; North America—36.0%; Asia—6.1%
1961–70	3,321,677	North America—43.9%; Europe—33.8%; Asia—12.9%
1971–80	4,493,314	North America—37.5%; Asia—35.3%; Europe—17.8%
1981–90	7,338,062	North America—43.0%; Asia—37.3%; Europe—10.4%
1991–98	7,606,068	North America—43.8%; Asia—30.9%; Europe—14.9%

Source: Immigration and Naturalization Service, *1998 Statistical Yearbook,* pp. 16, 19.

NOTE: In INS data, Mexico is included in "North America." In Census Bureau data Mexico is included in "Central America." One bureaucracy, two taxonomies.

FOREIGN BORN 1900–2000

YEAR	NUMBER IN MILLIONS	PERCENTAGE OF TOTAL POPULATION
1900	10.4	13.6%
1910	13.6	14.7%
1920	14.0	13.2%
1930	14.3	11.6%
1940	11.7	8.8%
1950	10.4	6.9%
1960	9.7	5.4%
1970	9.6	4.7%
1980	14.1	6.2%
1990	19.8	8.0%
2000	28.4	10.4%

Sources: Lisa Lollock. *The Foreign Born Population of the United States: March 2000.* Washington, D.C.: GPO, 2001; U.S. Census Bureau, Current Populations Reports, P20-534 (available at http://www.census.gov/prod/2000pubs/p20-534.pdf); U.S. Bureau of the Census. *We, The American Foreign Born.* Washington, D.C.: GPO, 1983 (available at http://www.census.gov/apsd/wepeople/we-7.pdf).

Chart 17.1
Immigration, 1901–1998

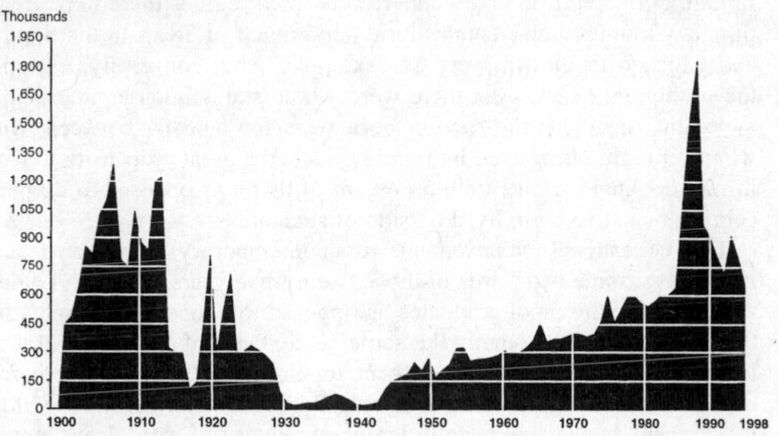

Source: Immigration and Naturalization Service. *1998 Statistical Yearbook.* Washington, D.C.: GPO, 2001, p. 15.

NOTE: The large spike in the data for 1989–91 represents, not a great surge of immigration in those years, but the major effects of the IRCA amesty of 1986. The vast majority of those persons had been in the country since 1982 or earlier but the INS bookkeeping records them in the year in which they were legalized. Source: Immigration and Naturalization Service. *1998 Statistical Yearbook.* Washington, D.C.: GPO, 2001, p. 15.

percent of the population (1910) and as little as 5 percent of the population (1970). The 28.4 million foreign born of the 2000 census was about halfway between those extremes: this group represented some 10.4 percent of the nation's population. Five-sixths (84 percent) of the foreign born of 2000 immigrated to the United States after 1970; almost two-fifths of them (39.5 percent) arrived in the 1990s.

The origins of the foreign born in 2000 reflected the sources of recent immigration. According to the Census Bureau, which does not differentiate between legal and illegal immigrants, just over a third (34.5 percent) came from Central America, a quarter (25.5 percent) from Asia, and nearly a sixth (15.5 percent) from Europe. They were almost evenly divided by gender. Very few were young children; only 7.2 percent of the foreign born were under fifteen years of age: that cohort accounted for 25.2 percent of all native-born Americans.

Immigrant settlement patterns in 2000 showed a heavy concentration in the western states. 39.9 percent of the foreign born had settled in that region which was home to only 20.8 percent of native-born Americans. 22.6 percent of the foreign born were attracted to the Northeast, which held 18.6 percent of the natives. Accordingly, there were relatively fewer immigrants in the South and Midwest. The former region was home to 35.9 percent of the natives and 26.8 percent of the foreign born, while the latter contained 24.6 percent of the natives and 10.7 percent of the foreign born. Yet even in the underrepresented regions there were areas of heavy immigration—south Florida and much of Texas in the South, and Chicago in the Midwest, for example. And, conversely, even in the immigrant-heavy West there were whole states, such as Idaho and Wyoming, in which the foreign born were few and far between. But whatever region they lived in, immigrants had a great propensity to live in metropolitan regions: only 5 percent of them, as opposed to 20 percent of the native born, lived outside of such areas.

The educational achievements of contemporary immigrants are impressive: some two-thirds of those twenty-five years of age and older are at least high school graduates as opposed to almost seven-eights of the native born. But roughly the same percentages of each group are at least college graduates (25.8 percent for immigrants, 25.6 percent for native born), and, although the Census Bureau has not yet released the most recent data, a significantly larger percentage of very recent immigrants have Ph.D.s than do native-born Americans. On the other hand, large numbers of foreign born have little or no formal education: 22.2 percent of them had less than a ninth-grade education, as opposed to only 4.7 percent of the natives. This lumping of all immigrants into one set of statistics obviously ignores very real differences between those from different regions and nations. To examine just one category as an example, 44.9 percent of all foreign-born Asians had bachelor's degrees or better, while only 5.5 percent of those from Central America, 19.1 percent of those from the Caribbean, and 25.9 percent of those from South America had made similar academic progress. Note that despite the stereotype of undereducated Hispanics the last figure is marginally above the achievement of native-born Americans. The very high figure for Asians, of course is well known, but every tenth foreign-born Asian immigrant had less than a ninth-grade education. And when we get data broken down by national origin there will be even more variation.

A 1995 study looked at United States residents with a Ph.D.—whether from American institutions or elsewhere—and found find that 23 percent of the science and engineering Ph.D.s, and 12 percent of the nonscience and engineering Ph.D.s were foreign born. The share of foreign-born Ph.D.s was highest in engineering (40 percent), math and computer science (33 percent), and physics, chemistry, and economics (31, 26, and 24 percent). About 55 percent of the immigrants with Ph.D.s in science and engineering were then naturalized U.S. citizens and a significant but unknown number were in the process of being naturalized.[2]

Immigrant income, like immigrant education, shows a distinct bipolarity. A Census Bureau survey of poverty status in 2000 found that 11.2 percent of native-born Americans and 13.3 percent of foreign-born Americans were below the poverty line. But when the foreign born were divided into naturalized citizens and aliens only 9.1 percent of the former and 21.3 percent of the latter were in poverty. Since naturalized citizens have been in the country at least five years and tend to have more education than most foreign born, this is congruent with many previous studies that have shown that the rate of poverty decreases as immigrants learn how to cope in their new society. The same kinds of disparities appear in the poverty data on immigrant families. 15.7 percent of all families headed by foreign-born householders were in poverty. When analyzed by region of origin of immigrant family head, 6.8 percent of Europeans, 11.8 percent of Asians, and 21.5 percent of Latin American families were below the line. Similarly, if families headed by Latin Americans are broken down by region, 23.5 percent of Central American families, 21.7 percent of Caribbean families, and only 10.3 percent of South American families were classified as poor. Although the 2000 Census data for immigrants from individual nations had not yet been released when this was written, it is obvious that, among the Caribbean families, for example, Haitian families would have a higher poverty level than, say, Dominican families.

What all these data tell us is that generalizations about the socioeconomic status of contemporary immigrants, when lumped together, are largely meaningless and misleading. Only when we are able to particularize and focus on individual groups, and on classes within those groups, is it possible to speak with any degree of precision. And, to be sure, there is also the experience of individuals to be examined. In the final analysis, as the great biologist Charles Darwin once noted, "there are no species, only individuals." But, as the data show, the experience of individuals is quite varied. By using individual experience, nativists can make the lives of contemporary immigrants seem like hell while immigrant advocates can paint those lives with rosy colors. The responsible historian must try to balance, somehow, these extremes and arrive at a reasonable estimate of immigrant status and expectations. It is difficult enough to do this for bygone eras: for the very recent past it is well nigh impossible, but in the pages to follow I will try to make a series of judicious assessments about the twelve million or so immigrants who came just before and after the turn of the current century. I will begin with the very latest immigrants for whom there is data.

Getting into America in 1998

In 1998, for the second year in a row, the number of legal immigrants reported by the INS decreased significantly: from 915,900 in 1996 to 798,378 in 1997 to 660,477 in 1998, an overall reduction of more than 25 percent. This figure did not, as the INS recognized, represent a lowering in the desire of eligible foreigners to come to America but was the result of the inability of an over-tasked agency to cope with the necessary paperwork. (The INS bureaucratese put it "due to an increase in pending adjustment of status applications.") The origins of 1998's immigrants were much as they had been: North America, largely Mexico, and Asia each contributed more than a third, with the rest divided among Europe, Africa, Oceania, and South America. (For some reason

Table 17.2

Immigrants Admitted by Major Category of Admission, 1996–1998

CATEGORY	1998 NUMBER	1997 NUMBER	1996 NUMBER
All Categories	**660,477**	**798,378**	**915,900**
New arrivals	357,037	380,719	421,405
Adjustments of status	303,440	417,659	494,495
Family-Sponsored Immigrants	**475,750**	**535,771**	**596,264**
Immediate relatives of U.S. citizens	283,368	322,440	302,090
Spouses	151,172	170,263	169,760
Children	70,472	76,631	63,971
Parents	61,742	74,114	66,699
Children of resident aliens born abroad	902	1,432	1,660
Other family-sponsored immigrants	191,480	213,331	294,274
Adult offspring, U.S. citizens	39,974	44,479	46,361
Siblings of U.S. citizens	63,018	55,171	64,979
Spouses and children of resident aliens	88,488	113,681	182,834
Employment-Based Preferences	**77,517**	**90,607**	**117,499**
Refugees and Asylees	**54,654**	**112,158**	**128,565**
Refugees	44,645	102,052	118,528
Asylees	10,000	10,106	10,037
Diversity Programs	**45,499**	**49,374**	**58,790**
Other	**5,809**	**9,822**	**14,782**

Source: INS, Adapted from Table 4, p. 28, *1998 Statistical Yearbook*

NOTE: Three small categories of persons, actually refugees, are counted elsewhere. Persons admitted under the Chinese Student Protection Act are tabulated as "Employment based" and both "Amerasians" and "Soviet/Indochinese Parolees" go under "Other." In 1997 these amounted to 142, 738, and 1,844 persons respectively.

or other the INS did not know, even by region, where 7,003 legal immigrants came from. This was 1 percent of the total and unprecedently high: the figures for 1996 and 1997 were 5 and 197 persons of unknown origin respectively.) The global breakdown by region and by major nations of emigration was as follows:

Table 17.3

Immigrants Admitted by Region of Birth, 1998 (Total = 660,477)

REGION	NUMBER	PERCENT
North America	252,996	38.3%
Asia	219,696	33.3%
Europe	90,793	13.7%
South America	45,394	6.9%
Africa	40,660	6.2%
Oceania	3,935	0.6%
Unknown	7,003	1.1%

Immigrants Admitted by Country of Birth, 1998

RANK	COUNTRY	NUMBER	PERCENT
1.	Mexico	131,575	19.9%
2.	China and Hong Kong	42,159	6.4%
3.	India	36,482	5.5%
4.	Philippines	34,466	5.2%
5.	Soviet Union (former)	30,163	4.6%
6.	Dominican Republic	20,387	3.1%
7.	Vietnam	17,649	2.7%
8.	Cuba	17,375	2.6%
9.	Jamaica	15,146	2.3%
10.	El Salvador	14,590	2.2%
Total		359,992	54.5%

Thus, just two countries accounted for just over a quarter of all immigrants and ten accounted for more than half of them.

Distribution of these immigrants throughout the United States was even more concentrated. More than 20 percent of all 1998 immigrants told the INS that they intended to live in one of two cities, New York and Los Angeles, while two-thirds of all immigrants planned to live in one of six states: California, New York, Florida, Texas, New Jersey, or Illinois.

Under what authority did these immigrants enter? Since the epochal 1965 immigration act was still the basic statute governing immigration the general preference system described in Table 13.2 (p. 342) remained in effect with minor modifications. Using the year 1998—the latest year for which detailed INS data were available in the summer of 2001—as an example, the INS statisticians came up with a total of 660,477 immigrants as the official total for that year. (See Table 17.3; for previous decades see Tables 13.3, 13.4, and 16.1 pp. 344, 346, and 404.) But only a little more than half of them—357,037 (54.1 percent)—were new arrivals. (In many recent years newcomers were a minority.) The rest were persons who had entered in some previous year and had had their status adjusted in 1998. Nearly three-quarters of the total—474,848 (71.9 percent)—were "family-sponsored" immigrants. Three-fifths of the family-sponsored group—283,368 (59.7 percent)—were "immediate relatives" (spouses, minor children, and parents) of U.S. citizens, whether naturalized or native born. These entered "without numerical limitation." The rest—191,480 (40.3 percent)—were either other relatives of U.S. citizens (adult sons and daughters plus brothers and sisters) or spouses and minor children of resident aliens. The latter group accounted for 88,488 (18.6 percent) of 1998's family-sponsored immigrants. Some of those who wish to reduce the number of immigrants admitted focus on these "other relatives" as a likely target. As will be noted later, when naturalization is discussed, each time an immigrant achieves resident alien status, there is a potential cohort of related individuals who have a privileged status in American immigration law. Similarly, each time a resident alien becomes a citizen, there is a potentially larger cohort of privileged individuals created.

About one immigrant in nine was admitted under what the INS calls employment based preferences. (But, as will be shown, there were nearly half a million temporary workers, trainees, and their families admitted during 1998 who are not counted as "immigrants.") These are of two general kinds. The larger category, some four-fifths of the total, is of skilled workers, mostly professionals whose work is needed: one thinks of nurses and other medical personnel and, more recently computer programmers and similarly skilled specialists in information sciences. The other fifth is of agricultural and service workers of various kinds, including sheepherders who were once almost exclusively Basque but are now largely Peruvians. The highly educated professionals tend to acculturate easily; the others tend to have little education and are less likely to acculturate quickly. In a third category, which has had much publicity in the press, are the so-called millionaire immigrants: persons who obtain preference visas on condition that they invest a million dollars in an American enterprise—or $500,000 if the investment is in an economically depressed area. The numbers of such persons have never been large—in our focus year of 1998 there were 824 of them, 0.1

percent of the year's crop of immigrants. The investor program—now categorized as "employment creation"—has been characterized by significant amounts of fraud, much of it on the part of a few former INS officials who have set up fee-charging agencies to help rich would-be immigrants gain admission.

Asylees and refugees continue to come in much greater numbers than anticipated. As we saw in chapter thirteen the 1980 refugee act envisaged a "normal" level of 50,000 refugees and 5,000 asylees annually, but those limits continue to be illusory. In 1998 54,645 persons, every twelfth legal immigrant admitted, was tabulated as a refugee: 10,006 of these were asylees. This was the first year in the 1990s when the adjusted refugee total fell below 100,000. Part of this decrease was caused by a reduction in arrivals; part by INS bookkeeping: according to the State Department's count, 76,181 refugees actually arrived during the year. Not counted in this total were Amerasians, "parolees" from Eastern Europe and Vietnam, and Cuban-Haitian "entrants": those three categories accounted for another 1,598 individuals in 1998. The courts have stretched the criteria for establishing both refugee and asylee status well beyond what Congress believed it was doing. I doubt if any congressperson in 1980 thought of battered women as asylum candidates, yet another example of "unintended consequences." The bulk of the refugees admitted in 1998—more than seven out of ten—came from two European regions, the former Yugoslavia (30,906, 40.6 percent) and the former Soviet Union (23,349, 30.6 percent). Another 10,288 (13.5 percent) came from Vietnam. Thus, despite much talk about a "new world order" refugee policy remained retro, a throwback to the Cold War. Africa, where the vast majority of the world's refugees are located, had been home to just 6,662 refugees, all but a thousand of them from three nations: Somalia, Liberia, and Sudan.

But not included in these totals are persons let into the country by the attorney general, but not counted as "admitted" by the INS. Parole authority, which as we saw was begun in wartime as a onetime event by FDR in 1944, became routine and statutory in the Eisenhower administration. At first parole was used only for refugees but it is now used for a wide variety of administrative purposes. In recent years more than 100,000 parolees have been let in annually; in 1998 a record 234,545 were let in. According to the INS more than 75 percent of these cases are port-of-entry paroles, most of which involve returning resident aliens with incomplete documents and are quickly resolved. However it is impossible to determine from the published data the disposition of the cases.

The asylum story is a bureaucratic nightmare. The INS reported that exactly 10,000 asylees arrived during 1998 and that 12,951 persons were granted asylum. But at the beginning of the year there were 397,809 asylum cases pending—that is, persons who were actually in the country and whose claims had not been adjudicated—and that "only" 360,247 cases were pending at year's end, which seems like a kind of progress. But a footnote explains that the number is lower "because of corrections to the database." The INS seems not to know whether it is gaining on its backlog or not, but perhaps it is just not telling.

Of the individuals granted asylum in 1998 nearly three-fifths, 7,581 (58.5 percent), came from seven countries: the former Soviet Union—3,336; Somalia—1,308; Iran—708; China—668; Ethiopia—543; former Yugoslavia, 527; and Liberia—491. More than a third of successful applicants for asylum were Africans, 4,915 or 38 percent. This indicates that those Africans—mostly elites—who managed to get to the United States were seen as deserving of admission, which underscores the obvious bias in granting refugee status in Africa. Other evidence of this bias is found in the fact that, at the end of 2000, there was just one refugee processing center in all of sub-Saharan Africa, as opposed to eight in Europe. The U.S. State Department, which almost never admits U.S. bias, attributes the imbalance to a "lack of infrastructure," as if this infrastructure were outside of the control of the government.

The "diversity" category was first introduced without using that term as a very small allocation—5,000 annually for two years—in the Immigration Reform and Control Act (IRCA) of 1986 under the rubric "making visas available to non-preference immigrants." This program was made much more significant—some 8 percent of allocated immigrant visas—in the 1990 legislation (see following). By 1998 it was actually functioning to diversify somewhat the immigrant mix although it had begun as a kind of backdoor affirmative action for white folks in general and Irish in particular. While the word "lottery" does not appear in the statute and is not used by the INS, the Department of State, which administers the program, calls it that and the term is used by the press and public. Diversity immigrants are now chosen from applicants who are nationals of countries that did *not* have a cumulative total of 50,000 immigrants in the previous five years. There may be as many as 55,000 in a given year, with a per country limit of 3,850. In 1998 there were 45,499 such immigrants, 6.9 percent of the total. Europe got the largest group of winners—21,765 (44.1 percent)—more than three-quarters of them from former east bloc nations. The next largest group was from Africa: 15,025 (30.0 percent), a particularly significant figure since there were "only" 44,668 immigrants from Africa if one includes every category. The other quarter of diversity admissions was distributed as follows: Asia, 8,631; North America, 2,014; South America, 1,044; and Oceania, 2,014.

"NONIMMIGRANT" IMMIGRANTS

In addition to the various groups of legal immigrants, the INS also records various categories of people whom it calls "nonimmigrants" but who might more reasonably be called "temporary immigrants": these are persons who enter the United States on various kinds of visas that do not entitle them to apply for permanent resident status and which have expiration dates. Their number had grown even more rapidly than that of immigrants: from two million annually in 1965 to thirty million in 1998. Much of that increase is very recent: between 1991 and 1998 the number of these arrivals almost tripled, growing from eleven million to thirty million. Note that these thirty million "nonimmigrant" arrivals outnumber recorded immigrant arrivals by almost ten to one. Of these thirty million incoming foreigners, slightly more than half came from just four nations: some 5.25 million came from Japan, with very large but smaller numbers from the United Kingdom, Mexico, and Germany.[3]

More than three-quarters of the "nonimmigrants" in 1998—some twenty-three million persons—were tourists, what the INS calls "visitors for pleasure." The vast majority of these return home or go somewhere else, but there is no real way to know exactly how many have done so. The incoming tabulation is based primarily on collecting I-94 forms: these are the forms airline flight attendants pass out on returning overseas flights. (The government seems to have lost or misplaced the cards for 1997: the INS has also reported the "unavailability" of these and related records for two or three months of 1979, all of 1980, and most of the data of 1981 and 1982.) The data on departing "nonimmigrants" are even sketchier.

The other major groups in this category are "visitors for business," students, the previously mentioned "temporary employees," and parolees. There were 4.4 million temporary visitors who came to the United States on business in 1988. Very few studies have been made of such visitors: many are employees of multinational corporations. We know a great deal about foreign students who have become a major component of American academic life, particularly at the graduate level in many scientific and technical disciplines. Colleges and universities often compete for them. While a substantial majority of foreign students return to their homelands, large numbers of them manage, in one way or another, to get the coveted green card. In 1998 564,683 foreign students were admitted along with 33,837 spouses and children.

A somewhat smaller number of persons, 371,653 "temporary workers and trainees," were admitted along with 86,866 spouses and children. The best-known segment of this group, some 240,000 workers—nearly two-thirds of the total—were "workers with specialty occupations" who entered with H-1B visas. Most of these were scientists and technicians to staff the various Silicon Valleys of America. Another 50,000 were at the other end of the economic spectrum, half of them agricultural workers on H-2A visas; most of the other half, on H-2B visas, were hired for seasonal employment in resorts and similar establishments. (Although most H-2A workers go to the sun belt, they are a national phenomenon. In 1993, for example, the New England Apple Council recruited 2,000 foreign workers and 400 American workers for its harvest.) The remaining fifth of 1998 "temporaries," nearly 80,000 workers, came in under nine discrete visa categories created by Congress.[4]

But in addition to the more than 450,000 workers and families listed above, there were another half million persons (plus 140,000 dependents) who were actually temporary workers in ordinary language but whom the INS called something else. 250,000 were "exchange visitors" on J-1 visas, many of them academics, along with 41,771 spouses and children; 200,000 were "intra-company transfers," along with 99,196 spouses and children; and nearly 60,000 were workers whose admittance was covered by the North American Free Trade Agreement (NAFTA).

Why are these thirty million nonimmigrants important? Simply because, as noted earlier, that despite all of the continuing uproar about illegal immigrants, most authorities continue to believe that it is visa overstayers—that is persons who entered legally as "nonimmigrants"—who comprise the major portion of what are called "illegal immigrants" or "undocumented persons" and not the so-called "wetbacks" and other surreptitious border crossers who get so much publicity in the media and from demagogues in and out of congress. Looking just at our focus year of 1988 there is no reason to believe that any significant number of the five million Japanese "nonimmigrants" have become overstayers. The same cannot be said for the 3.5 million Mexican visitors, the nearly one million Brazilian visitors, the 750,000 visitors from China, or the 500,000 Argentines. Because of the economic and social conditions in those and other similarly situated countries there is every reason to believe that significant numbers of the such 1998 visitors are still in the

United States, most of them without legal authority to be here. Not only will a significant number of the persons entering on the various special visa programs—students, workers, and their families—be able, by one means or another, to change their status to that of permanent resident, women in these programs who have children here become mothers of American citizens.

But the millions described above are only the tip of the iceberg in terms of legal entrance into the United States in 1988. Without citing any numbers the INS acknowledges that it has no data for most of Canadians and Mexicans who cross the borders in great numbers every day. The rules are openly discriminatory: Canadians may travel for business or pleasure without travel restrictions for six months without a visa. Mexicans who cross the border frequently may apply for border crossing cards, which can be used for admission to the United States as long as the entrant stays within twenty-five miles of the border and does not stay more than seventy-two hours. (In August 2001 a mild hue and cry was raised in the media about discrimination against Mexican trucks, when Congress applied different safety regulations for them as opposed to those regulating trucks from Canada, but little is ever said about this longstanding discrimination against Mexican as opposed to Canadian border crossers.) How many persons cross the border annually? No one knows, but one informed journalist in early 1999 put the figure at 400 million individual border crossings annually, a guess that would include multiple crossings by many individuals.[5]

ILLEGAL IMMIGRANTS

As noted on page 311 the problem of estimating the number of illegal immigrants in the United States at any given moment is highly problematic. The INS, under commissioner Doris Meissner (1993–2000) was more responsible and consistent in its estimates than it had been under previous commissioners, but they were still guesses. The INS devotes four pages of double-columned text in its *1998 Yearbook* to explicating its estimates of illegal immigrants, beginning with a large-type, bold-faced assertion about five million "undocumented immigrants" as of October 1996, but ending, in smaller type, with the admission that its estimates "should be used with caution because of the inherent limitations in the data."

"Data" is a pretty strong term for some of the things that the INS is dealing with, particularly given its own evidence of lost or missing information, etc. While it is clear that there are millions of illegal immigrants residing in the contemporary United States, there is no particular reason to accept the government's methods or its information, particularly when it gets down to estimates by country of origin and location within the United States. It recently had to adjust its figures for the Dominican Republic by 50 percent after discovering that some 25,000 presumed Dominican illegals had been attributed to the much smaller Caribbean island of Dominica. We must remember the cautious statistician's acronym, GIGO: garbage in, garbage out. *If,* however, we accept the five million figure we should remember that this would be 1.8 percent of the population, that the vast majority are hardworking, exploited toilers doing necessary jobs and not in any way a threat to the republic as some scaremongers would have us believe.

Immigration Legislation and Public Opinion: A Roller Coaster

The drastic changes in attitudes toward immigration in the 1990s and beyond make a curious counterpoint to the general continuity of the patterns of immigration itself. Yet this fluctuating attitude represents, in itself, a kind of continuity: ambivalence about immigration has long been a hallmark of what is sometimes called the American mind. The various major episodes of nativism in the American past—the Alien and Sedition Acts of the late Federalist Era, the Know-Nothing furor of the 1850s, the long buildup of immigration restriction leading up to the Immigration Act of 1924—were called to mind by many commentators who feared that the outburst of mean-spirited anti-immigrant legislation from the mid-1980s into the mid-1990s seemed to be headed toward some kind of a drastic climax. In a thoughtful book published in 1998 the historian David Reimers critically analyzed what he called "the turn against immigration" showing that most Americans had come to regard immigrants as "unwelcome strangers." And while he was careful not to predict the future, most readers surely put down his book feeling that the nativist trend would continue.[6] But by the time that Reimers's book appeared, it was beginning to seem as if the "turn against immigration" was one of those turning points of history at which history refused to turn, and most events of the few years since then have heightened that impression. However, the roller-coaster ride could restart at any time, and most students of American immigration policy wonder what would happen if the nation were to enter an economic recession, no less a depression.

Although it preceded the so-called "Gingrich revolution" of the mid-1990s, the demand for drastic immigration "reform" was not a part of the "contract with America"—now largely in default—that was its battle cry. But the Gingrich revolution quickly made immigration reform one of its objectives. That demand, of course, went back to the long and, for many, frustrating struggle over the various versions of the Mazzoli-Simpson bills (see pp. 391 ff.) which culminated in the passage of the IRCA in 1986. The basic—and foolish—promises of some of the law's supporters that it would cure or at least greatly lessen illegal immigration were soon seen by almost all specialists as unredeemable, and that disappointment accounts for some, but by no means all, of the venom of Gingrich and his allies in both parties.

IMMIGRATION "REFORM"?

It is now possible to arrive at a fuller balance sheet for the immediate results of IRCA, particularly its amnesty provisions, although as late as 1998 nearly a thousand immigrants were legalized under its provisions and additional legalizations will occur for some time. (For an official summary of IRCA provisions see Appendix I.)

Although IRCA seemed to be essentially an amnesty for recent illegal immigrants and a means of controlling illegal immigration, it contained other important provisions. These included a liberalization of "registry" and, as noted, the creation of what became know as "diversity visas." The concept of "registry" had entered American immigration law in 1929 establishing a kind of amnesty. If an otherwise eligible alien could show continuous residence in the United States before a certain date—originally 1924 but pushed back in the 1930s to 1921 and later advanced to 1948—he or she could qualify for adjustment to permanent status even if legal entry could not be demonstrated. IRCA advanced the registry date to 1972.

IRCA also began what has become the "diversity program," mentioned earlier. The 1986 law, in a provision sponsored by Brian J. Donnelly, a fourth-term Democratic congressman from greater Boston, set aside 5,000 visas in each of the two subsequent years for "qualified immigrants" who were "natives" of countries "the immigration of whose natives" was "adversely affected" by the 1965 immigration law. This, as noted earlier, was clearly affirmative action for whites. The use of the word "natives" rather than "citizens" was insurance against any of the visas going to naturalized third world immigrants in European nations. The winners were to be selected by a lottery conducted by the State Department. As Donnelly and his colleagues on the Massachusetts delegation—including, crucially Senator Edward M. Kennedy—expected, natives of Ireland garnered the lion's share, more than 40 percent of the visas in that two-year period.

Unlike an ordinary lottery, in the visa lottery it is not enough to win. To get a visa the winner has to be qualified for entry under immigration law and meet all kinds of administrative conditions. When I was teaching in Canada, in the summer of 1991, one of my Canadian students was a lottery winner who had qualified but had deferred his emigration until he finished his education. He told me that one of the first things he had to do was present himself to an American consulate: the closest one then was in Vancouver, 560 miles away.

The general amnesty provisions were supposed to legalize perhaps 3.1 million persons who had been residents in the United States since 1982. But Congress also created a new category of "special agricultural workers" who needed only to have done ninety days of agricultural labor and lived in the United States since May 1, 1985, and set aside up to 350,000 of the so-called amnesty places for such laborers. Subsequent Congresses increased the share for these agricultural workers significantly. Of the 2.68 million persons actually legalized under IRCA by the end of 1998 almost 1.25 million were "special agricultural workers" of one kind or another.

In the final analysis the amnesty provisions of IRCA not only increased significantly the number of legal immigrants in the United States but also created a well-publicized precedent for future liberalizations that would be impossible for Congress to resist. The legalization of persons already here did not contribute to the number of immigrants present, but each person legalized could become, in time, a naturalized American citizen whose relatives would be eligible for privileged admission status. Although IRCA and most of the sixteen statutes affecting immigration that were passed during the 1990s were full of anti-immigrant rhetoric and seemed to be getting "tough" with illegal immigrants, they were largely ineffectual, often intentionally so, a matter of thunder without lightning. (See the description of the enforcement provisions of IRCA at pp. 391–97; for a list of 1990s legislation see Appendix II).

Ronald Reagan, and, to a lesser extent, his two immediate successors, George H. W. Bush (Bush I) and Bill Clinton, all contributed to and exacerbated the anti-immigrant rhetoric, yet each signed legislation and was responsible for administrative actions that clearly contributed to the increase of legal immigration or other administrative admission to the United States. Reagan, for example, when he signed IRCA, said:

> Future generations of Americans will be thankful for our efforts to humanely regain control of our borders and thereby preserve the value of one of the most sacred possessions of our people: American citizenship.[7]

The enforcement provisions of IRCA, discussed below, created great fears in many nations of the circum-Caribbean, particularly the Dominican Republic and El Salvador, whose economies are greatly dependent on immigrant remittances. The fear was that large numbers of their citizens would be sent home and others excluded, thus reducing these remittances. To give just one example, President José Napoleón Duarte of El Salvador, an ally in the struggle against communism in Central America, wrote President Reagan in early 1987 requesting that Salvadorans in the United States illegally be given "extended voluntary departure" (EVD) status, an aspect of the attorney general's parole power enabling illegal immigrants to remain. While EVD had been usually granted on humanitarian grounds, Duarte stressed the economic loss if immigrant remittances ceased and pointed out that the remittances of $350 million to $600 million annually were larger than total United States aid and argued that "to eliminate remittances from the United States would be yet another blow that seems counterproductive to our joint aims of denying Central America to Marxist-Leninist regimes."[8]

Assistant Secretary of State Elliot Abrams, who had opposed EVD on humanitarian grounds, now supported it on Duarte's economic-ideological grounds, but the negative views of the Department of Justice and the congressional leadership prevailed, and EVD was not granted. As it turned out, these fears were chimerical; few were deported and U.S. borders remained porous. Eventually, for both humanitarian and economic reasons, forms of EVD or its equivalent were later put into place during the Bush I and Clinton presidencies for Salvadorans, Nicaraguans, and others by both the administration and Congress. The Immigration Act of 1990, for example, created Temporary Protected Status (TPS) as an extension of the parole power of the attorney general to be extended to aliens from a country in which there is warfare, or earthquakes and other environmental disasters, or when a country is temporarily unable to handle the return of its nationals and has requested such designation, or when the attorney general finds that other emergencies exist. Congress itself applied TPS to Salvadorans in the United States for eighteen months to July 1, 1992, but this was later extended by executive action until December 31, 1994. As these extensions were about to expire the Salvadoran congress urged El Salvadorans in the United States illegally to apply for asylum. In that year the country's gross domestic product was estimated at $8 billion and the $1.2 billion spent annually on imports came largely from remittances that were estimated at $800 million for 1994. The largest export, coffee, generated only $275 million.

Just four years after the passage of IRCA Congress approved a major adjustment of immigration law. The Immigration Act of 1990—a detailed summary is in Appendix I—was a complex measure that liberalized some aspects of immigration law and tightened others, but left the basic thrust of the law largely alone. TPS, noted previously, was a relatively minor provision unless you were its beneficiary. The "big news" was the increase in the putative "flexible cap" to 675,000 immigrants annually consisting of 480,000 "family sponsored" immigrants, 140,000 "employment based," and 55,000 "diversity" immigrants. Probably more significant was the increase in the "H" series of temporary employment visas. The 1990 statute also eliminated many of the antiradical shibboleths of the Cold War Era but greatly increased exclusion based on previous criminal activity.

This concern about crime largely replaced concern about communism. One subtitle of fifteen sections of the 1990 statute is titled "Criminal Aliens." While anyone still a Communist was banned, most Communists who had left the party as recently as two years before applying for a visa were no longer excludable. The same section made anyone who had engaged in prostitution in the ten years prior to applying excludable. Clearly communism was easier for a Congress in the grip of the Moral Majority to forgive than vice.

"CONTROLLING OUR BORDERS": MISSION IMPOSSIBLE

If the INS represents a "broken system," as many of its critics claim, the Border Patrol was established to set up a system that never worked. As noted earlier, the borders of the United States—the land borders of just the lower forty-eight extend 5,525 miles—have always been permeable for those determined to cross them, whether the purpose was warfare, Indian raids, immigration, the importation of liquor during prohibition, or in more recent years, the importation of illegal drugs. While reluctant to finance adequately the administrative functions of the INS, Congress has thrown money at the INS in general and at its enforcement arm, the Border Patrol, in particular. During the tenure of INS Commissioner Doris Meissner (1993–2000) the INS workforce doubled to 32,000 and the budget tripled to $4.3 billion. Reinforced fences, electronic sensors, and other technological devices were installed at heavy border crossing points, particularly in and around San Diego. Such operations were often given fancy pseudo-military names: the San Diego effort in 1994 was called Operation Gatekeeper. The results were predictable: large numbers of illegal migrants shifted their point of entry farther east, particularly in Arizona. Thus, in fiscal 1998, the San Diego Border Patrol sector reported a 13 percent drop in the number of illegals seized from the previous year, recording 248,000 apprehensions. In the same period the Tucson sector, where between 1994 and 1998 the number of border patrol agents increased almost 400 percent from 287 to 1,081, the number of apprehensions increased 42 percent to 420,000 in 1998. What the INS data deliberately do not show is how many of these apprehensions were of the same persons. And the INS's own guesstimates show that the number of illegal immigrants continued to grow despite the massive increments in enforcement expenditures and personnel. Thus although the Border Patrol spokesperson for the Tucson district could brag that its accomplishments demonstrate that "what has been done by Congress in putting additional resources, manpower and technology here in Arizona has paid off" the evidence indicates that the money is largely wasted.[9] In addition, as noted previously, many scholars of immigration argue that the tighter border control plus the hope of another amnesty have encouraged many who would otherwise have been circular migrants, returning regularly to Mexico, to remain in the United States once they get across the border.

A very different kind of operation was conducted in El Paso. Originally called Operation Blockade but changed to Operation Hold the Line, it had local rather than national goals. Immigration scholar Frank Bean who studied it for a federal commission, found four distinct types of illegal border crossers there. The kinds of crossers most troublesome to El Pasoans—teenagers crossing to party and street vendors—were stopped, but long-distance migrants headed for the interior United States and daily commuters were inconvenienced but not stopped by the massing of agents at the border.[10]

And, in general, the key thing to understand about immigration law is that its enforcement is highly selective. A few years ago when a history convention I was attending met in a hotel near Disneyland in California's Orange County, some fellow historians asked me during dinner if I could show them any illegal immigrants. I told them that, if they would meet me in the hotel lobby before breakfast, we would go for a walk and I would try to find some. As it turned out, there was a vacant corner lot within 500 yards of the hotel. Standing in it were large numbers of men, Hispanic in appearance, dressed in work clothes. We watched from across the street. Pickups and larger trucks drove up regularly, short conversations took place, and one or more of the men climbed into the trucks that then drove away. What we were watching was an informal, casual labor market, the most of whose "clients" were surely illegal aliens, operating for the convenience of their employers and without fear of the INS, which simply isn't interested.

A corollary of the increased border security was that, in addition to moving away from the fenced and heavily guarded areas, more and more border crossers and the "coyotes" who smuggled them chose to cross in very desolate and dangerous country, resulting in more and more deaths from heat exhaustion, extreme dehydration, and other causes. Although the fatalities represented only a tiny fraction of border crossers, the mounting total of deaths, and media interviews with survivors who had undergone harrowing experiences, led to increased humanitarian protests by members of the American public, and by the late 1990s the Border Patrol and the Mexican government were conducting educational campaigns about the dangers of desert and mountain crossings. In addition the Border Patrol also began to take positive steps such as making water supplies available in some dangerous areas and conducting regular searches by helicopter.

DEPORTATION

Interception at the border does not normally involve deportation, which can be a complex legal procedure. In 1998 for example, the INS made 1,679,439 "apprehensions," 1,555,776 of them by the Border Patrol. Ninety-seven percent of the latter were made along the border with Mexico. Although nationals of 186 countries were apprehended, 96.1 percent of them were recorded as Mexican nationals. That means that some 65,000 were of other nationalities. The overwhelming majority of these apprehensions result in "voluntary departure"—that is, aliens who do not claim the right to a hearing or asylum and simply are turned away at the border, or, if apprehended inland, agree to be escorted to the border. The following table shows total "removals" for 1998:

Table 17.4

Alien Removals—1998

Accepted offer of voluntary departure	1,569,817
Formally removed (with penalties)	172,547
	1,742,364

Source: *INS Yearbook*, 1998, p. 202.

Since the penalties referred to in the table involve "only" a denial of the right to enter the United States for a stated period, none of the persons listed in Table 17.4 went to jail. This does not represent leniency by the government. It is very expensive to keep a person in a federal prison: at least $35,000 a year and rising. It is cheaper to send a person to Harvard then to keep him or her in prison! If all those apprehended and removed were sent to prison the annual cost would be just below $60 billion annually. The INS budget for detention is "only" $800 million annually. About a third of this is spent to rent cells for the detained aliens in some 225 county jails, usually in rural areas where the costs of detaining persons is less than the $50 a day—$18,250 a year—that the INS pays. The country prisons rip off the inmates in a variety of ways, often, for example charging inmates as much as $5 for the first minute of telephone service at a time when standard rates were as low as 5–10 cents a minute.

The INS not only runs the largest—and least professional—federal law enforcement agency, the Border Patrol, but also is a large-scale custodian of federal prisoners. Toward the end of the year 2000 it was holding some 16,000 aliens in its own facilities. One of the best known and most troubled of these is the Krome Avenue detention center in Miami, which was opened in 1981 for Haitian boat people and quickly got much deserved criticism because of its conditions. It is supposed to hold four hundred detainees in fifty-inmate pods, each with bunk beds, chairs, one TV, showers, toilets, and pay telephones. The Miami facility received further notoriety in June 1995 when a visiting Congressional delegation discovered that local INS officials had hidden inmates to conceal overcrowding. Several INS employees were disciplined—but not dismissed—for covering up conditions there. In August 2000, Attorney General Janet Reno ordered an another investigation of Krome after some of those detained alleged that female inmates were promised freedom and other favors in exchange for sex. The government had so little confidence in the discipline of its custodial personnel there that three female inmates who had complained about sexual contact were released because Washington feared for their safety.[11]

The 172,000 "formal removals"—i.e., deportations—of 1998 were at an all-time high and up sharply—more than 400 percent from the early 1990s, largely because of new legislation and procedures. The vast majority of these deportations—80.6 percent—were of Mexicans and almost half of the rest came from three Central American nations—El Salvador, Guatemala, and Honduras—each of which was sent some 5,000 of its nationals.[12]

NATURALIZATION

But if deportations rose, so did naturalizations. Stimulated partly by fear, naturalizations increased spectacularly in the late 1990s. Except for a spike of some 440,000 in 1944 during World War II, there had never been as many as 400,000 naturalizations in any one year prior to 1994. In that year, 434,000 persons were naturalized, followed by 488,000 in 1995, 1,044,000 in 1996, 598,000 in 1997, and 463,000 in 1998. To put those numbers into perspective, since 1907—there are no reliable data before that—17.2 million persons have been naturalized. Just over five million were naturalized in the 1990s: that is, 29 percent of all of the naturalizations since 1907! And this number would have been considerably higher if the INS had been able to do its job properly. It had a growing backlog of more than a million qualified applicants to process at the end of 1997. This rise in the number of applications for naturalization is even more striking: of the 20.3 million petitions filed and fees paid for naturalization since 1907, almost 36 percent were in the 1990s as were almost two-thirds of the 1.5 million petitions denied. The fees have become quite expensive. In 1907 the filing fees were $5; as late as 1973 the cost was only $10; between then and 1995 it rose to $95 in four stages by 1995, and in 1999 it soared to $225. The justification for this is that, as an Office of Management and Budget directive puts it, a user fee "will be assessed against each identifiable recipient for special benefits derived from federal activities beyond those used by the general public," but Congress and the INS have stretched the rules so that the money is used for general operation and overhead. Forty separate fees are assessed for other immigration and naturalization services and a examination of the fee schedules makes it clear that there is no necessary relationship between the fee charged and the cost of that particular service. The INS's 1998 budget assumed that a total of some $789 million in various fees would be paid by immigrants. It seems clear that the ordinary, law-abiding immigrant applicant for naturalization is helping to defray costs for other services—apprehension, for example.

There are several reasons for the rise in the number of naturalizations that have gone up even more steeply than immigration. One special factor, of course, were the IRCA legalizations whose recipients first became eligible for citizenship in 1994. By the end of 1998 some 500,000 persons, 19 percent of the 2.68 million legalized under IRCA, had become American citizens, and an indeterminate number of others were among the million plus in the INS backlog.

Another was the spate of punitive federal and state laws and regulations enacted principally between 1994 and 1996—discussed below—which penalized legal resident aliens in a number of ways by limiting—or attempting to limit—their entitlement to benefits normally available to all. These, and fears of worse to come, provided real incentives for resident aliens to become U.S. citizens as did the continuing stress on family reunification in American immigration law. A third factor was the federally sponsored Citizenship USA initiative, which streamlined the naturalization process somewhat: Republicans insisted, not without reason, that this was a Clinton administration ploy designed to create more Democratic votes in the 1996 election.

Who is likely to be naturalized? INS data show that the young are more likely to do so than the old; that there is little difference in terms of gender; and that the more highly educated and skilled are also more likely to naturalize than those without such characteristics. Of those persons over sixteen years of age who entered as resident aliens in 1977, just over half—52.8 percent—had become citizens twenty years later. More than 60 percent of persons from China, the former USSR, the Philippines, Korea, India, Guyana, Cuba, and Colombia had naturalized, but fewer than 40 percent of those from Greece, Portugal, Dominican Republic, Mexico, the United Kingdom, and Canada. The last two are at the bottom, with 26.1 and 21.9 percent respectively, but the consistently low rate from Mexico—32.2 percent of the 1977 cohort—was and is a major factor in depressing the overall naturalization rate. On a regional basis the data for the years 1991–97 showed that 38.7 percent of naturalizations came from Asia, 38.1 percent from North America, 11.5 percent from Europe, 7.8 percent from South America, and 3.9 percent from the rest of the world.[13]

THE TURN AGAINST IMMIGRANTS

The immigration legislation of the early 1990s was largely minor and ad hoc. Two statutes passed in 1992 were in the Cold War mode and made minor expansions of eligibility. The Soviet Scientists Act authorized permanent resident status for up to 750 scientists from the former USSR as well as their spouses and children while the Chinese Student Protection Act—enacted 110 years after the Chinese Exclusion Act and intended to protect Chinese students who were in the United States at the time of the uprising and massacre in Tiananmen Square or who fled there after it—provided that Chinese nationals who had been in the United States between June 4, 1989, and April 11, 1990, were eligible for permanent resident status. Similarly, the adoption of NAFTA in

1993 liberalized somewhat temporary employment entries by Canadian and Mexican nationals and their families, with more generous terms for the northerners.

Then came what I call the turn against immigrants. It was a short-lived turn that produced punitive legislation primarily in the period between late 1994 and the election of 1996. The turn can be understood only in the context of the general climate of opinion in the post–Cold War United States and whose major manifestations were the so-called "Contract with America" and the spectacular but short-lived dominance of the Republican Speaker of the House of Representatives, Newton L. ("Newt") Gingrich (b. 1943). It was for immigrants a dark age, and, like an earlier "Dark Age," was nasty and brutish. Unable to agree on any effective measures that would curb both legal and illegal immigration—effective sanctions against employers of illegal immigrants probably would have been the most effective—Congress instead decided to punish immigrants who were legally in the United States but had not yet become citizens. Because of processing backlogs at the INS many immigrants who filed at or close to the five-year minimum waiting period had to wait an additional year or so—and occasionally longer—to complete the process of becoming a citizen.

Former Congressperson Barbara Jordan (1936–1996), speaking as chair of the U.S. Commission on Immigration Reform (CIR), a body created by the Immigration Act of 1990, put it well when she noted, in May 1994, that rather than a discussion, there was a "furor" about immigration. When she spoke there were 150 different pieces of immigration legislation pending in Congress, some of which called for a "moratorium" on all immigration reminiscent of the House of Representatives action in 1921. Jordan, and most of the other eight CIR members were mild restrictionists, mild at least, in the climate of the mid-1990s. She favored tightened border controls, better identification documents, ending welfare and other social services to illegal immigrants, and was sympathetic to claims by the states for reimbursement of costs caused by illegal immigrants but skeptical about the amounts claimed. But she was not opposed to the then current levels of immigration and insisted that the United States could and should improve opportunities for all Americans, whether natives or legal immigrants.[14]

The first tangible national results of the immigration furor Jordan discussed came in September 1994 with the passage of the Violent Crime Control and Law Enforcement Act, which reflected the alarm felt about alien criminals already in the United States and alleged "loss of control" of American borders. The act had no effect on the flow of legal immigration, but did authorize the establishment of a criminal alien center, made it easier and quicker to deport criminals and rejected asylum seekers, strengthened the penalties for passport and visa offenses, provided more money for the border patrol and other aspects of "border management." It also created a new type of nonimmigrant visa, the "S" visa, which allowed the Attorney General to bring in alien witnesses who could provide information about terrorism, some of them probably former terrorists. Such persons were presumably one of the sources for the "secret evidence" used in deportation proceedings against presumed terrorists. (See the discussion of the Kiareldeen case at p. 436.)

In October the first interim report of the Jordan Commission only added to the furor. The report could have been written by a modern day Dr. Pangloss, Voltaire's character who insisted, although surrounded by disasters, that everything was for the best because this was the best of all possible worlds. It expressed faith that money, technology, reorganization, and better management could fix everything: it recommended the "development of a simpler, more fraud-resistant system for verifying work authorization" and believed "that the most promising option for secure, nondiscriminatory verification is a computerized registry" using everyone's Social Security number. It also supported old-fashioned remedies like more border patrol agents and fences. If these didn't work, it gently threatened to break up the INS. The one thing it did *not* do was point a finger at a chief culprit in the illegal immigration dilemma, the Congress which, as we have seen, consistently refused to enact tough employer sanctions. But, of course, the Commission was Congress's creature, and such bodies only rarely turn on their creators.

But even more crucial to the growing turn against immigrants were events in California, where a successful political campaign for a drastic anti-immigrant ballot measure, Proposition 187, not only polarized politics in the largest and perhaps most volatile state but also had national implications. The campaign in California was orchestrated in part by ex-INS officials, including Alan C. Nelson (1933–1997), who had been commissioner from 1982–1989, and Harold Ezell (1937–1998) who had been western regional commissioner under Nelson, and who had performed as part of an ersatz mariachi, Trio Amnestía, that had serenaded in immigrant neighborhoods in an effort to promote the 1986 IRCA amnesty. Like the CIR report Prop 187 stressed barring the use of public services by illegal immigrants, but unlike it, included schooling in its ban. Although critics pointed out that a number of its propositions were clearly unconstitutional the initiative was clearly very popular among the California electorate.

Official 1994 California Voter Information
Proposition 187
Illegal Aliens

Makes illegal immigrants ineligible for public social services, public health care services (unless emergency under federal law), and public school education at elementary, secondary, and post-secondary levels.

Requires various state and local agencies to report persons who are suspected illegal aliens to the California Attorney General and the Immigration and Naturalization Service. Mandates California Attorney General to transmit reports to the Immigration and Naturalization Service and maintain records of such reports.

Makes it a felony to manufacture, sell or use false citizenship or residence documents.

Perhaps the crucial element in Prop 187's victory at the polls in November 1994 was its adoption by California's Republican governor, Pete Wilson (b. 1933), who had represented California in the United States Senate from 1982 until his election as governor in 1990. Early in the run-up to the election, after polls reported that Wilson was trailing his Democratic opponent, Kathleen Brown, by seventeen points, he made support for Prop 187 the linchpin of his campaign, calling it the "save our state" initiative. Few pointed out, then or later, that in his Senate years he had served as point man for one group of his constituents, California's growers, and consistently and effectively helped frustrate any attempts to legislate effective sanctions against the agricultural employers of illegal aliens. Wilson won reelection with a comfortable 55 percent of the vote while Prop 187's approval was close to the traditional landslide mark of 60 percent as it swept the state, carrying fifty of fifty-eight counties. According to exit polls, the initiative was favored by 64 percent of white voters, 57 percent of Asian American voters, 56 percent of African Americans, and even 31 percent of Hispanic voters. It was not a straight Republican vs. Democrat battle: some Democrats, like incumbent U.S. Senator Dianne Feinstein (b. 1933), backed Prop 187 and survived. President Bill Clinton, who had earlier endorsed the CIR findings, opposed Prop 187, but, of course, he was not up for reelection in 1994.[15]

The victory prompted many of the initiative's supporters to take the law into their own hands and California civil rights organizations had to set up "hot lines" to handle what *Time* magazine described as "thousands of calls from distraught victims reporting impromptu acts of discrimination that recalled the vigilante spirit of the old Wild West. Many of the callers were citizens or legal residents, wrongly suspected of being illegal." Some reported being turned away from emergency rooms, pharmacies, and being harassed by police, for not having immigration documents. "A customer at a Santa Paula restaurant demanded to see the cook's green card, declaring that it was every citizen's duty to report illegals."[16]

They were quickly relieved of that duty. Within days of the election U.S. District Judge Mariana R. Pfaelzer issued a preliminary injunction against the state's enforcement of most of Prop 187. In 1997 she made that injunction permanent, noting, for example, that the education ban conflicted with a 1982 Supreme Court ruling that required Texas to keep public schools available to all residents regardless of immigration status, and that the core of the initiative conflicted with the federal government's exclusive authority to regulate immigration. Finally, in September 1999, almost five years after it was approved by the electorate and after Wilson's successor, Gray Davis, a Democrat who had won the governorship in 1998 had dropped the state's appeal to the U.S. Supreme Court, Judge Pfaelzer issued a final order. All that remained of Prop 187 on the statute books were two laws that established state criminal penalties for the manufacture and use of false documents to conceal illegal immigration status, acts that had long been illegal under federal law. The legal director of the American Civil Liberties Union of Southern California called the ruling "the final shovel of dirt on the grave of Proposition 187."[17]

In the short run, however, Prop 187's anti-immigrant animus seemed the wave of the future and most Republicans, in and out of Congress, and many Democrats, including President Clinton, moved to align themselves with public opinion. Yet there were notable exceptions among the Republicans, none more so than then Texas governor, George W. Bush (b. 1946), who publicly denounced Prop 187 and similar measures. As the British journal *The Economist* put it in 1996, Texas and California represented "two states of mind" on the immigration issue. Texas not only had a more powerful Hispanic vote but its trade relations with Mexico were much more important, relatively, than California's.[18] In Congress, the one-term Republican Senator from Michigan, Spencer Abraham (b. 1952) was an effective advocate for immigrants in general and high-tech immigrants after his election in 1994.

The major federal legislative products of the immediate post–Prop 187 climate of opinion were four statutes passed in the run-up to the presidential election of 1996, not one of which had any direct effect on the numbers of legal immigrants permitted to enter. Instead all made life more difficult and dangerous for legal immigrants already in the country. Despite all the rhetoric about "controlling our borders" Congress simply could not agree about making basic changes in immigration law and instead enacted laws that were mean-spirited to demonstrate to the electorate that it was being "tough" on immigration.

The first of these, in April 1996, was the Antiterrorism and Effective Death Penalty Act. Terrorism, of course, was a real threat, and there had been two recent major atrocities in American cities. Immigrant Muslim terrorists had set off a bomb under New York City's World Trade Center in February 1993, killing six persons, and in April 1995, a native-born Christian terrorist blew up the federal office building in Oklahoma City, killing 168 persons. But most of the rhetoric in Congress and the stories in the media stressed foreign/Muslim/Arab terrorists, and, of course, the coupling of immigrants and crime was an old story in the history of American nativism.

The 1996 act did little or nothing to prevent terrorism. Title I of the act was entitled "Habeas Corpus Reform" and its purpose, in the spirit of more than a century of immigration law and regulations, was to remove or limit the constitutional protections allowed to aliens and, as much as possible, keep the courts out of immigration matters and to provide for expeditious deportation. As the legal historian Lucy Salyer had noted in 1995, "immigration law remains a resident alien, only partially integrated into American legal culture."[19] Other provisions were aimed to exclude or deport "terrorists" in ways that paralleled earlier legislation that had been aimed at excluding and deporting Communists. Many of those "terrorists"—largely Muslims or Arabs—were so identified by "secret evidence" that the government refused to reveal unless forced to do so by legal process. A number of "horror stories" were eventually revealed showing that immigrants had been kept in confinement for years on the basis of secret evidence later shown to be of no probative value.

In 1998, for example, the INS seized and tried to deport Hany Mahmoud Kiareldeen, a Palestinian who had lived legally in the U.S. for ten years, claiming that he was a member of a terrorist group, had been peripherally involved in the World Trade Center bombing conspiracy, and wanted to kill Attorney General Janet Reno. Even though seven immigration judges had ruled that the secret evidence in his case was unreliable, the government continued to hold him for a total of seventeen months until a federal district judge issued a writ of habeas corpus freeing him. In speaking of his and similar cases the *Washington Post* editorialized that "the use of secret evidence is antithetical to the premises of our judicial system." Bills outlawing the use of secret evidence in deportation cases were introduced into Congress in 1999, 2000, and 2001.[20]

In August 1998, as a parallel to ongoing welfare "reform" Congress passed the Personal Responsibility and Work Opportunity Reconciliation Act. Disregarding the advice of the CIR, this act clamped down on legally admitted resident aliens. As a preamble to the act, Congress, perhaps with a guilty conscience, found it necessary to make a statement of "national policy concerning welfare and immigration" and to resort to an imaginary history in an attempt to justify what seemed to many Americans as callous actions. Claiming, falsely, that "Self-sufficiency has been a basic principle of United States immigration law since this country's earliest immigration statutes," it went on to create a principle that "the availability of public benefits not constitute an incentive for immigration to the United States." Only in the fevered imaginations of nativists could the notion thrive that any significant number of immigrants came to the United States seeking welfare. But if any group of immigrants can be categorized as coming for "public benefits" it is surely those nineteenth century farm-seekers, before and after the Homestead Act of 1861, who came to get "free land" in the public domain. The law disqualified most legal immigrants from receiving food stamps, Supplemental Security Income, and barred any future legal immigrants from all means-tested federal programs for five years after entry, and encouraged the states to take similar actions.[21]

A month later further privileges were removed and more restrictions added by the Illegal Immigration Reform and Immigrant Responsibility Act of September 30, 1996. Most importantly it:

1. Denied earned Social Security benefits to any alien not legally present in the United States (A similar provision had been passed earlier in the Cold War forbidding payments to persons who were residing behind the so-called Iron Curtain, but it was struck down by the courts.)
2. Forbade illegal aliens from receiving college educational benefits.
3. Further increased funding for the Border Patrol and other border safeguards.
4. Tried to set up complex computer-assisted systems to exchange

information between the INS and other agencies to prevent ineligible aliens from receiving various forbidden benefits.

Apart from their meanness, each piece of legislation was incredibly complicated (some were parts of omnibus reconciliation measures) and, taken as a whole the legislation of the 1990s contributed significantly to making immigration law a major growth sector of the legal profession.

PUBLIC OPINION AND IMMIGRATION

The Gallup polls have been asking Americans about their views on immigration on a periodic basis for a long time. Chart 17.2 shows the changing responses of the general public to the same question, "In your view should immigration be kept at the present level, increased or decreased?" since the beginning of 1965. Just before the passage of the 1965 immigration act, 7 percent of Americans thought that immigration ought to be increased, while 33 percent thought that it should be decreased. Twenty-one years later, on the eve of the 1986 IRCA legislation, 49 percent thought that immigration should be decreased, while only 7 percent thought that it should be increased. Seven years later almost two-thirds of the public favored decrease, while the same 7 percent favored an increase. Just after the passage of Prop 187 those favoring a decrease remained at the same level—65 percent—but those favoring an increase had dropped slightly to 6 percent. Indicative of the way in which immigration had become more central to the whole politi-

Chart 17.2
American Attitudes Toward the Volume of Immigration, 1965–2001

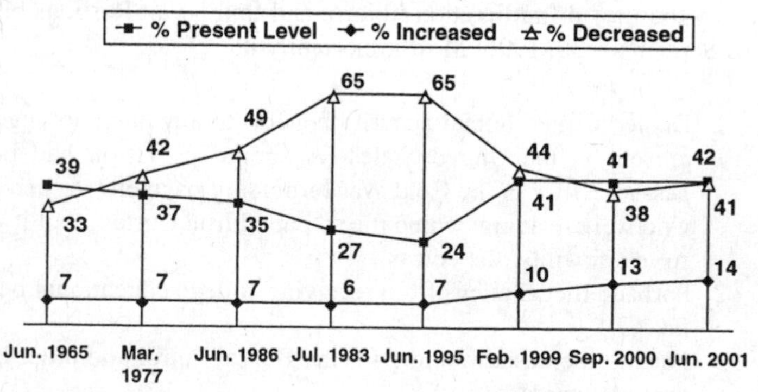

Source: Gallup Poll News Service, Poll Analysis, July 18, 2001.

cal and social process, the percentage answering "don't know" or not answering dropped steadily from 21 percent of all respondents in 1965 to only 2 percent in 1993, and has remained at less than 10 percent since then. But the views of those having opinions shifted sharply after Congress began passing the punitive anti-immigrant legislation of the mid-1990s. In three successive polls over a period of twenty-one months, the percentages of those favoring decreases or stability in the numbers of immigrants were, for statistical purposes, all but equal, while the percentage of those thinking that the current level of immigration should increase, which had remained virtually unchanged during five polls over thirty years, doubled in just over six years. Thus, in mid-1995 31 percent of Americans had a favorable view of the volume of immigration, but six years later a clear majority—56 percent—thought that way.

In 1995 at or close to the height of the anti-immigrant furor, a recent immigrant from Great Britain employed by *Forbes* magazine and a board member of *National Review,* published a tract, *Alien Nation: Common Sense about America's Immigration Disaster,* that was a clear throwback to the blatant racism of a Lothrop Stoddard or a Madison Grant. The great problem, according to Peter Brimelow, was that most contemporary immigrants were not white. His prescription for curing the "disaster" included repealing all immigration legislation of the past thirty years (i.e., from the 1965 act on), national identity cards, deporting and/or jailing all illegal immigrants, eliminating all refugee and asylee admissions, and suggested that the model for the U.S. labor policies ought to be Switzerland, which has many Gastarbeiter, guest workers who are not eligible for naturalization under normal circumstances. What was surprising about *Alien Nation* was not its content—such screeds are published regularly by obscure publishers—but that, as one reviewer noted, a mainstream publisher (Random House) "should feel comfortable publishing a reactionary ramble like this," and argued that it was evidence of "the ideological climate in which we are living." It, of course, got a rave notice from the *National Review:*

> After a decade and a half of confinement to fairly obscure pamphlets and technical monographs, the case against massive immigration has at last found a champion whose book . . . is likely to inform and shape the national discussion of immigration for years to come. . . . Mr. Brimelow's main case against immigration . . . is a cultural one rather than the usual economistic quibbling over how many jobs Mexicans take from black waiters and so forth.

> . . . Mr. Brimelow has ensured that his book is not only the main source for the strongest case for immigration restriction in decades but also an important contribution to American political thought.

But despite being a farrago of racist bias and half-truths, it was taken seriously by many people who should have know better, such as Richard Bernstein, one of the staff reviewers for the *New York Times,* who believed that it was "a highly cogent presentation of what is going to be the benchmark case against immigration" and praised Brimelow as a historian[!]:

> Mr. Brimelow also shows that America is not so much a country of immigration as it is one of "intermittent immigration." The periods when there was almost no immigration are more characteristic of our history than the briefer periods when the door was open.

Writing in the late summer of 2001 it is hard to remember that such nonsense could pass as serious criticism only a little over six years ago. It is inconceivable that such a review could appear in the *Times* today.[22]

How are we to account for these acute changes in public attitudes about immigration? For a time, the "furor" that Barbara Jordan had talked about, a furor that was stoked by politicians and most of the media, persuaded crucial elements of the public that "something" had to be done about immigration. But once the extreme legislation was on the books and a significant amount of opposition had developed, many Americans who considered themselves moderates reacted against what they perceived as unfair treatment. In addition, it now seems clear that Prop 187 and the anti-immigrant legislation that followed was a Pyrrhic victory for nativists. One of the factors in the 1994 California electoral victory was the traditional low participation of Latinos. In that election they were only 8–10 percent of the voters but were more than a quarter of the population, a goodly majority of them U.S. citizens. The Prop 187 victory not only energized Latinos to register to vote in future elections, but the subsequent federal legislation was a clear stimulus for many longtime resident aliens of all ethnicities, in California and elsewhere in the nation, to become citizens. In addition, other extreme acts, such as the impeachment and unsuccessful attempt to remove Bill Clinton after his reelection in 1996, further soured many American voters against extremism of all kinds. In California's 1998 election—immigration in general and Prop 187 in particular were major issues—Democrat Gray Davis defeated his Republican opponent for governor in a landslide, gaining 60 percent of the vote as Republican state officeholders became an endangered species.

By this time, many Republicans and most centrist Democrats, including President Clinton, who had voted for the anti-immigrant measures of the mid-1990s, were having second thoughts. In 1997 and 1998 provisions were inserted in two acts that had little to do with immigration, which restored most of the benefits that had been denied to legal immigrants by the 1996 Personal Responsibility legislation.[23] In addition, in late 1997 Congress receded from many of the restrictions legislated in the Illegal Immigration Reform Act and allowed large numbers of immigrants who had not entered with visas an opportunity to become regularized, a kind of amnesty *in petto* although the government did not use the term. According to the INS, these changes could enable approximately 150,000 Nicaraguans and 5,000 Cubans to achieve permanent resident status—and thus get green cards and the right to enter the naturalization process without being subjected to the so called "hardship rules" of 1996. Similarly, approximately 200,000 Salvadorans, 50,000 Guatemalans, and a smaller but unspecified number of aliens from the former Soviet Union were allowed to seek "hardship relief" under the rules that had prevailed before 1996.[24]

ELIÁN GONZALES AND OTHER CHILDREN

Cubans, once a major focus of American refugee policy, remain a small but continuing element. However, one tiny Cuban became, for a few months in 1999–2000, the most publicized refugee in the world. In November 1999, six-year-old Elián Gonzales was the sole and seemingly miraculous survivor of an ill-fated attempt of a group of Cuban boat people to reach Florida that included his mother, her lover, and an undetermined number of others. After his rescue by a fisherman, the INS placed him in the temporary custody of a cousin who was a member of the Miami Cuban community. His father, who was still in Cuba and had not given permission for the boy to be taken on the dangerous journey, wanted his son back. The Miami relatives, with the assistance of lawyers, claimed that the boy had filed for asylum and thus could not be sent back to Cuba against his will. Elián became the focus of a struggle between the Miami Cuban community, the boy's father, Fidel Castro, and the INS. The incident set off a media circus and the politics of the situation—the 2000 presidential election campaign was just heating up—perhaps caused Attorney General Janet Reno, who had final authority over the INS, to be slower to act than she might have been otherwise.

Although legal authorities were all but unanimous that a six-year-old child could have not legal standing in an asylum adjudication, there was an ugly Cold War precedent from the 1980s that had allowed ideology to trample normal parental rights and many in the Miami community undoubtedly remembered an even older governmental intervention for Cuban child immigrants. The 1980s case involved not Cubans but a twelve-year-old citizen of the USSR. Walter Polovchak, twelve years old, and two siblings had been brought to Chicago by their parents in 1980. After five months the elder Polovchaks decided to return to the Soviet Union. Walter and an older sister did not want to return and moved in with a cousin, an established Chicago resident. The parents went to the police, who took the runaway children into custody, but, on the advice of the INS and the State Department, did not return them to their parents. The government quickly arranged for the children to file for protection under the new asylum law, which had passed earlier that year. The parents got legal assistance and filed suit for custody of Walter, but not his sister, in state court but returned, as planned, to the USSR. An Illinois trial court made Walter its ward, frustrating the parents' attempt, but an Illinois appellate court ruled for them. The State Department countered this move by issuing a "departure control order" that declared that it was against the national interest for Walter to leave. The parents' lawyers filed suit in federal court.

The normal "law's delay" plus adroit government stalling kept the case from being heard for five years. The government's target date was October 3, 1985, when Walter would become eighteen and cease to be a child under both American and Soviet law, which would make the suit moot. But the case came to trial that summer and on July 17, 1985, the federal district court found for the parents. The government, naturally, appealed, which froze the district court's order. The case was argued before the Seventh Circuit Court of Appeals on September 9, which, the next day, found that the government had violated Walter's parents' rights to due process, but revoked the district court's order, which, because of Walter's imminent legal maturity, meant no action at all. Communist parents could not expect justice from the American legal system as long as the Cold War stayed hot.[25]

Although not in any way a precedent, between 1960–1962 the United States government supported a clandestine effort, named Operation Pedro Pan, which brought some 14,000 unaccompanied Cuban children, about 11,000 boys and 3,000 girls, to the United States and Costa Rica.

Although a few handfuls were toddlers Elián's age and younger, 60 percent were teenaged boys. The majority came by commercial airlines that were still able to fly the Havana-Miami route. Most were sent by their parents who believed false rumors, perhaps started by the Central Intelligence Agency, that Castro planned to nationalize children, take them from their parents, and indoctrinate them in special camps.

Reverend Bryan Walsh, an Irish-born priest who ran Miami's Catholic charities told a *New York Times* reporter in April 2000 that the U.S. government agreed to waive visas for the children as long as the program was administered by a nongovernmental agency, in this case the Catholic welfare apparatus. Most of the children went to live in Catholic family homes, but perhaps five hundred were either Protestant or Jewish and were placed by the United Way and the Hebrew Immigrant Aid Society in families of appropriate religious backgrounds. Of those who came to the United States, most were placed in middle-class Catholic homes in which Spanish was not spoken. Many parents eventually followed their children to the United States, but perhaps half the children never saw their parents again.[26]

But the political climate almost a decade after the collapse of the Soviet Union was different: both the local federal district court and the Sixth Circuit Court of Appeals upheld the INS view, and Elián and his father, who had come from Cuba to get him, returned to Cuba as national heroes. There was much bitterness and civil disorder in Miami: first when armed INS agents seized the boy from his relatives who had threatened to resist and later when the court decision went against the wishes of the majority of the Miami Cuban community. Wholly apart from the boy and his father ripples from the affair continued to have their effect. Inevitable generational rifts in the exile community between what Maria Cristina Garcia has styled "Cuban exiles" and "Cuban Americans" were increased, heightened hostility to Democrats may have lost Al Gore enough votes to have cost him Florida and the presidency in the 2000 election, and, as I write, Janet Reno has announced her candidacy for governor of Florida in 2002.[27]

GLOBALIZATION AND ITS EFFECTS

It now seems clear that "the turn against immigration" that Reimers so precisely delineated had quickly turned again. The causes of that turn are to be sought not only in the politics of ethnicity and fairness noted above, but also in the economic changes that have been subsumed by the shorthand term, globalization. Although, according to the *Oxford*

English Dictionary, the word came into the language in the early 1960s, but the concept really caught on only in the 1990s. A recent UN document put it this way:

> Since coming to the fore as one of the most talked-about issues of the late twentieth century and the new millennium, the phenomenon of globalization has captured world attention in various ways. From the information superhighway to the international trade in drugs and arms, to the phenomenal impact of MacWorld, Nike and the global media, the subject of globalization has come to concern all and sundry. At the core of most discussions of the issue is the extraordinary explosion of both technology and information, in ways that have considerably reduced the twin concepts of time and space. In particular, information and communications technology (ICT) has emerged as perhaps the most dominant force in the global system of production, albeit with significant ramifications in all other spheres of contemporary human existence.[28]

What that definition ignores is the phenomenon of expanded immigration on a worldwide basis wholly apart from immigration to the United States.

From the Vatican, Pope John Paul II provided an integrated view of globalization and immigration. John Paul, who had made many pro-immigrant statements, delivered a particularly strong plea for change in December 2000 as he integrated World Immigrant Day into the Great Jubilee, calling for solidarity with thousands of "desperate men and women, many of whom are young," who "every day face sometimes dramatic risks to escape from a life without a future." The pontifical message denounces those countries that, despite having relative abundance, tend to close their borders even tighter, "under the pressure of public opinion colored by the inconveniences implied in the phenomenon of immigration." This attitude amounts to tunnel vision on the part of the developed world, which fails to see the "immigrations of despair" that oblige "men and women without a future to abandon their country amidst a thousand dangers."

> "On one hand, globalization accelerates the movement of capital and the exchange of goods and services among men, inevitably influencing . . . human movements," the Pope explained. "The new generations go forward with the conviction that the planet

> today is a 'global village' and they are involved in friendships that overlook differences in language and culture."
>
> But, at the same time, "globalization causes new cleavages. In the context of unbridled freedom without adequate brakes, the difference between 'emerging' countries and 'loser' countries becomes more profound. The former have capital and technology that enables them to enjoy the planet's resources, a faculty that they do not always employ with a spirit of solidarity, by learning to share. The latter, on the contrary, have a difficult time accessing the necessary resources for adequate human development, and at times they even lack the means of subsistence. Crushed by debts and lacerated by internal divisions, they sometimes dissipate the little they have in war."

Given this picture, John Paul II believes that the challenge is very obvious: "to insure globalization in solidarity, a globalization without marginalization."[29]

A brief global survey can provide some specifics to flesh out the pontiff's eloquent generalizations. The more modern and industrialized nations of the entire world are facing tremendous pressures from would-be immigrants from lesser developed nations. In Europe this means the nations of the European Community (EEC), plus Norway and Switzerland; in Asia magnet countries include Japan, Singapore, and the Republic of Korea, plus Australia and New Zealand on its southern perimeter; while in Africa, the Republic of South Africa is the chief attractor of labor-seeking migrants. (Refugees, of course, are another matter. Most of the world's millions of refugees are in Africa and they tend to be in the nations of first asylum, which are usually proximate to the nations from which they fled.) In late 2001 Pakistan had perhaps two million refugees within its borders, more than any other state. The policies towards both economic and refugee migrants of members of the EEC are governed, in part, by the community-wide standards adopted by the European Parliament, which tend to be more welcoming than those of its individual members.

In addition, many of the developed nations are, somewhat belatedly, becoming aware of the economic implications of a population pattern affected by modernity, which results in, among other things, a longer life span, later age of marriage, fewer children, and a larger percentage of the population beyond traditional retirement age. This means that such countries have very low rates of population growth, which, in terms of natural increase alone, is at or below the zero population growth

(ZPG) point. Germany and Japan—the chief industrial competitors of the United States—have each, belatedly, come to realize that immigration is vital to their industrial success and at the century's end had begun to try to change both policies and attitudes. Germany has long had a large immigrant population. Before the unification of the two Germanys there was a slightly higher percentage of foreigners in the German Federal Republic (West Germany) than in the United States, even though its then chancellor, Helmut Kohl, insisted that Germany was not and never had been a nation of immigration. It was very difficult for anyone not of "German blood" to become a German citizen, even if born on German soil. Rather than making a special group, "aliens ineligible to citizenship," as the United States did until 1952, Germany simply refused to give rights to the vast majority of its post–World War II immigrants or their descendents, the largest group of whom were Turks. Even the multiculturally oriented European Union describes the 2.1 million persons of Turkish ethnicity in Germany, about 30 percent of its foreign residents, as "the largest foreign community in a European OECD country [accounting] for almost 30 percent of the country's foreign residents," even though large numbers of them are German-born. By the beginning of the new century, Germany is not only in the process of liberalizing its naturalization statutes but also has begun to emulate American policy by importing high-tech personnel from India and other nations.

Such trends were general throughout western Europe. A 2001 European Union report pointed out that "both the number of foreigners and foreign-born inhabitants and their share in the total population have increased in the last decade. Between 1988 and 1998, the share of foreigners in Finland's population multiplied by four, although their absolute number remains low. In Austria, Denmark, Italy, Portugal and Spain the number of foreigners doubled in absolute terms, and in Germany it increased by 63 percent." It also noted that the number of immigrants from outside the community was also rising. "Chinese immigrants now count among the top 10 nationalities settling in France, Italy and Spain. In the Nordic countries, immigration from neighbouring countries is gradually giving way to immigration by Asian nationals and citizens of the former Yugoslavia." Even more striking was the number of immigrants from northern Africa. "The distribution in Europe of nationals from the Maghreb countries (Algeria, Morocco and Tunisia) . . . has changed in the last fifteen years. Although France remains the main host country for Maghreb nationals, increasing numbers of Moroccans and Tunisians are settling in Italy, Spain and Germany."[30]

Japan, with a largely homogenous population, has one large group of "foreigners"—ethnic Koreans, most of them now second- or third-generation inhabitants of Japan—whom it treats abominably and to whom almost no civic rights are granted.[31] Japan's initial search for immigrant labor turned to Brazil, where many poor, largely third-generation Japanese Brazilians were willing to emigrate. The Brazilian Nikkei experience in Japan has not been ideal. As a Toyota executive put it to me: "They are Japanese on the outside, but on the inside they work to a samba beat." Brazilian sojourners complain that, despite their appearance, they are discriminated against. One woman reported that when she boarded a bus the driver announced: "Please watch your bags. There is a foreigner onboard." The current Asian economic crisis has reduced entries and accelerated departures, but the unemployment rate in São Paulo—perhaps 20 percent—and Japan's chronic labor shortages make it seem likely that this westerly labor migration will continue.[32] United Nations and Japanese government projections are that the island nation will need perhaps 600,000 immigrants annually to avoid drastic population shrinkage. One current projection in that without such an importation of labor Japan's current population of 127 million—only 1.7 million of them foreigners—will shrink to 100 million by 2050 and to 67 million by the year 3000.

In the former USSR population decline has already begun. The disintegration of the Soviet state has meant that the old barriers against emigration are largely down and many of its best and brightest are leaving even without encouragement such as the previously mentioned Soviet Scientists Act. This population outflow plus a very low birth rate, the collapse of the health services system, and epidemics of tuberculosis, AIDS, and alcoholism have resulted in a *lowered* life expectancy, making Russia the only major nation in the late twentieth-century world to face the problem of long-term actual population decline. Both Russian and western demographers expect the decline to continue for decades.[33] Yet there is immigration to Russia from various other parts of the former Soviet Union, particularly from what the Russians now call the "near abroad" and, in the Russian Far East, from China. The most desired immigrants are ethnic Russians, and, after them, other Europeans.

This "graying" of the population has had less effect on the United States, where, as we have seen, the relatively large influx of young, adult immigrants has kept the average age much lower than it would have been otherwise. But even with current relatively high levels of immigration there the population does continue to age with all of the problems that this creates for the Social Security system. Although many emphasize the restrictiveness of current American immigration policy—and it is restrictive—the view from overseas is often quite different. A Kenyan journalist writing in *The Nation* (Nairobi) argued that:

> While Europeans are moving mountains to keep illegal immigrants out, it is increasingly becoming clear that their aging, and sometimes declining populations, cannot support a vibrant, competitive economy. They are forced to compete with the US, which has perfected the art of keeping its labour force young, energetic and relatively cheap, thanks to regular infusion of immigrants. . . . What to do? Well, if Europe believes that it can be an island of riches in a sea of starvation and poverty, it is in for more surprises. A lot of the Africans swimming the Mediterranean would not risk their lives, the stigma of racism and the hardships of sweatshops were there any chance of keeping body and soul together at home. They are taking the high risks because there is nothing to eat at home.[34]

At the Turn of the Century

As the calendar turned to the twenty-first century, the United States continued to expand ways for some kinds of people to enter the country. For example, listed below are the limits established, as of September 2001, for the numbers of temporary workers and trainees who would be allowed in as "non-immigrants."[35]

1. 65,000 in each fiscal year before fiscal year 1999;
2. 115,000 in fiscal year 1999;
3. 115,000 in fiscal year 2000;
4. 195,000 in fiscal year 2001;
5. 195,000 in fiscal year 2002;
6. 195,000 in fiscal year 2003; and
7. 65,000 in each succeeding fiscal year.

No one familiar with the pattern of American immigration legislation will be surprised if, before fiscal year 2004 rolls around, the drastic reduction scheduled to go into effect in that year is modified upward. The authorization of almost 600,000 temporary workers in three years does not include their spouses and children who may come in without numerical limitation, and many of whom, once here, will enter the labor force legally or illegally.

Much more surprising, even shocking, was the 180-degree turn in policies toward immigrants made by the American labor movement. At its winter meeting in New Orleans on February 16, 2000, the Executive Council of the AFL-CIO, representing the vast majority of organized American workers, called for blanket amnesty for illegal immigrants and an end to most sanctions against employers who hire them. This reversed the anti-immigrant policies of American labor federations that had existed since post–Civil War America and had been, as we have seen, emphasized by such giants of the labor movement as Samuel Gompers, who dominated the AFL from its birth until his death in 1924. In the recent past organized labor had campaigned and lobbied, without much success, for strict enforcement of employer sanctions, and had vigorously opposed NAFTA partly because it feared that the agreement would increase the number of border-crossing laborers. Why did it change? Because the America labor movement was one of the great negative success stories of the later twentieth century. In the ten or fifteen years after the end of World War II, American trade unions had organized about one American worker in three. Since 1960 that figure had spiraled steadily downward so that by 1999, according to the Department of Labor's Bureau of Labor Statistics, it had organized fewer than one worker in seven. Since 1980 absolute numbers of trade union members had fallen: there were 19.8 million organized workers in that year but only 16.4 million in 1999, even though the labor force had grown from 90 million to 119 million in the same period. The manufacturing sector, a traditional source of members for labor, was losing jobs steadily, and the burgeoning high-tech sector had proved all but impossible to organize. Thus the only possible growth areas for the unions were government jobs and the low-paying, often dirty jobs that were increasingly filled with immigrants, legal and illegal. Since illegal immigrants could not effectively organize—at the first sign of protest the boss would call in the INS—a call for legalization by the unions was logical, and had been so for some time.

To be sure, the unions and their spokespersons insisted that their action was a traditional trade union call for workers' rights, but that was a transparent fiction, although many of the union organizers were earnest strivers for justice. "Immigrants are not only the history of the union movement, they are its future, its soul, its spirit," argued Linda Chavez-Thomson, the AFL-CIO's secretary-treasurer, whose gender and ethnicity were both unprecedented in the federation's hierarchy. She added that "we are on the side of working people everywhere, and that means we are on the side of immigrants."

Steven Greenhouse, the ace labor reporter for the *New York Times,* noted that "For decades, many unions and union members looked askance at newcomers, viewing them as interlopers stealing jobs and driving down wages. But as the AFL-CIO has focused increasingly on unionizing workers on the lowest rungs, like waiters, janitors and hotel maids, union leaders have recognized that they have to improve labor's image among immigrants." He reported on a labor forum held in New York City in April 2000, at which immigrant activists told their stories.

Herbert Jean-Baptiste, an immigrant from Haiti, complained that 90 percent of immigrants who work at nursing homes lack health insurance. Syed Armughan, a limousine driver from Pakistan, said that many drivers had to work seventy-hour weeks, received no health insurance or retirement benefits, and earned less than $30,000 a year. And Zonia Villanueva, a housekeeper from El Salvador, complained that many employment agencies charged exorbitant fees and that many homeowners cheated their maids out of wages. "We are workers who contribute to the economy," Villanueva said. "People depend on us to clean their houses and care for their children. We have been in the country for many years and we pay taxes. We want to stand up for our rights. We want to live and work free from intimidation. We only want the respect that every worker deserves." The unions' new stand put labor and business, traditionally antagonistic on this and many other issues, on the same side. For example, Randy Johnson, vice president for labor policy of the United States Chamber of Commerce, welcomed labor's new policy. "It's certainly a departure from organized labor's traditional position, which was: an increased number of immigrant workers is bad for domestic workers," he said. "I think this is an area where the business community and organized labor can work together."[36]

Conversely some traditional friends of organized labor saw the change of policy as a betrayal of all that labor had stood for and a disaster for "American" workers. For example, Vernon M. Briggs, Jr., a senior professor at the Cornell University School of Labor and Industrial Relations and a favorite of the anti-immigration Washington think tank, the Center for Immigration Studies, deplored the policy change. "Immigration is not the source of all evil, but it is disparately affecting our low-skilled workers. I think the unions have made a terrible mistake in putting the interests of immigrant workers first."[37]

The first seven and a half months of the administration of George W. Bush (Bush II) represented "Texas-style" rather than "California-style" Republicanism. The administration saw Mexico as a partner as opposed to a threat; the new president of Mexico, Vincente Fox, addressed the American Congress, and some kind of a general amnesty seemed in the works. President Bush's initial notions about immigration, as articulated in his first budget message, are printed as Appendix III. But the terrorist attacks on the World Trade Center and the Pentagon of September 11, 2001, put immigration reform on hold and may well have triggered or exacerbated a recession. In the short run, at least, the pace of not only new legal immigration initiatives but also immigration would surely slow.

In the long run, however, there was every reason to believe that the economic and social forces that had produced the burgeoning of American immigration in the last half of the twentieth century would continue into the twenty-first.

Appendix I

INS Summary of IRCA (1986) and Immigration Act of 1990

Immigration Reform and Control Act of November 6, 1986 (IRCA)
(*100 Statutes-at-Large 3359*)

Comprehensive immigration legislation:

A. Authorized legalization (i.e., temporary and then permanent resident status) for aliens who had resided in the United States in an unlawful status since January 1, 1982 (entering illegally or as temporary visitors with authorized stay expiring before that date or with the Government's knowledge of their unlawful status before that date) and are not excludable.

B. Created sanctions prohibiting employers from knowingly hiring, recruiting, or referring for a fee aliens not authorized to work in the United States.

C. Increased enforcement at U.S. borders.

D. Created a new classification of seasonal agricultural worker and provisions for the legalization of certain such workers.

E. Extended the registry date (i.e., the date from which an alien has resided illegally and continuously in the United States and thus qualifies for adjustment to permanent resident status) from June 30, 1948, to January 1, 1972.

F. Authorized adjustment to permanent resident status for Cubans and Haitians who entered the United States without inspection and had continuously resided in country since January 1, 1982.

G. Increased the numerical limitation for immigrants admitted under the preference system for dependent areas from 600 to 5,000 beginning in fiscal year 1988.

H. Created a new special immigrant category for certain retired employees of international organizations and their families and a new nonimmigrant status for parents and children of such immigrants.

I. Created a nonimmigrant Visa Waiver Pilot program allowing certain aliens to visit the United States without applying for a nonimmigrant visa.

J. Allocated 5,000 nonpreference visas in each of fiscal years 1987 and 1988 for aliens born in countries from which immigration was adversely affected by the 1965 act.

Immigration Act of November 29, 1990
(104 Statutes-at-Large 4978)

A major overhaul of immigration law:

A. Increased total immigration under an overall flexible cap of 675,000 immigrants beginning in fiscal year 1995, preceded by a 700,000 level during fiscal years 1992 through 1994. The 675,000 level to consist of: 480,000 family sponsored; 140,000 employment-based; and 55,000 "diversity immigrants."

B. Revised all grounds for exclusion and deportation, significantly rewriting the political and ideological grounds. For example, repealed the bar against the admission of Communists as nonimmigrants and limited the exclusion of aliens on foreign policy grounds.

C. Authorized the Attorney General to grant temporary protected status to undocumented alien nationals of designated countries subject to armed conflict or natural disasters.

D. Revised and established new nonimmigrant admission categories:

1. Redefined the H-1(b) temporary worker category and limited number of aliens who may be issued visas or otherwise provided nonimmigrant status under this category to 65,000 annually.

2. Limited number of H-2(b) temporary worker category aliens who may be issued visas or otherwise provided nonimmigrant status to 66,000 annually.

3. Created new temporary worker admission categories (O, P, Q, and R), some with annual caps on number of aliens who may be issued visas or otherwise provided nonimmigrant status.

E. Revised, and extended the Visa Waiver Pilot Program through fiscal year 1994.

F. Revised naturalization authority and requirements:

1. Transferred the exclusive jurisdiction to naturalize aliens from the Federal and State courts to the Attorney General.

2. Amended the substantive requirements for naturalization: State residency requirements revised and reduced to three months; added another ground for waiving the English language requirement; lifted the permanent bar to naturalization for aliens who applied to be relieved from U.S. military service on grounds of alienage who previously served in the service of the country of the alien's nationality.

G. Revised enforcement activities. For example:

1. Broadened the definition of "aggravated felony" and imposed new legal restrictions on aliens convicted of such crimes.

2. Revised employer sanctions provisions of the Immigration Reform and Control Act of 1986.

3. Authorized funds to increase Border Patrol personnel by 1,000.

4. Revised criminal and deportation provisions.

H. Recodified the thirty-two grounds for exclusion into nine categories, including revising and repealing some of the grounds (especially health grounds).

Source: INS, *1988 Statistical Yearbook,* Washington, D.C.: GPO, 2000, Appendix 1.

Appendix II

Other Legislation Affecting Immigration in the 1990s (Title and Citation Only)

Armed Forces Immigration Adjustment Act of October 1, 1991
(105 Statutes-at-Large 555)

Act of December 12, 1991
(105 Statutes-at-Large 1733)
Miscellaneous and Technical Immigration and Naturalization Amendments Act [to 1990 Act].

Chinese Student Protection Act of October 9, 1992
(106 Statutes-at-Large 1969)

Soviet Scientists Immigration Act of October 10, 1992
(106 Statutes-at-Large 3316)

North American Free-Trade Agreement Implementation Act of December 8, 1993
(107 Statutes-at-Large 2057)

Violent Crime Control and Law Enforcement Act of September 13, 1994
(108 Statutes-at-Large 1796)

Antiterrorism and Effective Death Penalty Act of April 24, 1996
(110 Statutes-at-Large 1214)

Personal Responsibility and Work Opportunity Reconciliation Act of August 22, 1996
(110 Statutes-at-Large 2105)

Illegal Immigration Reform and Immigrant Responsibility Act of September 30, 1996
(110 Statutes-at-Large 3009)

Balanced Budget Act of August 5, 1997
(111 Statutes-at-Large 270)

Nicaraguan Adjustment and Central American Relief Act (NACARA) of November 19, 1997
(111 Statutes-at-Large 2193)

Agricultural Research Reform Act of February 11, 1998
(112 Statutes-at-Large 575)

Visa Waiver Pilot Program Reauthorization Act of April 27, 1998
(112 Statutes-at-Large 56)

American Competitiveness and Workforce Improvement Act of October 21, 1998
(112 Statutes-at-Large 2681)
Part of the Omnibus Appropriations Act, 1999.

Non-Citizen Benefit Clarification Act of October 28, 1998
(112 Statutes-at-Large 2926)

Irish Peace Process Cultural and Training Program Act of October 30, 1998
(112 Statutes-at-Large 3013)

Agriculture, Rural Development, Food and Drug Administration, and Related Agencies Appropriations Act of October 22, 1999
(113 Statutes-at-Large 1135)

Nursing Relief for Disadvantaged Areas Act of November 12, 1999
(113 Statutes-at-Large 1312)

Act of November 13, 1999
(113 Statutes-at-Large 1483)

Consolidated Appropriations Act of November 29, 1999
(113 Statutes-at-Large 1501A-233)

Act of December 7, 1999
(113 Statutes-at-Large 1696)

Source: INS, *1988 Statistical Yearbook*, Washington, D.C.: GPO, 2000, Appendix 1

Appendix III

George W. Bush, Budget Message: Immigration Proposals February 28, 2001

14. Reform the Immigration System

The United States is a nation of immigrants. Unfortunately, today when new immigrants arrive on our shores, their first experience is often one of frustration and anxiety. The Administration believes that legal immigrants should be greeted with open arms, rather than endless lines. We must be responsive to those who seek to immigrate to this country by legal means, and to those who have emigrated and now seek to become U.S. citizens.

While we seek to improve the system that welcomes legal immigrants, the United States is a nation of laws and must act to combat illegal immigration. Working through the Immigration and Naturalization Service (INS), the Federal Government should take additional steps to defend the security and stability of our nation against the threats of organized crime, drug traffickers, and terrorist groups. The Administration is committed to improving U.S. immigration law enforcement and ensuring the safety of our borders.

Accelerating INS Processes Pertaining to Citizenship and Benefits: The Administration is committed to building and maintaining an immigration services system that ensures integrity, provides services accurately and in a timely manner, and emphasizes a culture of respect. It often takes three years or more for INS to process immigration applications and/or petitions. At times, in some areas of California, Federal delays in processing adjustment of status applications have averaged fifty-two months; in some areas of Texas, delays have averaged sixty-nine months; in some areas of Arizona, forty-nine months; in some areas of Illinois, thirty-seven months.

To improve INS's focus on service and to reduce the delays in INS processing of immigration applications, the Administration proposes a universal six-month standard for processing all immigration applications. To meet this standard, the Administration supports a five-year, $500 million, initiative to fund new personnel, introduce employee performance incentives to process cases quickly and accurately, and make customer satisfaction a priority. The $100 million proposed in 2002 is the first installment in this effort to provide quality service to all legal immigrants, citizens, businesses, and other INS customers.

Strengthening Border Control and Enforcement: The budget provides funds to support additional Border Patrol agents, as well as technology to supplement the new agents. Congress authorized hiring 5,000 new Border Patrol agents between 1997 and 2001. To date, INS has received funding for roughly 3,860 new agents. In order to hire and train the remaining agents that are needed, the President's budget requests $75 million to fund 570 new agents per year in each of 2002 and 2003. With the new agents to be added, it is estimated that about 11,000 agents will be deployed along the nation's Northern and Southern borders by the end of 2003. This is 12 percent more than 2001 and represents more than 175 percent growth in agent staffing since 1993. In addition, $20 million is requested for 2002 to fund intrusion detection technology including high-resolution color and infrared cameras and state-of-the-art command centers as force multipliers to supplement the new agents and provide continuous monitoring of the border from remote sites. The proposed combination of intrusion detection technology, and a substantial number of new Border Patrol agents will permit INS to enforce the rule of law and enhance border management over larger portions of the border.

The 2002 Budget also provides an additional $7 million to establish intelligence units along the Northern and Southwest borders. The units will collect, analyze, and disseminate information to identify and interdict illegal entrants to the United States; monitor potential terrorist activity and smuggling operations; and track the movement of illicit narcotics, weapons, and other contraband across our borders.

Ensuring Detention and Removal of Illegal Aliens: The Administration is committed to removing those who have entered the country illegally and to detain criminal aliens. The budget funds INS detention and deportation staff and provides resources to remove criminal and illegal aliens swiftly. The 2002 Budget provides $89 million to support an additional 1,607 average daily detention bed spaces for a total level of more than 21,000 bed spaces. INS will continue to target its efforts primarily on removing deportable aliens held in Federal, State, and local facilities to ensure that these criminal aliens are not allowed back on the street. The budget also continues funding to fully implement detention standards to ensure those detained, particularly those who have pending asylum cases, are treated fairly.

Creating an INS Structure for the Future: The INS has suffered from systemic problems the last few years, particularly those related to INS's dual missions of service and enforcement. These systemic problems include: competing priorities; insufficient accountability between field offices and headquarters; overlapping organizational relationships; and, lack of consistent operations and policies. The Administration believes that it is critical to address these problems. The Administration proposes restructuring and splitting the INS into two agencies with separate chains of command and accountability, reporting to a single policy leader in the Department of Justice. One agency will be focused exclusively on service and the other will be focused exclusively on law enforcement. The Administration will work with Congress in a bipartisan manner to enact legislation that fundamentally improves the way the nation's immigration system is administered.

Source: http://www.whitehouse.gov/news/usbudget/blueprint/budyoc.html (pp. 85–86)
The printed version is ISBN 0-16-050683-2

Notes

The notes that follow have a dual purpose. They are designed to show the sources of direct quotations and to guide the reader to further information. As even a casual glance will show, I am heavily in debt to the editors (Stephen Thernstrom et al.) of and contributors to the *Harvard Encyclopedia of American Ethnic Groups* (Cambridge, Mass., 1980), which will be cited as *HEAEG*. For the strengths and weaknesses of this invaluable tool see my review essay, "The Melting Pot: A Content Analysis," in *Reviews in American History* (December 1981): 428–33. In addition, two textbooks were of great assistance. Maldwyn A. Jones's *American Immigration* (Chicago, 1960), is the book that most shaped my total view of the immigrant past, while Thomas Archdeacon's *Becoming American* (New York, 1983), is the best recent text and contains a good bibliography. All works cited in the notes except for newspapers and popular periodicals, appear in the Selected Bibliography, which follows the notes.

CHAPTER 1 OVERSEAS MIGRATION FROM EUROPE

1. Donald C. Johnson and Mailand A. Edey, *Lucy: The Beginnings of Humankind* (New York, 1981). On Aleut languages see Don E. Dumond, *The Eskimos and Aleuts* (New York, 1987).

2. Lynn White, Jr., *Medieval Religion and Technology: Collected Essays* (Los Angeles, 1978).

3. Edward Eggleston, *The Transit of Civilization* (New York, 1901).

4. Alfred J. Crosby, *The Columbian Exchange* (Westport, Conn., 1972), and *Ecological Imperialism: The Biological Expansion of Europe* (Cambridge, England, 1986). For Winthrop, see Lyle Koehler, "Red-White Power Relationships and Justice in the Courts of Seventeenth-Century New England," *American Indian Culture and Research Journal* 3:4 (1979): 4.

5. Magnus Mörner, *Race Mixture in the History of Latin America* (Boston, 1967). For the Berkeley school itself, see Woodrow Borah and S. F. Cook,

The Aboriginal Population of Central Mexico on the Eve of the Spanish Conquest (Berkeley and Los Angeles, 1963).

6. For colonial identity, see Nicholas Canny and Anthony Pagden, eds., *Colonial Identity in the New World, 1500–1800* (Princeton, 1988).

7. Walter Goffart, *Barbarians and Romans, A.D. 418–584. The Techniques of Accommodation* (Princeton, 1980).

8. Amerigo Castro, *The Spaniards: An Introduction to Their History* (Berkeley and Los Angeles, 1980).

9. Vilhjalmur Stefansson, ed., *Great Adventures and Explorations* (New York, 1949).

10. Erik Wahlgren, *The Kensington Stone. A Mystery Solved* (Madison, Wis., 1958).

11. Samuel Eliot Morison, *The European Discovery of America,* vol. 1, *The Northern Voyages, A.D. 500–1600* (New York, 1971).

12. John Larner, "The Certainty of Columbus: Some Recent Studies," *History* (February 1988): 3–23, is a scholarly and witty account of recent Columbus scholarship and controversy.

13. Ravenstein's major essay is "The Laws of Migration," *Journal of the Royal Statistical Society* (1889): 241–301.

14. Nathan Glazer and Daniel P. Moynihan, *Beyond the Melting Pot* (New York, 1963).

15. Robert Harney, "The Italian Experience in America," in Anthony Mollica, ed., *Handbook for Teachers of Italian* (Ontario, Canada, 1976), pp. 219–41. William Chazanof, *Valledolmo-Fredonia* (Fredonia, N.Y., 1961).

16. Thomas J. Archdeacon, *Becoming American: An Ethnic History* (New York, 1983), p. 139.

17. Thistlewaite's essay, "Migration from Europe Overseas in the Nineteenth and Twentieth Centuries," often reprinted, first appeared in *XI Congrès International des Sciences Historiques, Rapports* (Stockholm, 1960), vol. 5, pp. 32–60.

18. For Argentina, see Carl Solberg, *Immigration and Nationalism: Argentina and Chile, 1890–1914* (Austin, Tex., 1970); for Brazil, Thomas H. Holloway, *Immigrants on the Land* (Chapel Hill, N.C., 1980).

19. Friedrich Edding, "Intra-European Migration and the Prospects of Integration," in Brinley Thomas, ed., *Economics of International Migration* (London, 1958), pp. 238–48; Dudley Kirk, *Europe's Population in the Interwar Years* (Princeton, 1946), p. 126.

20. Archdeacon, *Becoming American,* p. 139.

21. Richard A. Esterlin, "Immigration: Economic and Social Characteristics," in *HEAEG,* pp. 476–86, is a good summary of the data.

CHAPTER 2 ENGLISH IMMIGRANTS IN AMERICA

1. Ulrich Bonnell Phillips, "The Central Theme of Southern History," *American Historical Review* 33 (1928): 30–50.

2. Adam Smith, *The Wealth of Nations,* (New York, 1937), pp. 70–71.

3. Quotations from Hakluyt and most of the other pamphleteers in this chapter are from David Cressy's fine book, *Coming Over* (Cambridge, England, 1988).

4. Edmund S. Morgan, *American Slavery; American Freedom: The Ordeal of Colonial Virginia* (New York, 1975) is the source for most of the statements about Virginia.

5. Karen O. Kupperman, "Apathy and Death in Early Jamestown," *Journal of American History* 66 (1979): 24–40, at 24–25.

6. A. L. Rowse, *The Elizabethans and America* (New York, 1959), p. 82.

7. Campbell, *The English Yeoman under Elizabeth and the Early Stuarts* (New Haven, 1942). See also Abbot Smith, *Colonists in Bondage* (Chapel Hill, N.C., 1947) and David Galenson, *White Servitude in Colonial America* (Cambridge, England, 1981).

8. E. A. Wrigley, *Population and History* (New York, 1969).

9. David Souden, "Rogues, Whores and Vagabonds," *Social History* 4 (1978): 23–41.

10. Bernard Bailyn, *Voyagers to the West* (New York, 1986).

11. Günter Moltmann, "The Migration of German Redemptioners to North America, 1720–1820," in P. C. Emmer, ed., *Colonialism and Migration: Indentured Labour Before and After Slavery* (Dordrecht, the Netherlands, 1986), pp. 105–72.

12. Russell R. Menard, *Economy and Society in Early Colonial Maryland* (New York, 1985), is the source for most of the statements about Maryland.

13. Cressy, *Coming Over,* is the source of most of the statements about seventeenth-century migration to New England.

14. Compare several statements in a collection of Morison's early essays, *By Land and by Sea* (New York, 1953), with his *Oxford History of the American People* (New York, 1965).

15. Galenson, *White Servitude.*

16. Oscar Handlin, *Boston's Immigrants,* 2nd ed. (Cambridge, Mass., 1959).

CHAPTER 3 SLAVERY AND IMMIGRANTS FROM AFRICA

1. Sidney Mintz, *Sweetness and Power* (New York, 1986).

2. Maldwyn A. Jones, *American Imigration* (Chicago, 1960); Thomas J. Archdeacon, *Becoming American: An Ethnic History* (New York, 1983).

3. August Meier and Elliott Rudwick, *Black History and the Historical Profession* (Urbana, Ill., 1986).

4. John Hope Franklin, *From Slavery to Freedom* 6th ed. (New York, 1988); August Meier and Elliott Rudwick, *From Plantation to Ghetto* (New York, 1976).

5. Melville J. Herskovits, *The Myth of the Negro Past* (New York, 1941).

6. Lorenzo Turner, *Africanisms in the Gullah Dialect* (New York, 1949).

7. E. Franklin Frazier, *The Negro Family in the United States* (New York, 1939).

8. Philip D. Curtin, *The African Slave Trade: A Census* (Madison, Wis., 1969).

9. W. E. B. DuBois, *The Souls of Black Folk* (New York, 1904).

CHAPTER 4 OTHER EUROPEANS IN COLONIAL AMERICA

1. The essay "Germans" by Kathleen Neils Conzen in *HEAEG* is the best possible beginning for a study of Germans. See also the essays on other German-speaking groups noted in the text. For German redemptioners, see Günter Moltmann's "The Migration of German Redemptioners to North America, 1720–1820," in P. C. Emmer, ed., *Colonialism and Migration: Indentured Labour Before and After Slavery,* (Dordrecht, the Netherlands, 1986), pp. 105–22. For Pennsylvania Germans, the brief essay in *HEAEG* by folklorist Don Yoder is a good start. Settlement patterns and farming practices of the Germans are discussed in geographer James Lemon's useful study, *The Best Poor Man's Country: A Geographical Study of Southeastern Pennsylvania* (Baltimore, 1972). The best recent general history is LaVern J. Rippley's *The German-Americans* (Boston, 1976).

2. The most recent full account is James G. Leyburn's *The Scotch-Irish: A Social History* (Chapel Hill, 1962). Quotations from it are at p. 169, pp. 171–72 (Archbishop Boulter's letters), and pp. 217–72. The essay by Maldwyn A. Jones in *HEAEG* is most useful, as is his contribution to E. E. R. Green, ed., *Essays in Scotch Irish History* (London, 1969). John William Ward, *Andrew Jackson: Symbol for an Age* (New York, 1955) explores and analyzes the mythic element that Jackson personified. There are illuminating comments in Lemon's *The Best Poor Man's Country,* even though most of Pennsylvania's Scotch Irish lived outside his area of focus.

3. Gordon Donaldson, author of both the *HEAEG* essay and *The Scots Overseas* (London, 1966), is the best guide. For the colonial period, Ian C. C. Graham's *Colonists from Scotland: Emigration to North America, 1707–1783* (Ithaca, N.Y., 1965) supplies many details. For a good example of what genealogists can do, Donald Whyte's *Dictionary of Scottish Emigrants to the USA* (Baltimore, 1972) should be browsed in. Its cutoff date is 1854.

4. Kerby P. Miller's *Emigrants and Exiles: Ireland and the Irish Exodus to North America* (New York, 1985) is the best single book on Irish immigration to America. For a survey of Catholicism in America, John Tracy Ellis's *American Catholicism* (Chicago, 1956, 1969) is the place to start, but the more

recent works of Jay P. Dolan, for example, *The Immigrant Church: New York's Irish and German Catholics, 1815–1865* (Baltimore, 1975), and *The American Catholic Experience* (New York, 1985), are more detailed and analytical.

5. Rowland Berthoff's essay in *HEAEG* is first-rate. Edward George Hartmann's *Americans from Wales* (Boston, 1967) is the best survey. James Lemon's *The Best Poor Man's Country* analyzes Welsh settlement in Pennsylvania.

6. Robert P. Swierenga's excellent essay in *HEAEG* should be supplemented by Ernst van den Boogaart, "The Servant Migration to New Netherland, 1624–664," in P. C. Emmer, ed., *Colonialism and Migration,* pp. 124–44, which is broader than its title indicates.

7. For the French empire in North America see W. J. Eccles, *France in America* (New York, 1973). Jon Butler's *The Huguenots in America* (Cambridge, Mass., 1983) is, as indicated in the text, indispensable. For Acadians, Marietta M. LeBreton's essay in *HEAEG* is most useful. For Martin E. Marty's remark, see his "Ethnicity: The Skeleton of Religion in America," *Church History* 41 (1972): 5–21.

8. Two essays in *HEAEG,* Carlos E. Cortés's long one, "Mexicans," and Frances Leon Quintana's brief one, "Spanish," have useful information about the colonial period. For more detail see David J. Weber's *Foreigners in Their Native Land: Historical Roots of the Mexican Americans* (Albuquerque, 1973). For the Spanish empire see Charles Gibson, *Spain in America* (New York, 1967).

9. Ulf Beijbom's essay in *HEAEG* is a good beginning. Boogaart, "The Servant Migration to New Netherland, 1624–1664," has information on New Sweden and its fate.

10. Jacob Rader Marcus's *The Colonial American Jew,* 3 vols. (Detroit, 1970), and *Early American Jewry,* 2 vols. (Philadelphia, 1951–53) are monuments by the man who "invented" the field. Nathan Glazer's *American Judaism,* rev. ed. (Chicago, 1972), is a good general survey by a sociologist. Arthur A. Goren's essay in *HEAEG* is thorough.

CHAPTER 5 ETHNICITY AND RACE IN AMERICAN LIFE

1. Marcus Eli Ravage, *An American in the Making: The Life Story of an Immigrant* (New York, 1917), p. 60 ff.

2. Nicholas Canny and Anthony Pagden, eds., *Colonial Identity in the Atlantic World, 1500–1800* (Princeton, 1988).

3. Maldwyn A. Jones, *American Immigration* (Chicago, 1960).

4. Robert R. Berkhofer, Jr., *Salvation and the Savage* (Lexington, Ky., 1965).

5. John Donald Duncan, "Indian Slavery," in Bruce A. Glasrud and Alan M. Smith, eds., *Race Relations in British North America, 1607–1783* (Chicago, 1982), pp. 85–106.

6. *Somerset* case in Helen Catterall, ed., *Judicial Cases Concerning American Slavery and the Negro* (Washington, D.C., 1926), pp. 4–5.

7. George Washington Williams, *History of the Negro Race in America* (New York, 1882).

8. The best account of the New York slave revolt is Ferenc M. Szasz's "The New York Slave Revolt of 1741: A Re-Examination," *New York History* 48 (1967): 215–30.

9. Henry M. Muhlenberg, as quoted in Gary B. Nash, *The Urban Crucible* (Cambridge, Mass., 1979), p. 509, n. 54.

10. Franklin's pamphlet from Leonard W. Labaree, ed., *The Papers of Benjamin Franklin,* vol. 4 (New Haven, 1961), p. 234.

11. Nash, *Urban Crucible,* p. 284.

12. George M. Frederickson and Dale T. Knobel, "History of Prejudice and Discrimination," *HEAEG,* p. 840.

13. For American loyalism, see Wallace Brown's *The King's Friends* (Providence, R.I., 1965) and Mary Beth Norton's *The British-Americans: The Loyalist Exiles in England, 1774–1789* (New York, 1972).

14. For naturalization, see James H. Kettner's *The Development of American Citizenship, 1607–1870* (Chapel Hill, N.C., 1978).

15. For the Alien and Sedition Acts, see John C. Miller's *Crisis in Freedom* (Boston, 1952).

16. For Bishop Carroll, see John Tracy Ellis's *American Catholicism,* 2nd ed. (Chicago, 1969), pp. 36–37.

17. Philip Gleason, "American Identity and Americanization", *HEAEG.*

18. John Quincy Adams, letter of June 4, 1819, cited in Moses Rischin, ed., *Immigration and the American Tradition* (Indianapolis, 1976), pp. 48–49.

CHAPTER 6 PIONEERS OF THE CENTURY OF IMMIGRATION

1. The most recent and comprehensive book on Irish immigration to America is Kerby Miller, *Exiles and Emigrants* (New York, 1986). Arnold Schrier, *Ireland and the Emigration* (Minneapolis, 1958), is a pioneering interpretation. Miller and Schrier are preparing for publication a large collection of the immigrant letters that have enriched both their books.

2. Hansen and the master of the *Ocean* quoted from "The Second Colonization of New England," in Marcus Lee Hansen, *The Immigrant in American History* (Cambridge, Mass., 1940).

3. Oscar Handlin, *Boston's Immigrants,* 2nd ed. (Cambridge, Mass., 1969).

4. Schrier, *Ireland and the Emigration,* p. 24.

5. Reverend John Francis Maguire, *The Irish in America* (New York, 1868), pp. 319–20.

6. James P. Shannon, *Catholic Colonization on the Western Frontier* (New Haven, Conn., 1967.)

7. Handlin, *Boston's Immigrants,* p. 216. See also Handlin, *The Uprooted* (Boston, 1951).

8. Cecil Woodham-Smith, *The Great Hunger* (New York, 1962), is a classic.

9. Robert E. Kennedy, *The Irish: Emigration, Marriage and Fertility* (Berkeley and Los Angeles, 1973).

10. Earl Niehaus, *The Irish in New Orleans, 1800–1865* (Baton Rouge, 1956), p. 123.

11. Dennis Clark, *The Irish in Philadelphia* (Philadelphia, 1974).

12. R. A. Birchall, *The San Francisco Irish, 1848–1880* (Berkeley, 1980).

13. Robert E. Hennings, *James D. Phelan and the Wilson Progressives of California* (New York, 1985), p. 3.

14. John Tracy Ellis, *American Catholicism* (Chicago, 1956).

15. Jay P. Dolan, *The Immigrant Church* (Baltimore, 1977).

16. David M. Emmons, *The Butte Irish: Class and Ethnicity in An American Mining Town, 1875–1925* (Urbana, Ill., 1989), and Brian C. Mitchell, *The Paddy Camps: The Irish of Lowell, 1821–61* (Urbana, Ill., 1988), are two recent local studies.

17. William L. Riordan, *Plunkitt of Tammany Hall* (New York, 1963).

18. Mack Walker, *Germany and the Emigration, 1816–1885* (Cambridge, Mass., 1964).

19. Kathleen Neils Conzen, *Immigrant Milwaukee, 1836–1860* (Cambridge, Mass., 1976).

20. Carl Wittke, "Ohio's Germans, 1840–1875," *Ohio Historical Quarterly* 62 (1957): 339–54.

21. Conzen, "Germans," *HEAEG,* p. 415.

22. Arthur A. Goren, "Jews," *HEAEG,* p. 579.

23. Two works by Don Heinrich Tolzmann, *German-American Literature* (Metuchen, N.J., 1977), and *German-Americana* (Metuchen, N.J., 1975), are excellent surveys.

24. Kristian Hvidt, *Flight to America: The Social Background of 300,000 Danish Emigrants* (New York, 1977), although chiefly about Danes, contains a superb comparative analysis upon which I draw extensively.

25. Ulf Beijbom, "Swedes," *HEAEG,* is the source for much that follows.

26. Ulf Beijbom, *Swedes in Chicago* (Vaxjo, Sweden, 1971).

27. Roger Daniels, *Asian America: Chinese and Japanese in the United States since 1850* (Seattle, 1988), p. 70.

28. I have relied heavily on Peter A. Munch, "Norwegians," *HEAEG.*

29. Theodore C. Blegen, *Norwegian Migration to America, 1825–1860* (Northfield, Minn., 1931).

30. Rølvaag cited in Dorothy Burton Skårdal, *The Divided Heart* (Lincoln, Neb., 1974).

31. Hvidt, as noted, is a major source for what follows. See also Skårdal, "Danes," *HEAEG.*

32. Bernard Bailyn, *Voyagers to the West* (New York, 1986), p. xix.

33. William Mulder, *Homeward to Zion* (Minneapolis, 1957).

34. Marcus Lee Hansen, "Immigration as a Field for Historical Research," *American Historical Review* 32 (1927): 500–518; John Higham, *Strangers in the Land* (New Brunswick, N.J., 1955).

CHAPTER 7 FROM THE MEDITERRANEAN

1. Maldwyn A. Jones, *Destination America* (London, 1976), has a good account of immigrant ships and their disasters. A detailed analysis is in an essay by Günter Moltmann, "Steamship Transport of Emigrants from Europe to the United States, 1850–1914: Social, Commercial, and Legislative Aspects" (in press).

2. Robert F. Forster, *The Italian Emigration of Our Times* (Cambridge, Mass., 1919).

3. Andrew Rolle. *The American Italians* (Wadsworth, Calif., 1972), pp. 23–24. That and his *The Immigrant Upraised* (Norman, Okla., 1968) are particularly good on Italians in the American West. For Italians generally I follow Humbert Nelli, "Italians," *HEAEG.*

4. Fiorello La Guardia. *The Making of an Insurgent* (Philadelphia, 1948), pp. 27–28.

5. Rudolph J. Vecoli, "Contadini in Chicago: A Critique of *The Uprooted,*" *Journal of American History* 51 (December, 1963): 404–17. William Chazanof, *Valledolmo-Fredonia* (Fredonia, N.Y., 1961).

6. Daniel Bell, *The End of Ideology* (Glencoe, Ill., 1960).

7. The careers of March and Kelly are from Victor R. Greene, *American Immigrant Leaders, 1800–1910* (Baltimore, 1987).

8. Theodore Saloutos's *A History of Greeks in the United States* (Cambridge, Mass., 1964) and his "Greeks," *HEAEG,* are the definitive authorities.

9. Alixa Naff is the author of a pathbreaking monograph, *Becoming American: The Early Arab Immigrant Experience* (Carbondale, Ill., 1985), and the essay "Arabs," *HEAEG.*

10. Robert Mirak's "Armenians," in *HEAEG,* is the source for most of the material on Armenian Americans. For the modern historical background, see Richard G. Hovannisian, *Armenia on the Road to Independence* (Berkeley, 1967).

CHAPTER 8 EASTERN EUROPEANS

1. Charlotte Erickson, *Invisible Immigrants* (London, 1972).

2. Frederick Jackson Turner, "The Significance of the Frontier in American History," American Historical Association, *Annual Report* (1893).

3. Ewa Morwaska, *For Bread, With Butter* (Cambridge, England, 1985).

4. Oscar Handlin, *The Uprooted* (Boston, 1951); John Bodnar, *The Transplanted* (Bloomington, Ind., 1985).

5. Victor P. Greene's "Poles," in *HEAEG,* is the source for much that follows.

6. John J. Bukowczyk. *And My Children Did Not Know Me* (Bloomington, Ind., 1987), is the source of the folksong and the letter quoted below.

7. John Tracy Ellis, *American Catholicism* (Chicago, 1956).

8. Arthur Goren's "Jews," in *HEAEG,* is a good place to begin reading about modern American Jewry.

9. Lamar Cecil, *Albert Ballin* (Princeton, N.J., 1967); Richard J. Evans, *Death in Hamburg* (London, 1981).

10. Moses Rischin's *The Promised City* (Cambridge, Mass., 1962) is the standard work on the Jews of the Lower East Side.

11. I owe the information about the *Forward* building to David Reimers.

12. Elizabeth Israels Perry, *Belle Moscowitz* (New York, 1984).

13. Much of the general information about Hungarian immigrants comes from Paula Benkart's "Hungarians," in *HEAEG.*

14. Julianna Puskás, "Hungarian Migration Patterns," in Ira Glazier and Luigi De Rosa, eds., *Migration Across Time and Nations* (New York, 1986).

15. Morawska, *For Bread With Butter,* p. 157.

CHAPTER 9 MINORITIES FROM OTHER REGIONS

1. For Chinese I have relied heavily on Henry Shih-shan Tsai's *The Chinese Experience in America* (Bloomington, Ind., 1986); Him Mark Lai's "Chinese," in *HEAEG,* and my own *Asian America* (Seattle, 1988).

2. For the transportation of other Asians, see Hugh Tinker's *A New System of Slavery* (London, 1974).

3. U.S. Congress, House, *Coolie Trade* (Washington, 1856).

4. Sucheng Chan, *This Bittersweet Soil* (Berkeley, 1986).

5. Victor and Brett de Bary Nee, *Longtime Californ'* (New York, 1973), p. 63.

6. Yung Wing, *My Life in China and America* (New York, 1909).

7. Sue Fawn Chung, "The Chinese American Citizens' Alliance," honors thesis, UCLA, 1965.

8. Rose Hum Lee, *The Growth and Decline of Chinese Communities in the Rocky Mountain Region* (New York, 1978), pp. 252–53.

9. Stanford L. Lyman, *The Asian in North America* (Santa Barbara, Calif., 1977).

10. For Japanese I have relied primarily on my own work, much of which is summarized in *Asian America.*

11. John Modell, *The Economics and Politics of Racial Accommodation* (Urbana, Ill., 1977).

12. Delber L. McKee, *Chinese Exclusion Versus the Open Door Policy* (Detroit, 1977), p. 51.

13. S. Frank Miyamoto, *Social Solidarity Among the Japanese in Seattle* (Seattle, 1984), p. 102.

14. Massachusetts Bureau of Statistics of Labor, *Twelfth Annual Report* (Boston, 1881), p. 469.

15. For French Canadians I have used chiefly Gerard J. Brault, *The French Canadian Heritage in New England* (Hanover, N.H., 1986), and Elliott Robert Barkan's "French Canadians," in *HEAEG*.

CHAPTER 10 THE TRIUMPH OF NATIVISM

1. Ray Allen Billington's *The Protestant Crusade* (New York, 1938) is the best account of pre–Civil War nativism. For nativism up to 1924, John Higham's *Strangers in the Land* (New Brunswick, N.J., 1955) is seminal. For an overview of nativism and more, see George M. Frederickson and Dale T. Knobel's "Prejudice and Discrimination, History of," in *HEAEG*.

2. For Chinese Exclusion and anti-Chinese activities generally, see Elmer C. Sandmeyer's *The Anti-Chinese Movement in California* (Urbana, Ill., 1939); Stuart Creighton Miller's *The Unwelcome Immigrant* (Berkeley, 1969); Alexander Saxton's *The Indispensable Enemy* (Berkeley, 1971); and my own *Asian America* (Seattle, 1988).

3. Fiorello La Guardia, *The Making of an Insurgent* (Philadelphia, 1948), pp. 64–65.

4. Barbara Miller Solomon's *Ancestors and Immigrants* (Cambridge, Mass., 1956) is the standard account of the Immigration Restriction League.

5. I have analyzed the anti-Japanese movement in *The Politics of Prejudice* (Berkeley, 1962).

6. Roger Daniels, *Racism and Immigration Restriction* (St. Charles, Mo., 1974).

CHAPTER 11 MIGRATION IN PROSPERITY

1. Henry James, *The American Scene* (New York, 1907).

2. *HEAEG,* p. 493.

3. Roger Daniels, "American Refugee Policy in Historical Perspective," in J. C. Jackman and C. M. Borden, eds., *The Muses Flee Hitler* (Washington, D.C., 1983), pp. 61–77.

4. Madison Grant, as cited in John Higham, *Strangers in the Land* (New Brunswick, N.J., 1955), p. 306.

5. Michael A. Meyer, "The Refugee Scholars Project of the Hebrew Union College," in B. W. Korn, ed., *A Bicentennial Festschrift for Jacob Rader Marcus* (New York, 1976) pp. 359–75.

6. Roger Daniels, *Asian America* (Seattle, 1988).

7. Commission on Wartime Relocation and Internment of Civilians, *Personal Justice Denied* (Washington, D.C., 1982).

8. Philip Gleason, "American Identity and Americanization," in *HEAEG.*

CHAPTER 12 FROM THE NEW WORLD

1. David J. Weber, *Foreigners in Their Native Land* (Albuquerque, N.M., 1973).

2. Carlos E. Cortes, "Mexicans," in *HEAEG.*

3. For further details of discrimination against Mexicans in California see Roger Daniels and Harry H. L. Kitano's *American Racism* (Englewood Cliffs, N.J., 1970), pp. 68–71, 73–78.

4. Carey McWilliam's *North from Mexico* (Boston, 1948) is a pioneering account.

5. Joseph P. Fitzgerald, "Puerto Ricans," in *HEAEG.*

CHAPTER 13 CHANGING THE RULES

1. Robert Shaplen, "One-Man Lobby," *The New Yorker* (March 24, 1951), pp. 35–55; Fred Riggs, *Pressures on Congress* (New York, 1950). Harry H. L. Kitano and Roger Daniels, *Asian Americans* (Englewood Cliffs, N.J., 1988).

2. Leonard Dinnerstein, *America and the Survivors of the Holocaust* (New York, 1982).

3. Roger Daniels, "Changes in Immigration Law and Nativism since 1924," *American Jewish History* (December 1986), pp. 159–80.

4. Commission on Immigration and Naturalization, *Whom We Shall Welcome* (Washington, 1952).

5. V. S. Naipaul, *The Enigma of Arrival* (New York, 1987).

6. Leon W. Bouvier and Robert W. Gardner, *Immigration to the U.S.: The Unfinished Story* (Washington, D.C., 1986).

7. Nathan Glazer, ed., *Clamor at the Gates* (San Francisco, 1985).

8. David M. Reimers's *Still the Golden Door* (New York, 1985) is the best account of post-1965 immigration.

CHAPTER 14 THE NEW ASIAN IMMIGRANTS

1. Harry H. L. Kitano and Roger Daniels. *Asian Americans* (Englewood Cliffs, N.J., 1988); Roger Daniels, *Asian America* (Seattle, 1988). Robert W. Gardner et al., *Asian Americans: Growth, Change, and Diversity* (Washington, D.C., 1985).

2. Carlos Bulosan, *America Is in the Heart* (New York, 1943); Mary Dorita Clifford, "The Hawaiian Sugar Planter Association and Filipino Exclusion," and Roger Daniels, "Filipino Immigration in Historical Perspective" in

J. Saniel, ed., *The Filipino Exclusion Movement* (Quezon City, Philippines, 1967).

3. Joan M. Jensen, *Passage from India* (New Haven, 1988), and "East Indians," in *HEAEG.*

4. Roger Daniels, *A History of Indian Immigration to the United States* (New York, 1989).

5. Dalip Singh Saund, *Congressman from India* (New York, 1960).

6. Kitano and Daniels, *Asian Americans.*

7. David M. Reimers, *Still the Golden Door* (New York, 1985).

CHAPTER 15 CARIBBEANS, CENTRAL AMERICANS, AND SOVIET JEWS

1. Lisandro Perez, "Cubans," in *HEAEG.*

2. George Pozzetta and Gary Mormino, *The Immigrant World of Ybor City* (Urbana, Ill., 1986).

3. Eleanor M. Rogg, *The Assimilation of Cuban Exiles* (New York, 1974).

4. David M. Reimers, *Still the Golden Door* (New York, 1985).

5. Ibid.

6. Michel S. Laguerre, "Haitians," in *HEAEG.*

7. Walter LaFeber, *Inevitable Revolutions: The United States in Central America* (New York, 1984).

8. U.S. Select Commission on Immigration and Refugee Policy, U.S. *Immigration Policy and the National Interest: Staff Report and Final Report* (Washington, D.C., 1981).

9. Roger Boyes, "Locked Out of the Promised Land," London *Times,* February 7, 1989.

CHAPTER 16 THE 1980S AND BEYOND

1. U.S. Select Commission on Immigration and Refugee Policy, *U.S. Immigration Policy and the National Interest: Staff Report and Final Report* (Washington, D.C., 1981).

2. Theodore Hesburgh, "Enough Delay on Immigration," *New York Times,* March 20, 1986. For a defense of the commission by its staff director, see Lawrence H. Fuchs's "Immigration Reform in 1911 and 1981: The Role of Select Commissions," *Journal of American Ethnic History* 3 (1983): 58–89.

3. Carl Rowan, "Marshall Plan for Mexico?" *Cincinnati Enquirer,* June 21, 1986. For Galbraith see his *A Life in Our Times* (Boston, 1981), pp. 280–81. Council of Economic Advisors and F. Ray Marshall quoted in Leon F. Bouvier and Robert W. Gardner, *Immigration to the United States: The Unfinished Story* (Washington, D.C., 1986), pp. 28, 30.

4. Alan Simpson quoted in David M. Reimers, *Still the Golden Door* (New York, 1985), pp. 152–54.

5. *New York Times* story by Roberto Suro, June 18, 1989.

6. INS Form I-9 (01/07/87).

7. *New York Times,* June 18, 1989.

8. Gardner and Bouvier, *Immigration to the United States,* p. 40. For US English and the English-only movement see Ana Celia Zentella, "English-Only Laws Will Foster Divisiveness, Not Unity: They Are Anti-Hispanic, Anti-Elderly, and Anti-Female," *Chronicle of Higher Education,* November 23, 1988, p. B-1; *Time,* December 5, 1988; *Newsweek,* February 20, 1988; Eric Schmitt, "As the Suburbs Speak More Spanish, English Becomes a Cause," *New York Times,* February 26, 1989, p. E6; and Christopher Hitchens, "Deporting the Native Tongues," *Times Literary Supplement,* June 30–July 6, 1989, which contains the quotations from John Tanton.

9. Kingsley Davis, "The Migrations of Human Populations," *Scientific American,* 231, no. 3 (September 1974), p. 105; Ben J. Wattenberg, *The Birth Dearth* (New York, 1987).

10. Francis X. Clines, "The New Illegals: The Irish," *New York Times Magazine,* November 20, 1988.

11. *International Herald Tribune,* July 15–16, 1989.

12. For an informed look at the whole Statue of Liberty-Ellis Island Project see F. Ross Holland, *Idealists, Scoundrels, and the Lady: An Insider's View of the Statue of Liberty-Ellis Island Project* (Urbana, 1993).

13. Nathan Glazer, ed., *Clamor at the Gates* (San Francisco, 1985), p. 3.

CHAPTER 17 IMMIGRATION IN AN AGE OF GLOBALIZATION

1. The Maxwell School Report may be found online at: http://www.govexec.com/gpp/reportcard.htm.

2. David North, 1995. "Soothing the Establishment: The Impact of Foreign-Born Scientists and Engineers on America," as cited, *Migration News,* February 1996.

3. The following categories of "nonimmigrants" not otherwise discussed in the text came in the numbers indicated in 1998: Transit aliens—365,607; foreign government officials—126,543; international representatives—86,129; representatives of foreign media—28,888; NATO officials—12,176; Other and unknown—87; and fiancés and fiancées of U.S. citizens—13,748. Persons in the last group become "immigrants" as soon as the marriage ceremony is performed. INS, *1998 Statistical Yearbook*, Washington, D.C.: GPO, 2000, Table 38.

4. The categories are and respective numbers in 1998 are: 1. H-1A, registered nurses—555; 2. H-3, industrial trainees—3,157; 3. O-1—workers with "extraordinary ability or achievement"—12,221; 4. O-2—helpers for O-1s—2,802; 5. P-1—"internationally recognized" athletes or entertainers—34,477; 6. P-2—Artists or entertainers in reciprocal exchange programs—3,089; 7. P-3—Artists or entertainers in culturally unique programs—9, 452; 8. Q-1—Workers in international cultural exchange programs—1,921; and 9. R-1—Workers in religious occupations—10,863. *Ibid.*, Tables 38 and 40.

5. "Plan to check ID at border under fire," *Tucson Citizen*, March 30, 1999.

6. David M. Reimers, *Unwelcome Strangers: American Identity and the Turn Against Immigration* (New York: Columbia University Press, 1998).

7. *Public Papers of the Presidents: Ronald Reagan, 1986*, p. 1521. Washington, D.C.: GPO, 1989.

8. Christopher Mitchell, "Changing the Rules: The Impact of the Simpson-Rodino Act on Inter-American Diplomacy," pp. 177–189 in Georges Vernez, ed., *Immigration and International Relations*. (Santa Monica, CA: Rand Corporation, 1990). Quotation at p. 183. Simpson-Rodino is another name for IRCA.

9. Graciela Sevilla, "Border Patrol captures increase: Record year due to more agents and funding." *Arizona Republic*, Oct. 7, 1988.

10. "Operation Hold the Line in Texas," *Migration News*, (March 1994).

11. *Migration News*, December 2000.

12. *INS Yearbook*, 1998, p. 203.

13. *INS Yearbook*, 1998, pp. 134–41. For a good analysis of earlier naturalization data see Elliott R. Barkan, "Whom Shall We Integrate? A Comparative Analysis of the Immigration and Naturalization Trends of Asians Before and After the 1965 Immigration Act (1951–1978)." *Journal of American Ethnic History* (Fall 1983) 29–57.

14. For Jordan's comments, see *Migration News*, May 1994, Vol. 1, No. 5. For the reports and other materials generated by the U.S. Commission on Immigration Reform (1990–1997) see http://www.utexas.edu/lbj/uscir/

15. There is a good summary in the perceptive essay by the German scholar Herbert Dittgen, "The American Debate about Immigration in the 1990s: A New Nationalism After the Cold War?," pp. 197–225 in Knud Krakau, ed., *The American Nation, National Identity, Nationalism*. Münster, Germany, 1997.

16. "Hot Lines and Hot Tempers." *Time* magazine (domestic edition), Nov. 28, 1994.

17. *Los Angeles Times*, Sep. 14, 1999.

18. *The Economist* (London), July 13, 1996.

19. Lucy E. Salyer, *Laws Harsh as Tigers: Chinese Immigrants and the Shaping of Modern Immigration Law* (Chapel Hill, 1995).

20. *Washington Post*, November 3, 1999. The legal authority for the use of secret evidence is the Classified Information Procedures Act (CIPA) to Immigration Proceedings, Chapter 9 of Title II of the Immigration and Nationality Act (8 U.S.C. 1351 et seq.)

21. The immigration statute is Title IV of PL 104–193.

22. Peter Brimelow, *Alien Nation: Common Sense about America's Immigration Disaster* (New York, 1995.) Reviews cited: Samuel Francis in *National Review* v. 47 (May 1, 1995), p. 76; Lawrence Chua in *Voice Literary Supplement* v. 134 (April 1995), p. 17; and Richard Bernstein, *New York Times* 1995. For a contrast see the review by Nicholas Lemann in *The New York Times Book Review*, April 16, 1995, p. 3.

23. The statutes were the Balanced Budget Act of August 5, 1977 (111 *Stat.* 270), and the Agricultural Research Reform Act of February 11, 1998 (112 *Stat.* 575).

24. Nicaraguan Adjustment and Central American Relief Act of November 19, 1997 (111 *Stat.* 2193).

25. *Polovchak v. Meese* 774 *F. 2d* 731 (1985).

26. Almost nothing appeared about "Pedro Pan" until shortly before the Elián episode; see Victor Andres Triay, *Fleeing Castro: Operation Pedro Pan and the Cuban Children's Program* (Gainesville, 1998), and Yvonne M. Conde, *Operation Pedro Pan: The Untold Exodus of 14,048 Cuban Children* (New York, 1999). After Elián there was a flurry of stories in the press. Dirk Johnson, "Children of 'Operation Peter Pan' Recall Painful Separations from Parents" *New York Times*, April 22, 2000.

27. Maria Cristina Garcia, *Havana USA: Cuban Exiles and Cuban Americans in South Florida, 1959–1994* (Berkeley, 1996).

28. UN Press Release E/CN.4/Sub.2/2000/13, June 15, 2000.

29. "Pope Calls for Change in Mentality Toward Immigration Contrasts Immigrants' Desperate Plight with Tunnel Vision of Developed Countries," Vatican City: ZENIT, December 8, 1999.

30. *Trends in International Migration*. Paris: OECD, 2001.

31. Michael Weiner, "Japan in the Age of Migration," pp. 52–69 in M. Douglass and G.S. Roberts, eds., *Japan and Global Migration* (New York, 2000). For historical background see Weiner's definitive *The Origins of the Korean Community in Japan, 1910–1923* (Manchester, England, 1989).

32. Material on contemporary migration is from several interviews and the following newspaper articles: "Job Conditions Severer for Japanese Latin Americans," *Japan Weekly Monitor*, October 5, 1998, as cited *Asian Migration News*, December 1998; Howard LaFranchi, "The Revolving Door Connecting Brazil and Japan for 90 Years is Taking Another Turn," *Christian Science Monitor*, November 19, 1998; Matt Moffett, "Brazil's Japanese Community Wonders Where the Boys Are," *Wall Street Journal*, February 11, 1998.

33. Sarah Kurush, "Russian government addresses population decline," Associated Press story datelined Moscow, Feb. 15, 2001.

34. Mutuma Mathiu in *The Nation* (Nairobi), September 9, 2001.

35. 8 *United States Code* 1184(g). Title 8 governs "Aliens and Nationality"; Section 1184 governs "admission of nonimmigrants."

36. Steven Greenhouse, "Labor Urges Amnesty for Illegal Immigrants," *New York Times*, February 17, 2000, and "Labor Warms to Immigrant Workers," *New York Times*, April 2, 2000.

37. Joann Kelly, "Study: Immigrant amnesty hurts unions," United Press International, September 14, 2001.

Selected Bibliography

Alexander, June G. *The Immigrant Church and Community: Pittsburgh's Slovak Catholics and Lutherans, 1880–1915.* Pittsburgh, 1987.

Altschuler, Glenn C. *Race, Ethnicity and Class in American Social Thought,* 1865–1919. Arlington Heights, Ill., 1982.

Archdeacon, Thomas J. *Becoming American: An Ethnic History.* New York, 1983.

Bailyn, Bernard. *Voyagers to the West.* New York, 1986.

Barkan, Elliott Robert. "French Canadians." In *HEAEG,* pp. 388–401.

Barton, Josef J. *Peasants and Strangers.* Cambridge, Mass., 1975.

Bayor, Ronald. *Neighbors in Conflict.* Baltimore, 1978.

Beijbom, Ulf. "Swedes." In *HEAEG,* pp. 971–81.

———. *Swedes in Chicago.* Vaxjo, Sweden, 1971.

Bell, Daniel. *The End of Ideology.* Glencoe, Ill., 1960.

Benkart, Paula. "Hungarians." In *HEAEG,* pp. 462–71.

Berkhofer, Robert R., Jr. *Salvation and the Savage.* Lexington, Ky., 1965.

Berthoff, Rowland T. *British Immigrants in Industrial America 1790–1950.* Cambridge, Mass., 1953.

———. "Welsh." In *HEAEG,* pp. 1011–17.

Billington, Ray Allen. *The Protestant Crusade.* New York, 1938.

Birchall, R. A. *The San Francisco Irish, 1848–1880.* Berkeley and Los Angeles, 1980.

Blegen, Theodore C. *Norwegian Migration to America, 1825–1860.* Northfield, Minn., 1931.

Boogaart, Ernst van den. "The Servant Migration to New Netherland, 1624–1664." In P. C. Emmer, ed., *Colonialism and Migration.* Dordrecht, the Netherlands, 1986.

Borah, Woodrow, and S. F. Cook. *The Aboriginal Population of Central Mexico on the Eve of the Spanish Conquest.* Berkeley, 1963.

Bouvier, Leon W., and Robert W. Gardner. *Immigration to the U.S.: The Unfinished Story.* Washington, D.C., 1986.

Brault, Gerard. *The French Canadian Heritage in New England.* Hanover, N.H., 1986.

Brown, Wallace. *The King's Friends.* Providence, R.I., 1965.

Bukowczyk, John J. *And My Children Did Not Know Me.* Bloomington, Ind., 1987.

Bulosan, Carlos. *America Is in the Heart.* New York, 1943.

Butler, Jon. *The Huguenots in America.* Cambridge, Mass., 1983.

Camarillo, Albert. *Chicanos in a Changing Society.* Cambridge, Mass., 1979.

Campbell, Mildred. *The English Yeoman under Elizabeth and the Early Stuarts.* New Haven, 1942.

Canny, Nicholas, and Anthony Pagden, eds. *Colonial Identity in the New World, 1500–1800.* Princeton, 1988.

Cardoso, Lawrence. *Mexican Emigration to the United States.* Tucson, Ariz., 1980.

Castro, Amerigo. *The Spaniards: An Introduction to Their History.* Berkeley and Los Angeles, 1980.

Catterall, Helen, ed. *Judicial Cases Concerning American Slavery and the Negro.* Washington, D.C., 1926.

Cecil, Lamar. *Albert Ballin.* Princeton, N.J., 1967.

Chan, Sucheng. *This Bittersweet Soil.* Berkeley and Los Angeles, 1986.

Chazanof, William. *Valledolmo-Fredonia.* Fredonia, N.Y., 1961.

Chung, Sue Fawn. "The Chinese American Citizens' Alliance." Honors thesis, UCLA, 1965.

Cinel, Dino. *From Italy to San Francisco.* Stanford, Calif., 1982.

Clark, Dennis. *The Irish in Philadelphia.* Philadelphia, 1974.

Clifford, Mary Dorita. "The Hawaiian Sugar Planter's Association and Filipino Exclusion." In J. Saniel, ed., *The Filipino Exclusion Movement.* Quezon City, Philippines, 1967.

Commission on Immigration and Naturalization. *Whom We Shall Welcome.* Washington, D.C., 1952.

Commission on the Wartime Relocation and Internment of Civilians. *Personal Justice Denied.* Washington, D.C., 1982.

Conzen, Kathleen Neils. *Immigrant Milwaukee, 1836–1860.* Cambridge, Mass., 1976.

———. "Germans." In *HEAEG,* pp. 405–25.

Cortes, Carlos E. "Mexicans." In *HEAEG,* pp. 697–719.

Cressy, David. *Coming Over.* Cambridge, England, 1988.

Crosby, Alfred J. *The Columbian Exchange.* Westport, Conn., 1972.

———. *Ecological Imperialism: The Biological Expansion of Europe.* Cambridge, England, 1986.

Curtin. Philip D. *The African Slave Trade: A Census.* Madison, Wis., 1969.

Daniels, Roger. "American Refugee Policy in Historical Perspective." In J. C. Jackman and C. M. Borden, eds., *The Muses Flee Hitler,* pp. 61–77. Washington, D.C., 1983.

———. *Asian America: Chinese and Japanese in the United States since 1850.* Seattle, 1989.

———. "Changes in Immigration Law and Nativism since 1924." *American Jewish History* (December, 1986), pp. 159–80.

———. "Filipino Immigration in Historical Perspective." In J. Saniel, ed., *The Filipino Exclusion Movement.* Quezon City, Philippines, 1967.

———. *A History of Indian Immigration to the United States.* New York, 1989.

———. *The Politics of Prejudice.* Berkeley and Los Angeles, 1962.

———. *Racism and Immigration Restriction.* St. Charles, Mo., 1974.

Daniels, Roger, and Harry H. L. Kitano. *American Racism: Exploration of the Nature of Prejudice.* Englewood Cliffs, N.J., 1970.

Davis, Kingsley. "The Migrations of Human Populations." *Scientific American* 231, no. 3 (September 1974): 105.

Diner, Hasia R. *Erin's Daughters in America.* Baltimore, 1983.

Dinnerstein, Leonard. *America and the Survivors of the Holocaust.* New York, 1982.

Dolan, Jay P. *The American Catholic Experience.* New York, 1985.

———. *The Immigrant Church: New York's Irish and German Catholics, 1815–1865.* Baltimore, 1975.

Donaldson, Gordon. "Scots." In *HEAEG,* pp. 908–16.

———. *The Scots Overseas.* London, 1966.

DuBois, W. E. B. *The Souls of Black Folk.* New York, 1904.

Dumond, Don E. *The Eskimos and the Aleuts.* New York, 1987.

Duncan, John Donald. "Indian Slavery." In Bruce A. Glasrud and Alan M. Smith, eds., *Race Relations in British North America, 1607–1783.* Chicago, 1982.

Eccles, W. J. *France in America.* New York, 1973.

Edding, Friedrich. "Intra-European Migration and the Prospects of Integration." In Brinley Thomas, ed., *Economics of International Migration.* London, 1958.

Eggleston, Edward. *The Transit of Civilization.* New York, 1901.

Ellis, John Tracy. *American Catholicism,* 2nd. ed. Chicago, 1969.

Emmons, David M. *The Butte Irish: Class and Ethnicity in an American Mining Town, 1875–1925.* Urbana, Ill., 1989.

Erickson, Charlotte. *Invisible Immigrants.* London, 1972.

Esterlin, Richard A. "Immigration: Economic and Social Characteristics." In *HEAEG,* pp. 476–86.

Evans, Richard J. *Death in Hamburg.* New York, 1987.

Fitzgerald, Joseph P. "Puerto Ricans." In *HEAEG,* pp. 858–67.

Foerster, Robert F. *The Italian Emigration of Our Times.* Cambridge, 1919.

Franklin, John Hope. *From Slavery to Freedom,* 6th ed. New York, 1988.

Frazier, E. Franklin. *The Negro Family in the United States.* New York, 1939.

Fredrickson, George M., and Dale T. Knobel. "Prejudice and Discrimination, History of." In *HEAEG*, pp. 820–29.

Fuchs, Lawrence H. "Immigration Reform in 1911 and 1981: The Role of Select Commissions." *Journal of American Ethnic History* 3 (1983): 58–89.

Gabaccia, Donna R. *From Italy to Elizabeth Street.* Albany, N.Y., 1983.

Galenson, David. *White Servitude in Colonial America.* Cambridge, England, 1981.

Gardner, Robert W., et al. *Asian Americans: Growth, Change, and Diversity.* Washington, D.C., 1985.

Gerber, David A. *The Making of an American Pluralism: Buffalo, New York, 1825–60.* Urbana, Ill., 1989.

Gibson, Charles. *Spain in America.* New York, 1967.

Glazer, Nathan. *American Judaism,* rev. ed. Chicago, 1972.

———, ed. *Clamor at the Gates.* San Francisco, 1985.

Glazer, Nathan, and Daniel P. Moynihan. *Beyond the Melting Pot.* New York, 1963.

Gleason, Philip. "American Identity and Americanization." In *HEAEG,* pp. 31–58.

Goffart, Walter. *Barbarians and Romans, A.D. 418–584: The Techniques of Accommodation.* Princeton, 1988.

Goren, Arthur A. "Jews." In *HEAEG,* pp. 571–98.

Graham, Ian C. C. *Colonists from Scotland: Emigration to North North America, 1707–1783.* Ithaca, N.Y., 1965.

Green, E. E. R., ed. *Essays in Scotch-Irish History.* London, 1969.

Greene, Victor R. *American Immigrant Leaders, 1800–1910.* Baltimore, 1987.

———. "Poles." In *HEAEG,* pp. 787–803.

Handlin, Oscar. *Boston's Immigrants,* 2nd ed. Cambridge, Mass., 1959.

———. *The Uprooted.* Boston, 1951.

Hansen, Marcus Lee. *The Atlantic Migration, 1607–1860.* Cambridge, Mass., 1940.

———. "Immigration as a Field for Historical Research." *American Historical Review* 32 (1927): 500–18.

———. *The Immigrant in American History.* Cambridge, Mass., 1940.

Harney, Robert F. "The Italian Experience in America." In Anthony Mollica, ed., *Handbook for Teachers of Italian.* Ontario, Canada, 1976.

Harney, Robert F., and J. Vincenza Scarpaci, eds. *Little Italies in North America.* Toronto, Canada, 1981.

Hartmann, Edward George. *Americans from Wales.* Boston, 1967.

Hennings, Robert E. *James D. Phelan and the Wilson Progressives of California.* New York, 1985.

Herskovits, Melville J. *The Myth of the Negro Past.* New York, 1941.

Higham, John. *Send These to Me: Immigrants in Urban America,* rev. ed. Baltimore, 1984.

———. *Strangers in the Land.* New Brunswick, N.J., 1955.
Holloway, Thomas H. *Immigrants on the Land.* Chapel Hill, N.C., 1980.
Hovannisian, Richard G. *Armenia on the Road to Independence.* Berkeley, 1967.
Hvidt, Kristian. *Flight to America: The Social Background of 300,000 Danish Immigrants.* New York, 1977.
James, Henry. *The American Scene.* New York, 1907.
Jensen, Joan M. "East Indians." In *HEAEG,* pp. 296–302.
———. *Passage from India.* New Haven, 1988.
Johansen, Donald J., and Mailand A. Edey. *Lucy: The Beginnings of Humankind.* New York, 1981.
Jones, Maldwyn A. *American Immigration.* Chicago, 1960.
———. "The Background of Emigration from Great Britain in the Nineteenth Century." *Perspectives in American History* 7 (1973): 3–92.
———. *Destination America.* London, 1976.
———. "Scotch-Irish." In *HEAEG,* pp. 893–908.
Kennedy, Robert W. *The Irish: Emigration, Marriage and Fertility.* Berkeley and Los Angeles, 1973.
Kettner, James H. *The Development of American Citizenship, 1607–1870.* Chapel Hill, N.C., 1978.
Kirk, Dudley. *Europe's Population in the Interwar Years.* Princeton, 1946.
Kitano, Harry H. L., *Japanese Americans: The Evolution of a Subculture.* Englewood Cliffs, N.J., 1969.
Kitano, Harry H. L., and Roger Daniels. *Asian Americans: Emerging Minorities.* Englewood Cliffs, N.J., 1988.
Koehler, Lyle. "Red-White Power Relationships and Justice in the Courts of Seventeenth-Century New England." *American Indian Culture and Research Journal* 3:4 (1979): 1–31.
Kraut, Alan M. *The Huddled Masses: The Immigrant in American Society, 1880–1921.* Arlington Heights, Ill., 1982.
Kupperman, Karen O. "Apathy and Death in Early Jamestown." *Journal of American History.* 66 (1979): 24–40.
La Guardia, Fiorello. *The Making of an Insurgent.* Philadelphia, 1948.
Labaree, Leonard W., ed. *The Papers of Benjamin Franklin,* vol. 4. New Haven, 1961.
LaFeber, Walter. *Inevitable Revolutions: The United States in Central America.* New York: 1984.
Laguerre, Michel S. "Haitians." In *HEAEG,* pp. 446–49.
Lai, Him Mark. "Chinese." In *HEAEG,* pp. 217–34.
Larner, John. "The Certainty of Columbus: Some Recent Studies." *History* (February 1988): 3–23.
LeBreton, Marietta M. "Acadians." In *HEAEG,* pp. 1–3.
Lee, Rose Hum. *The Growth and Decline of Chinese Communities in the Rocky Mountain Region.* New York, 1978.

Lemon, James. *The Best Poor Man's Country: A Geographical Study of Southeastern Pennsylvania.* Baltimore, 1972.

Leyburn, James G. *The Scotch-Irish: A Social History.* Chapel Hill, N.C., 1962.

Luebke, Frederick C. *Bonds of Loyalty: German-Americans and World War I.* De Kalb, Ill., 1974.

Lyman, Stanford L. *The Asian in North America.* Santa Barbara, Calif., 1977.

Maguire, Reverend John Francis. *The Irish in America.* New York, 1868.

Marcus, Jacob Rader. *The Colonial American Jew,* 3 vols. Detroit, 1970.

———. *Early American Jewry,* 2 vols. Philadelphia, 1951–53.

Marty, Martin E. "Ethnicity: The Skeleton of Religion in America." *Church History* 41 (1972): 15–21.

Massachusetts Bureau of Statistics of Labor. *Twelfth Annual Report.* Boston, 1881.

McKee, Delber L. *Chinese Exclusion versus the Open Door Policy.* Detroit, 1977.

McWilliams, Carey. *North from Mexico.* Boston, 1948.

Meier, August, and Elliott Rudwick. *Black History and the Historical Profession.* Urbana, Ill., 1986.

———. *From Plantation to Ghetto.* New York, 1976.

Menard, Russell. *Economy and Society in Early Colonial Maryland.* New York, 1985.

Meyer, Michael A. "The Refugee Scholars Project of the Hebrew Union College." In B. W. Korn, ed., *A Bicentennial Festschrift for Jacob Rader Marcus,* pp. 359–75. New York, 1976.

Miller, John C. *Crisis in Freedom.* Boston, 1952.

Miller, Kerby A. *Emigrants and Exiles: Ireland and the Irish Exodus to North America.* New York, 1985.

Miller, Stuart Creighton. *The Unwelcome Immigrant.* Berkeley, Calif., 1969.

Mintz, Sidney. *Sweetness and Power.* New York, 1986.

Mirak, Robert. "Armenians." In *HEAEG,* pp. 136–49.

Mitchell, Brian C. *The Paddy Camps: The Irish of Lowell, 1821–61.* Urbana, Ill., 1988.

Miyamoto, S. Frank. *Social Solidarity Among the Japanese in Seattle.* Seattle, 1984.

Modell, John. *The Economics and Politics of Racial Accommodation.* Urbana, Ill., 1977.

Moltmann, Günter. "The Migration of German Redemptioners to North America, 1720–1820." In P.C. Emmer, ed., *Colonialism and Migration.* Dordrecht, the Netherlands, 1986.

———. "Steamship Transport of Immigrants from Europe to the United States, 1850–1914." (In press.)

Morgan, Edmund S. *American Slavery; American Freedom: The Ordeal of Colonial Virginia.* New York, 1975.

Morison, Samuel Eliot. *By Land and By Sea.* New York, 1953.

———. *The European Discovery of America.* Vol. 1, *The Northern Voyages, A.D. 500–1600.* New York, 1971.

———. *Oxford History of the American People.* New York, 1965.

Morner, Magnus. *Race Mixture in the History of Latin America.* Boston, 1967.

Morwaska, Eva. *For Bread, With Butter.* Cambridge, England, 1985.

Mulder, William. *Homeward to Zion.* Minneapolis, Minn., 1957.

Munch, Peter A. "Norwegians." In *HEAEG,* pp. 750–61.

Naff, Alixa. "Arabs." In *HEAEG,* pp. 128–36.

———. *Becoming American: The Early Arab Immigrant Experience.* Carbondale, Ill., 1985.

Naipaul, V. S. *The Enigma of Arrival.* New York, 1987.

Nash, Gary B. *The Urban Crucible.* Cambridge, Mass., 1979.

Nee, Victor, and Brett de Bary. *Longtime Californ'.* New York, 1973.

Nelli, Humbert. "Italians." In *HEAEG,* pp. 545–60.

Niehaus, Earl. *The Irish in New Orleans, 1800–1865.* Baton Rouge, La., 1956.

Norton, Mary Beth, *The British-Americans: The Loyalist Exiles in England, 1774–1789.* New York, 1972.

Perez, Lisandro. "Cubans." In *HEAEG,* pp. 256–61.

Perry, Elizabeth Israels. *Belle Moscowitz.* New York, 1984.

Phillips, Ulrich Bonnell. "The Central Theme of Southern History." *American Historical Review* 33 (1928): 30–50.

Pozzetta, George, and Gary Mormino. *The Immigrant World of Ybor City.* Urbana, Ill., 1986.

Puskás, Julianna. "Hungarian Migration Patterns." In Ira Glazier and Luigi De Rosa, eds., *Migration Across Time and Nations.* New York, 1986.

Quintana, Frances Leon. "Spanish." In *HEAEG,* pp. 950–53.

Ravage, Marcus Eli. *An American in the Making: The Life Story of an Immigrant.* New York, 1917.

Ravenstein, E. G. "The Laws of Migration." *Journal of the Royal Statistical Society* (1889): 241–301.

Reimers, David M. *Still the Golden Door: The Third World Comes to America.* New York, 1985.

Riggs, Fred. *Pressures on Congress.* New York, 1950.

Riordan, William L. *Plunkitt of Tammany Hall.* New York, 1963.

Rippley, LaVern J. *The German-Americans.* Boston, 1976.

Rischin, Moses, ed. *Immigration and the American Tradition.* Indianapolis, 1976.

Rischin, Moses. *The Promised City.* Cambridge, Mass., 1962.

Rogg, Eleanor. *The Assimilation of Cuban Exiles.* New York, 1974.

Rolle, Andrew. *The American Italians.* Wadsworth, Calif., 1972.

———. *The Immigrant Upraised.* Norman, Okla., 1968.

Rowse, A. L. *The Elizabethans and America.* New York, 1959.

Saloutos, Theodore. "Greeks." In *HEAEG,* pp. 430–40.

———. *A History of the Greeks in the United States.* Cambridge, Mass., 1964.

Sandmeyer, Elmer C. *The Anti-Chinese Movement in California.* Urbana, Ill., 1939.

Saund, Dalip Singh. *Congressman from India.* New York, 1960.

Saxton, Alexander. *The Indispensable Enemy.* Berkeley, Calif., 1971.

Schrier, Arnold. *Ireland and the Emigration.* Minneapolis, 1958.

Seller, Maxine, ed. *Immigrant Women.* Philadelphia, 1981.

Shannon, James P. *Catholic Colonization on the Western Frontier.* New Haven, 1967.

Shaplen, Robert. "One-Man Lobby." *The New Yorker,* March 24, 1951.

Skardal, Dorothy Burton. *The Divided Heart.* Lincoln, Neb., 1974.

Smith, Abbot. *Colonists in Bondage.* Chapel Hill, N.C., 1947.

Smith, Adam. *The Wealth of Nations.* New York, 1937.

Solberg, Carl. *Immigration and Nationalism: Argentina and Chile, 1890–1914.* Austin, Tex., 1970.

Solomon, Barbara Miller. *Ancestors and Immigrants.* Cambridge, Mass., 1956.

Souden, David. "Rogues, Whores and Vagabonds." *Social History* 4 (1978): 23–41.

Stefansson, Vilhjalmur, ed. *Great Adventures and Explorations.* New York, 1949.

Swierenga, Robert P. "Dutch." In *HEAEG,* pp. 284–95.

Szasz, Ferenc M. "The New York Slave Revolt of 1741: A Re-Examination." *New York History* 48 (1967): 215–30.

Taylor, Philip. *The Distant Magnet: European Emigration to the U.S.A.* New York, 1971.

Thistlewaite, Frank. "Migration from Europe Overseas in the Nineteenth and Twentieth Centuries." *XI Congrès International des Sciences Historiques, Rapports,* vol. 5, pp. 32–60. Stockholm, 1960.

Tinker, Hugh. *A New System of Slavery.* London, 1974.

Tolzmann, Don Heinrich. *German-Americana.* Metuchen, N.J., 1975.

———. *German-American Literature.* Metuchen, N.J., 1975.

Tsai, Henry Shih-shan. *The Chinese Experience in America.* Bloomington, Ind., 1986.

Turner, Frederick Jackson. "The Significance of the Frontier in American History." In American Historical Association, *Annual Report,* 1893.

U.S. Congress. House. . . . *Coolie Trade.* Washington, D.C., 1860.

U.S. Select Commission on Immigration and Refugee Policy. *U.S. Immigration Policy and the National Interest: Staff Report and Final Report.* Washington, D.C., 1981.

Vecoli, Rudolph. "Contadini in Chicago: A Critique of *The Uprooted.*" *Journal of American History* 51 (December 1963): 404–17.

———. "Prelates and Peasants: Italian Immigrants and the Catholic Church." *Journal of Social History* 2 (1969): 217–68.

Wade, Richard C. *Slavery in the Cities.* New York, 1964.

Wahlgren, Erik. *The Kensington Stone. A Mystery Solved.* Madison, Wis., 1958.

Walker, Mack. *Germany and the Emigration, 1816–1885.* Cambridge, Mass., 1964.

Ward, David. *Cities and Immigrants.* New York, 1971.

Ward, John William. *Andrew Jackson: Symbol for an Age.* New York, 1955.

Wattenberg, Ben J. *The Birth Dearth.* New York, 1987.

Weber, David J. *Foreigners in Their Native Land.* Albuquerque, N.M., 1973.

White, Lynn T., Jr. *Medieval Religion and Technology: Collected Essays.* Los Angeles, 1978.

Whyte, Donald. *Dictionary of the Scottish Emigrants to the USA.* Baltimore, 1972.

Williams, George Washington. *History of the Negro Race in America.* New York, 1882.

Wing, Yung. *My Life in China and America.* New York, 1909.

Wittke, Carl. "Ohio's Germans, 1840–1875." *Historical Quarterly* 66 (1957): 339–54.

Woodham-Smith, Cecil. *The Great Hunger.* London, 1962.

Wrigley, E. A.. *Population and History.* New York, 1969.

Yans-McLaughlin, Virginia. *Family and Community: Italian Immigrants in Buffalo, 1880–1930.* Ithaca, 1977.

Yoder, Don. "Pennsylvania Germans." In *HEAEG,* pp. 770–72.

[illegible] *[illegible]*. Revised ed. [illegible] 1966.

W[illegible] [illegible]. Cambridge, Mass., 19[illegible].

Ward, [illegible] *[illegible] and Anthropology*. New York, 19[illegible].

[illegible] William [illegible]. [illegible] 193[illegible].

[illegible] New York, 19[illegible].

White, David [illegible]. Albuquerque, N.M., 1973.

White, [illegible] *[illegible] Collected Essays*. Los Angeles, 19[illegible].

White, [illegible] *[illegible] of the Second Empire [illegible]*. Baltimore, [illegible].

Williams, [illegible] Washington [illegible]. New York, 19[illegible].

Wolf, [illegible] *[illegible] in America*. New York, 1969.

Wolfe, [illegible] 1870 [illegible] *Quarterly* [illegible] 1970.

Woodham-Smith, Cecil. *[illegible]*. London, 1950.

Wright, [illegible]. New York, 1960.

[illegible] Indian [illegible] Philadelphia, 19[illegible].

[illegible] 1967.

Additional Bibliography

There has been a veritable explosion of scholarship on immigration and ethnicity since the publication of the first edition. The bibliography below is chiefly devoted to a selection of that scholarship but also includes several volumes that should have been in the initial bibliography.

Anbinder, Tyler. *Five Points*. New York, 2001.

———. *Nativism and Slavery*. New York, 1992.

Ansari, Maboud. *The Iranians in the United States*. New York, 1988.

Axtell, James. *After Columbus: Essays in the Ethnohistory of Colonial North America*. New York, 1990.

Baily, Samuel I. and Franco Ramella, eds. *One Family, Two Worlds: An Italian Family's Correspondence Across the Atlantic, 1901–1922*. New Brunswick, N.J., 1989.

Baines, Dudley. *Emigration from Europe, 1815–1930*. Basingstoke, U.K., 1991.

Barkan, Elliott Robert, ed. *A Nation of Peoples: A Sourcebook on America's Multicultural Heritage*. Westport, Conn., 1999.

Baseler, Marilyn C. *"Asylum for Mankind": America, 1607–1800*. Ithaca, N.Y., 1998.

Bayor, Ronald H. and Thomas J. Meagher, eds. *The New York Irish*. Baltimore, 1996.

Beiter, John and Mark. *An Enduring Legacy: The Story of the Basques in Idaho*. Reno, 2000.

Beltman, Brian W. *Dutch Farmer in the Missouri Valley: The Life and Letters of Ulbe Eringa, 1866–1950*. Urbana, 1995.

Berrol, Selma. *Growing Up American: Immigrant Children in America*. New York, 1995.

Bao, Xiaolan. *Holding Up More than Half the Sky: Chinese Women Garment Workers in New York City, 1948–1992*. Urbana, 2001.

Brasseaux, Carl A. *The Founding of New Acadia*. Baton Rouge, 1987.

———. *Acadian to Cajun*. Jackson, Miss., 1992.

Brault, Gerard J. *The French Canadian Heritage in New England.* Hanover, N.H., 1986.

Butler, Jon. *Becoming America: The Revolution before 1776.* Cambridge, Mass., 2000.

Chan, Sucheng. *Asian Americans: An Interpretative History.* Boston, 1991.

———, ed. *Hmong Means Free.* Philadelphia, 1994.

Cohen, Miriam. *Workshop to Office: Two Generations of Italian Women in New York City, 1900–1950.* Ithaca, N.Y., 1993.

Cooper, Patricia. *Once a Cigar Maker: Men, Women and Work Culture in American Cigar Factories.* Urbana, 1987.

Crosby, Alfred W., *Ecological Imperialism: The Biological Expansion of Europe, 900–1900.* Cambridge, U.K., 1986.

Daniels, Roger. *Not Like Us: Immigrants and Minorities in America, 1890–1924.* Chicago, 1997.

———. *Prisoners Without Trial: Japanese Americans in World War II.* New York, 1993.

Din, Gilbert C. *The Canary Islanders of Louisiana.* Baton Rouge, 1988.

Douglass, William A. *Emigration in a South Italian Town: An Anthropological History.* New Brunswick, N.J., 1984.

Ekirch, A. Roger. *Bound for America: The Transportation of British Convicts to the Colonies, 1718–1775.* New York, 1987.

Emmons, David M. *The Butte Irish.* Urbana, 1989.

Erickson, Charlotte. *Leaving England.* Ithaca, N.Y., 1994.

Feingold, Henry L., ed. *The Jewish People in America*, 5 vols. Baltimore, 1992.

Fogleman, Aaron S. *Hopeful Journeys: German Immigration, Settlement, and Political Culture in Colonial America.* Philadelphia, 1996.

Foner, Nancy, ed. *Islands in the City: West Indian Migration to New York.* Berkeley, 2001.

———. *New Immigrants in New York.* New York, 1987.

Gabaccia, Donna. *From the Other Side: Women, Gender, and Immigrant Life in the U.S., 1820–1990.* Bloomington, Ind., 1995.

———. *Militants and Migrants: Rural Sicilians Become American Workers.* New Brunswick, N.J., 1988.

Games, Alison F. *Migration and the Origins of the English Atlantic World.* Cambridge, Mass., 1999.

Garcia, Maria Cristina. *Havana USA: Cuban Exiles and Cuban Americans in South Florida, 1959–1994.* Berkeley, 1996.

Garcia, Mario T. *Desert Immigrants: The Mexicans of El Paso, 1880–1920.* New Haven, 1981.

———. *Mexican Americans: Leadership, Ideology and Identity, 1930–1960.* New Haven, 1990.

Gerber, David A. *The Making of An American Pluralism: Buffalo, New York, 1825–60.* Urbana, 1989.

Gjerde, Jon. *The Minds of the West: Ethnocultural Evolution in the Rural Middle West, 1830–1917.* Chapel Hill, 1997.

———. *From Peasants to Farmers: The Migration from Balestrand, Norway to the Middle West*. Cambridge, Eng., 1985.

Glenn, Susan A. *Daughters of the Shtetl: Life and Labor in the Immigrant Generation*. Ithaca, N.Y., 1990.

Gonzalez, Gilbert C. *Labor and Community: Mexican Citrus Worker Villages in a Southern California County, 1900–1950*. Urbana, 1994.

Goodfriend, Joyce D. *Before the Melting Pot: Society and Culture in Colonial New York City, 1624–1730*. Princeton, N.J., 1992.

Gordon, Milton. *Assimilation in American Life: The Role of Race, Religion, and National Origins*. New York, 1964.

Gutiérrez, David G. *Walls and Mirrors: Mexican Immigrants, Mexican Americans and the Politics of Ethnicity*. Berkeley, 1995.

Halter, Marilyn. *Between Race and Ethnicity: Cape Verdean American Immigrants, 1860–1965*. Urbana, 1993.

Hein, Jeremy. *From Vietnam, Laos, and Cambodia: A Refugee Experience in the United States*. New York, 1995.

Henige, David. *Numbers from Nowhere: The American Indian Contact Debate*. Norman, Okla., 1998.

Hoerder, Dirk and Horst Rössler, eds. *Distant Magnets: Expectations and Realities in the Immigrant Experience, 1840–1930*. New York, 1993.

Jacobson, Matthew Frye. *Special Sorrows: The Disaporic Imagination of Irish, Polish and Jewish Immigrants in the United States*. Cambridge, Mass., 1995.

Kamphoefner, Walter D., *The Westfalians: From Germany to Missouri*. Princeton, N.J., 1987.

———. Wolfgang Helbich and Ulrike Sommer, eds. *News from the Land of Freedom: German Immigrants Write Home*. Ithaca, N.Y., 1991.

Kasinitz, Philip. *Caribbean New York: Black Immigrants and the Politics of Race*. Ithaca, N.Y., 1992.

Kroes, Rob. *The Persistence of Ethnicity: Dutch Calvinist Pioneers in Amsterdam, Montana*. Urbana, 1992.

Kuropas, Myron B. *The Ukrainian Americans: Roots and Aspirations, 1884–1954*. Toronto, 1991.

Laguerre, Michel S. *American Odyssey: Haitians in New York City*. Ithaca, N.Y., 1984.

Lamphere, Louise. *From Working Mothers to Working Daughters: Immigrant Women in a New England Industrial Community*. Ithaca, N.Y., 1987.

Leonard, Karen I. *Making Ethnic Choices: California's Punjabi Mexican Americans*. Philadelphia, 1992.

Levine, Bruce C. *The Spirit of 1848: German Immigrants, Labor Conflict, and the Coming of the Civil War*. Urbana, 1992.

Loewen, Royden. *Family, Church, and Market: A Mennonite Community in the Old and New Worlds, 1850–1930*. Urbana, 1993.

Luebke, Frederick C. *Germans in the New World: Essays in the History of Immigration*. Urbana, 1990.

Morawaska, Ewa. *Insecure Prosperity: Small-Town Jews in Industrial America, 1890–1940*. Princeton, N.J., 1996.

Mormino, Gary and George Pozzetta. *The Immigrant World of Ybor City: Italians and Their Latin Neighbors in Tampa, 1885–1985*. Urbana, 1987.

Nugent, Walter. *Crossings: The Great Transatlantic Migrations, 1870–1914*. Bloomington, Ind., 1992.

Øverland, Orm. *Immigrant Minds, American Identities: Making the United States Home, 1870–1930*. Urbana, 2000.

Pacyga, Dominic. *Polish Immigrants and Industrial Chicago: Workers on the South Side, 1890–1922*. Columbus, Ohio, 1991.

Patterson, Wayne. *The Korean Frontier in America: Korean Immigration to Hawaii, 1896–1910*. Honolulu, 1988.

Pickle, Linda S. *Contented Among Strangers: Rural German-Speaking Women and Their Families in the Nineteenth Century American Midwest*. Urbana, 1996.

Posadas, Barbara M. *The Filipino Americans*. Westport, Conn., 1999.

Pula, James S. *Polish Americans: An Ethnic Community*. New York, 1995.

Puskas, Juliana, ed. *Overseas Migration from East-Central and Southeastern Europe*. Budapest, 1990.

Ramirez, Bruno. *On the Move: French-Canadian and Italian Migrants in the Atlantic Economy, 1860–1914*. Toronto, 1991.

Rose, Peter I. *Mainstream and Margins: Jews, Blacks and Other Americans*. New Brunswick, N.J., 1983.

Ruiz, Vicki. *Cannery Women, Cannery Lives: Mexican Women, Unionization, and the California Food Processing Industry, 1930–1950*. New Mexico, 1987.

Rutledge, Paul J. *The Vietnamese Experience in America*. Bloomington, Ind., 1992.

Salinger, Sharon V. *"To Serve Well and Faithfully": Labor and Indentured Servants in Pennsylvania, 1682–1800*. Cambridge, Eng., 1987.

Sanchez, George J. *Becoming Mexican American: Ethnicity and Acculturation in Chicano Los Angeles, 1900–1943*. New York, 1992.

Schneider, Dorothee. *Trade Unions and Community: The German Working Class in New York City, 1870–1900*. Urbana, 1994.

Sollors, Werner, ed. *The Invention of Ethnicity*. New York, 1989.

Semmingsen, Ingrid. *Norway to America: A History of the Migration*. Minneapolis, 1982.

Strauss, Herbert A. ed. *Jewish Immigrants of the Nazi Period in the U.S.A.* 6 vols., New York, 1979–87.

Swierenga, Robert P. *Dutch Jewry in the North American Diaspora*. Detroit, 1994.

———. *Faith and Family: Dutch Immigration and Settlement in the United States*. New York, 2000.

Takaki, Ronald. *Strangers from a Different Shore: A History of Asian Americans*. Boston, 1989.

Tamura, Eileen. *Americanization, Acculturation and Ethnic Identity: The Nisei Generation in Hawaii*. Urbana, 1994.

Taylor, Robert M. et al., eds. *Peopleing Indiana: The Ethnic Experience*. Indianapolis, 1996.

Ungar, Sanford J. *Fresh Blood: The New American Immigrants*. New York, 1995.

Valdez, Dennis Nodin. *Al Norte: Agricultural Workers in the Great Lakes Region, 1917–1970*. Austin, 1991.

Van Sant, John E. *Pacific Pioneers: Japanese Journeys to America and Hawaii, 1850–80*. Urbana, 2000.

Van Vugt, William E. *Britain to America: Mid-Nineteenth Century Immigrants to the United States*. Urbana, 1999.

Vargas, Zaragosa. *Proletarians of the North: Mexican Industrial Workers in Detroit and the Midwest, 1917–1933*. Berkeley, 1993.

Vecoli, Rudolph J. and Suzanne Sinke, eds. *A Century of European Migrations, 1830–1930*. Urbana, 1992.

Vinyard, JoEllen. *For Faith and Fortune: The Education of Urban Catholic Immigrants, Detroit, 1805–1925*. Urbana, 1998.

Virden, Jenel. *Goodbye Piccadilly: The American Immigrant Experience of British War Brides of World War II*. Urbana, 1996

Weber, David J. *The Spanish Frontier in North America*. New Haven, 1992.

Welaratna, Usha. *Beyond the Killing Fields: Voices of Nine Cambodian Survivors in America*. Stanford, 1993.

White, Richard. *The Middle Ground: Indians, Empires and Republics in the Great Lakes Region, 1650–1815*. Cambridge, Eng., 1991.

Wokeck, Marianne S. *A Trade in Strangers: The Beginnings of Mass Migration to North America*. State University, Pa., 1999.

Wyman, Mark. *DP: Europe's Displaced Persons, 1945–1951*. Philadelphia, 1989.

———. *Round-Trip to America: The Immigrants Return to Europe, 1880–1930*. Ithaca, N.Y., 1993.

Yans-McLaughlin, Virginia, ed. *Immigration Reconsidered: History, Sociology, and Politics*. New York, 1990.

Yung, Judy. *Unbound Feet: A Social History of Chinese Women in San Francisco*. Berkeley, 1995.

Zucci, John. *The Little Slaves of the Harp: Italian Child Street Musicians in Nineteenth Century Paris, London, and New York*. Montreal, 1992.

Acknowledgments

This book was written during a year's sojourn in Germany where I was a Visiting and Fulbright professor at the University of Hamburg. I like to think that residence in one of the greatest ports of emigration and visits to Bremen, Genoa, Liverpool, and London somehow enriched this work. My greatest debts are to my fellow students of immigration and ethnicity, whose work I have quoted and paraphrased throughout. They are acknowledged in the traditional way in the text, notes, and bibliography. My colleagues and friends at the University of Hamburg gave me a light teaching load and provided an atmosphere in which I could write. My sponsor, Professor Dr. Günter Moltmann, a specialist in German emigration and perhaps the Federal Republic's leading Americanist, was unfailingly generous of his time and resources. The director of the Historisches Seminar that year, Professor Dr. Horst Pietschmann, went out of his way to make me feel welcome, and Dr. Hans-Jürgen Grabbe was an unfailing source of good advice on all things German, from high culture to haircuts. Dr. Andreas Brinck, then in the final stages of writing his Ph.D. thesis, helped me cope with the bewildering array of separate libraries in Hamburg. Frau Renate Daumann provided secretarial and logistical support and much more, including expanding my German vocabulary with such terms as *gummibands* and *super!* In addition, the students in my Hamburg seminars often gave me very different European perspectives on immigration matters, as did students in other European universities I visited. In the United States, my editor, Paula McGuire of Visual Education Corporation, emended the text with skill, unfailing good sense, and enthusiasm; and Cynthia Cappa, Picture Research Coordinator, fulfilled the picture specifications with imagination and perseverance. Susan H. Llewellyn, the copyeditor at HarperCollins, contributed substantially to the coherence and clarity of the book.

My graduate student and assistant in Cincinnati, Kriste Lindenmeyer, performed a wide range of chores cheerfully and with great discretion. My colleague John K. Alexander saved me from a grievous error.

The picture research was greatly assisted by cooperation and courtesy from Phyllis Montgomery of the Metaform consortium in New York, Barry Moreno of the Ellis Island Archives of the National Park Service, Joachim Frank of the Hamburg Staatsarchiv, and Astrid Knopke.

I must also acknowledge the financial assistance of the American and German Fulbright commissions, which helped support my stay in Germany, and those of Britain, France, Italy, and Austria, which financed travels in those countries. Important support was provided by the University of Cincinnati, which gave me leave and funded some of my year abroad. I particularly wish to thank Dean Joseph Caruso and my chair, Otis C. Mitchell. Last but not at all least, I thank my fellow traveler, Judith M. Daniels, who was constantly interrupted in her own work in American Jewish history to listen, suggest, and counsel, and who put up with, almost without complaint, the various noises of literary creation in very close quarters.

Index